D1797504

Mathematics in Physics Education

Gesche Pospiech • Marisa Michelini
Bat-Sheva Eylon

Editors

Mathematics in Physics Education

 Springer

Editors
Gesche Pospiech
Technische Universität Dresden
Dresden, Germany

Marisa Michelini
DCFA - Section of Mathematics and
Physics
University of Udine
Udine, Italy

Bat-Sheva Eylon
The Weizmann Institute of Science
Rehovot, Israel

ISBN 978-3-030-04626-2   ISBN 978-3-030-04627-9   (eBook)
https://doi.org/10.1007/978-3-030-04627-9

© Springer Nature Switzerland AG 2019
This work is subject to copyright. All rights are reserved by the Publisher, whether the whole or part of the material is concerned, specifically the rights of translation, reprinting, reuse of illustrations, recitation, broadcasting, reproduction on microfilms or in any other physical way, and transmission or information storage and retrieval, electronic adaptation, computer software, or by similar or dissimilar methodology now known or hereafter developed.
The use of general descriptive names, registered names, trademarks, service marks, etc. in this publication does not imply, even in the absence of a specific statement, that such names are exempt from the relevant protective laws and regulations and therefore free for general use.
The publisher, the authors and the editors are safe to assume that the advice and information in this book are believed to be true and accurate at the date of publication. Neither the publisher nor the authors or the editors give a warranty, express or implied, with respect to the material contained herein or for any errors or omissions that may have been made. The publisher remains neutral with regard to jurisdictional claims in published maps and institutional affiliations.

This Springer imprint is published by the registered company Springer Nature Switzerland AG.
The registered company address is: Gewerbestrasse 11, 6330 Cham, Switzerland

Preface

In recent years, the interest in the role of mathematics in physics education has risen steadily. This can be seen in connection with the efforts of showing students the nature of physics which has gained increasing importance in physics education. Whereas the conceptual understanding of physics is an important goal of physics education, it is equally undoubted that mathematics is inevitably inherent in physics and its methods. As the physical method lives on the mathematical description of the reality, the mathematical aspects have to be taken into account from the perspective of teaching and learning physics itself as well as learning about physics. This book is devoted to the struggle to make the interplay of physics and mathematics insightful to students. Besides these fundamental considerations, there are practical aspects: the lecturers of first year university science courses all over the world complain increasingly about the difficulties of students in applying mathematical instruments to physical problems. Therefore, the questions arise: what are the deeper reasons of this deficiency, and what can be done about it during secondary school? The research about this topic has become more intense during the last 15 years and has reached a certain state that makes it desirable to gather the most important results from people working in the field in a book. We collected results of research on teaching and learning about the role of mathematics in physics gained so far and cover the whole spectrum of research, the theoretical foundation, as well as empirical results. The book should also raise the awareness of the role of mathematics in physics education and induce further research projects.

The research areas addressed in this book cover a broad range. Therefore, the book is divided into four parts, each concentrating on a specific aspect. In the first part, "Perspectives on Mathematics in Physics Education," theoretical viewpoints are treated, enlightening the interplay of physics and mathematics and also including historical developments. In the second part, "Learning Mathematization," with the most contributions, we delve into the learners' perspective. In this part, not only aspects of the learning by secondary school students but also by students just entering university or teacher students are considered as far as they could shed light onto learning in secondary school. The third part "Teaching Mathematization" includes a broad range of subjects from teachers' views and knowledge, the analysis of classroom discourse, and an evaluated teaching proposal. In the last part, "Facilitating Mathematization by Visual Means," approaches are described that take up mathematization in a broader interpretation. These contributions show that formal thinking might start before the use of mathematical elements in a narrower sense and thus prepare mathematization.

We hope this book will be a valuable source for researchers in the field, teacher educators, advanced students, as well as interested teachers.

We have to thank all the contributors of the book for their patience and, especially, Marisa Michelini for her constant support.

Dresden, Germany   Gesche Pospiech
July 2018

Contents

Contributors

Esther Bagno The Weizmann Institute of Science, Rehovot, Israel

H. Berger The Weizmann Institute of Science, Rehovot, Israel

Onne van Buuren Faculty of Behavioural and Movement Sciences, Vrije Universiteit Amsterdam, Amsterdam, The Netherlands

Mark Eichenlaub Department of Physics, University of Maryland, College Park, MD, USA

Elias Euler Disciplinary Domain of Science and Technology, Physics, Uppsala University, Uppsala, Sweden

Bat-Sheva Eylon The Weizmann Institute of Science, Rehovot, Israel

Marie-Annette Geyer Faculty of Physics, TU Dresden, Dresden, Germany

Ileana M. Greca Department of Specifics Didactics, Faculty of Education, Universidad de Burgos, Burgos, Spain

Bor Gregorcic Disciplinary Domain of Science and Technology, Physics, Uppsala University, Uppsala, Sweden

Lena Hansson School of Education and Environment, Kristianstad University, Kristianstad, Sweden

Örjan Hansson School of Education and Environment, Kristianstad University, Kristianstad, Sweden

André Heck Korteweg-de Vries Institute for Mathematics, University of Amsterdam, Amsterdam, The Netherlands

Dietmar Höttecke Faculty of Education, Physics Education, University of Hamburg, Hamburg, Germany

Lana Ivanjek Austrian Competence Center Physics, University of Vienna, Vienna, Austria

Kristina Juter School of Education and Environment, Kristianstad University, Kristianstad, Sweden

Ricardo Karam Department of Science Education, University of Copenhagen, København, Denmark

Olaf Krey Universität Augsburg, Augsburg, Germany

Wiebke Kuske-Janßen Faculty of Physics, TU Dresden, Dresden, Germany

Yaron Lehavi The Weizmann Institute of Science, Rehovot, Israel
The David Yellin Academic College of Education, Jerusalem, Israel

E. Magen The Weizmann Institute of Science, Rehovot, Israel
Ostrovsky High School, Raanana, Israel

Terhi Mäntylä Faculty of Education and Culture, Tampere University, Tampere, Finland

Željka Milin Šipuš Department of Mathematics, Faculty of Science, University of Zagreb, Zagreb, Croatia

Roni Mualem The Weizmann Institute of Science, Rehovot, Israel

Ana Raquel Pereira de Ataíde Department of Physics, Universidade Estadual da Paraíba, Paraíba, Brazil

Maja Planinic Department of Physics, Faculty of Science, University of Zagreb, Zagreb, Croatia

C. Polingher The Weizmann Institute of Science, Rehovot, Israel
Hemda Schwartz-Reisman Science Education Center, Tel Aviv, Israel

Jaska Poranen Faculty of Education and Culture, Tampere University, Tampere, Finland

Gesche Pospiech Technische Universität Dresden, Dresden, Germany

Andreas Redfors School of Education and Environment, Kristianstad University, Kristianstad, Sweden

Edward F. Redish Department of Physics, University of Maryland, College Park, MD, USA

Alberto Stefanel Physics Education Research Unit, Department of Chemistry, Physics and Environment, University of Udine, Udine, Italy

Ana Susac Department of Physics, Faculty of Science, University of Zagreb, Zagreb, Croatia

Department of Applied Physics, Faculty of Electrical Engineering and Computing, University of Zagreb

Olaf Uhden Faculty of Education, Physics Education, University of Hamburg, Hamburg, Germany

Chapter 1
Framework of Mathematization in Physics from a Teaching Perspective

Gesche Pospiech

1.1 Introduction

This book roots in the perception of physics as an empirical science which cannot be thought without a mathematical description of fundamental physical structures, this being one of the characteristic traits and most powerful tools of physics. Therefore at the heart of this book lies the idea of mathematics and physics as being involved in a constant interplay starting from the very beginning of scientific thought. Therefore we do not think of mathematics as an "application" in physics but more as a "mutual interaction between mathematics and physics that develops and shapes both disciplines" (Kjeldsen and Lützen 2015). This disciplinary view is not without relevance for physics education.

Physics education at school tries to achieve a variety of aims at different levels. The overarching goal is to provide students with a basis for further learning about science and the prerequisites for participation in the scientifically oriented societal discourse in shaping a scientific world view. That means that students should acquire scientific literacy encompassing physics knowledge, skills and insight into the nature of physics. This implies learning the concepts, principles and the structure of physics as well as the use of scientific methods including mathematical elements as an intrinsic feature of doing physics. The importance of teaching this complete picture of physics is underlined as most "national educational standards call for pupils to connect mathematics and science to real world phenomena" (Carrejo and Marshall 2007). Therefore, if scientific literacy as a goal of physics education is taken seriously, physics should be taught not only relying on the experiment as an empirical basis but also applying mathematics. It is the task of the teachers

G. Pospiech (✉)
Technische Universität Dresden, Dresden, Germany
e-mail: gesche.pospiech@tu-dresden.de

© Springer Nature Switzerland AG 2019
G. Pospiech et al. (eds.), *Mathematics in Physics Education*,
https://doi.org/10.1007/978-3-030-04627-9_1

to use both aspects in a balanced way for achieving an adequate understanding of physics with the students. However, experience shows this is not an easy task. Often students may be able to solve quantitative problems by certain techniques but have not reached an "understanding" of physical concepts or their relation with mathematics. This indicates that understanding the meaning of mathematical tools and their interrelation with the physical description of the world seems to be one of the most difficult and time-consuming steps in physics learning. Therefore, the basis for awareness and competences in this interplay has to be laid early in the educational career, starting from lower secondary school.

Besides the fundamental analysis of the interplay of mathematics and physics and underlying theoretical frameworks, we give an overview of empirical research on competences and views or attitudes of students and teachers and aspects of its teaching and learning. Mostly, the contributions in this book focus on secondary school students and their teachers. This does not exclude some results from physics education in college or university if we think them instructive with respect to secondary school. As the views and competences of teachers play a central role for the successful learning of students, also the teachers come into focus.

1.2   An Educational Perspective on the Interplay

In this section we will highlight the peculiarities of mathematics and physics with respect to their interplay. The role of mathematics in physics has multiple aspects: it serves as a tool (pragmatic perspective), it acts as a language (communicative function, see also Sect. 1.4), and it provides a logical and structural framework for describing, ordering and classifying physical processes and theories (Krey 2012). This will be discussed in this section and enriched by the historical perspective.

1.2.1   Historical-Philosophical Perspective on the Interplay

The interplay was intensely studied from the perspectives of history of science and its philosophy (see, e.g. Pask 2003; Brush 2015 and many others) as well as from physicists themselves (see, e.g. Dirac 1939; Wigner 1960; Einstein 1921).[1] An insight into the developing role of mathematics for physics and doing physics during the past centuries and its consequences is indicated by Gingras (2001). Even if we do not expand on this literature, all the contributors in this book are aware of

---

[1]The reference to the broad and deep literature on this topic would strongly go beyond the scope of this book, concentrating on the teaching perspective. Therefore we refer the interested reader to the relevant literature.

the deep, manifold and fruitful interrelation showing a rich variety, implying that the interplay is by no means "one-way":

- mathematics is seen as a tool in physics: e.g. formula and equations serve for making quantitative predictions
- physics could be seen as a field of application of mathematics: e.g. the construct of Hilbert spaces is applied in quantum theory
- there are physical concepts that developed from mathematics: e.g. the concept of curved space time was derived from Riemannian manifolds
- there are mathematical elements or structures that were inspired by physics: e.g. the problem of the three-body system in gravitational physics boosted the development of the theory of nonlinear systems or the development of the theory of distributions (Kjeldsen and Lützen 2015)
- there are concepts where the contribution of mathematics and physics cannot be separated: e.g. the development of the concept of derivative, as used in describing velocity and acceleration

This listing spans the range from mathematics as a technical tool to an inseparable mathematical-physical reasoning. A core feature of physics confirming its image as a rigorous and powerful science is the ability to make precise quantitative predictions enabled by the precise formulation of physical laws by equations and formula. But the power of mathematics goes far beyond such calculation as can be seen with many examples from the history of physics. Even with relatively simple instances such as the ideal gas law, it can be argued that the mathematical description combined with the power of algebraic and analytic manipulation allows for logical deduction of consequences that would not be possible only on the basis of qualitative physical arguments (de Berg 1992). This shows that physics inherits from the mathematical formulation also a deductive power enabling the development of theories. In a deeper sense, it is this structural significance of mathematical elements for physics that seems to be a central component of an adequate approach to science. For example, central physical concepts such as acceleration (Basson 2002) or force or the principle of least action cannot be thought of in a meaningful way without mathematics. Hestenes (1986) is even advocating the advantages of a unified language, based on geometric structures such as Clifford algebras.

This story of success relies on a long (historical) process of giving physical meaning to mathematical constructs. However, some steps that seem to be easy nowadays required a long development. Karam, Uhden and Höttecke in their contribution (*"The 'math as prerequisite' illusion: Historical considerations and implications for physics teaching"*) derive possible learning difficulties from the historical pathway. Their analysis leads to the insight that it would be too short-sighted to assume that mathematical techniques could be applied straightforward in the domain of physics. There are specific differences between the scientific cultures of mathematics and physics, implying different foci and therefore specific approaches. This becomes obvious in different conventions and interpretations. It is important that also teachers are aware of this additional difficulty for appropriately shaping their instruction.

1.2.2   Mathematical Perspective on the Interplay

As mathematics serves as a "language of physics", it provides mathematical elements and symbols such as numbers, tables, diagrams and algebraic notation for representing physical constructs. But beyond this representational role, it gives structural insights where the mathematical structures may even be richer than the physical world (Quale 2010). So in stressing the interplay, it may not be forgotten that physics and mathematics have their specific viewpoints and methods. This general statement can be clarified even with the most simple mathematical elements. Handling numbers in mathematics requires knowledge of their properties and the calculation rules, but there are no additional meanings behind. In physics numbers often are connected to units, expressing a physical quantity and implying that a big or small number has a special meaning in the real world. Because of the units, even the numbers have to be interpreted from a physical point of view. In addition, in experiments the numerical values have to be considered together with the measuring deviation. Hence even a number is laden with meaning in a physical context, even more so algebraic terms such as, e.g. a formula or a function (see Sect. 1.4.2). Indeed, mathematics provides many structures with specific meaning in a physics context. These structures contribute to the precise formulation of physical laws and allow for recognizing analogies, classifying, noticing patterns and so on. De Berg (1995) writes:

> .. algebraic expressions are primarily useful for theoretical development which leads not only to new data but new concepts. This development takes place through the laws of mathematics.

From a mathematical point of view, the existing differences are elaborated on in the chapter by *Heck and van Buuren ("Students' understanding of algebraic concepts")*. Their analysis shows the broadness of certain concepts in mathematics, e.g. of a variable, even in mathematics itself. The different possible meanings of variables and algebraic expressions which are not always communicated clearly add to the learning difficulties of students (Redish 2005). Therefore even in the introduction of a mathematical construct, it has to be made explicit for the students how to embed it into a physics context.

Another example is the notion of "function". Its mathematical definition focusses on the aspect of pointwise relation between the independent and the dependent variables. In physics however, the functional dependence is the most important aspect describing how one physical quantity depends on others. Furthermore in physics, there is a certain choice which variable is the dependent and which the independent variable. Sometimes their roles can be interchanged depending on the concrete situation. Also the distinction between parameters and variables often is difficult for students.

1.2.3 Mathematics and Conceptual Physics Understanding

From a school perspective, the role of mathematics sometimes is equated with solving end-of-chapter problems. Such a narrow focus might indeed lead to the impression that using mathematics could be opposed to understanding the concepts of physics. That the reconciliation of formal mathematics and physics concepts is not an easy task also on university level was shown by many studies, (e.g. Hu and Rebello 2013; Kuo et al. 2013; Nguyen and Rebello 2011). Even Hund (1975, p11), from the perspective of a physicist, pointed out a dichotomy between "handling physics" and "understanding physics", which were not easy to bridge. So it is no surprise that there is a struggle between the emphasis on conceptual understanding and the use of mathematics. Experience of many teachers shows that the merging of both is not easy to achieve (e.g. Monk 1994), especially as often the students' mathematical understanding itself is instrumental (Richland et al. 2012). However, that both aspects can support each other was advocated by Hewitt (1983, 2011). The mathematical, mostly algebraic, formulation of physical laws requires the ability to translate or transfer physical and mathematical elements onto each other.

A conceptual mathematical understanding is the more important as the mathematical elements cannot be simply transported into physics, but their meaning has to be framed by physical concepts (Bing and Redish 2007). The interpretation of mathematical symbols in terms of physics requires that mathematical and physical meanings have to be blended (Sherin 2001). In order to capture this aspect, the notion of a "symbolic form" was introduced. These symbolic forms cover quite a range of possible meanings of mathematical operations and hence could promote the understanding of physics equations. Sherin's capturing of understanding-driven processes of problem solving by symbolic forms gives valuable ideas about the interrelated role of mathematical experiences, semantic understanding of an equation and physical intuition.

Mathematics does not provide a direct description but even more includes an abstraction or idealization connected with a conceptual understanding. From a theoretical perspective, this aspect was analysed with the example of Coulomb's law (Kneubil and Robilotta 2015). They show that mathematics can be used as an epistemological tool in physics teaching in the sense that the interpretation of the interplay may not be unique and can hint to several aspects of a physical concept, in their case the concept of charge. Another study asked how prospective teachers connect experimental situations with mathematical models in an inquiry-based approach to kinematics (Carrejo and Marshall 2007). This study shows the complexity of the pathway from the phenomenon to a mathematical description. The complexity of the physics-mathematics interrelation and its strong dependence on the concrete context is discussed with the field of electricity and the learning gains of students by Meltzer (2002). A related example in the field of optics is presented in this book by *Krey ("What is learned about the role of mathematics in physics while learning physics concepts? A mathematics sensitive look at physics teaching and learning")*. He focusses on the added value of a mathematical formulation of physical examples going hand in hand with a reduction of complexity. Krey's example at the same time refers to the importance of varied representations. These different representations are at the core of the communication role of mathematics in physics and hence serve also as facilitating insight into physics. How the representations support each other and in which way the transformation between them could support conceptual understanding is discussed in the contribution of Geyer/Kuske-Janßen (*"Mathematical Representations in Physics Lessons"*, see also Sect. 1.4).

From these studies one can infer that the interplay of mathematics and physics can be seen from different perspectives and is by no means straightforward but needs closer analysis with the help of models.

1.3 Modelling the Interplay from a Physics Education Perspective

In order to capture details of the described complex interplay, models are needed that allow for ordering and classifying selected aspects. Because of its many facets, it will not be possible to devise a single model incorporating all possibilities and serving all research purposes. Accordingly there is quite a range of models of the interplay, each one specific for the intended research framework or instructional goal. The models highlight certain aspects, provide focus and allow for detailed analysis. In the end they should – from a teaching perspective – contribute to shaping the interplay in the actual lesson. In this section we will present a selection of models.

First there are theoretical models of mathematization in physics. By *mathematization* we mean the process of gradually transferring (with focus on conceptual considerations) and translating (focus on mathematics as language of physics) physical processes or phenomena into mathematical elements and structures. The most basic model is shown in Fig. 1.1 with a very schematic principle of the mathematization process. However, it displays a central element present in all other models, namely, processes starting from a physical situation going to mathematics and vice versa, thus including the processes of mathematization and interpretation or validation, respectively.

In order to proceed through this process, we will describe in more detail which aspects of the interplay mathematics and physics should be considered.

1.3.1 Technical and Structural Role of Mathematics in Physics

Describing physical processes or solving physical problems requires that the steps of the aforementioned model are done successfully. In the light of Sect. 1.2, it can

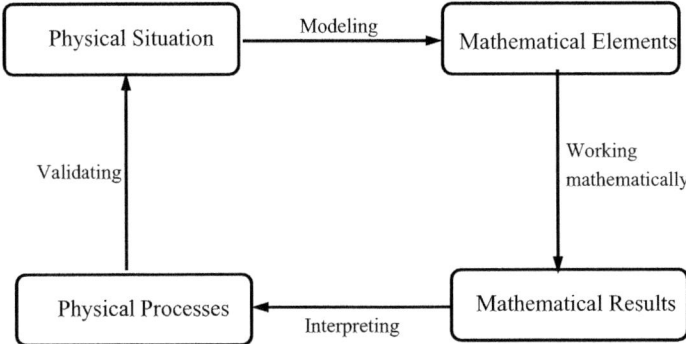

Fig. 1.1 Basic model of interplay mathematics and physics. (Modified according to Redish and Kuo 2015)

be concluded that this requires to understand the physical content of mathematical representations and to reach conceptual understanding in mathematics as well as in physics. However, many students appear to have difficulties to be aware of the meaning of formulas (perceived as the predominant feature of physics) and tend to rote application of mathematical techniques. Therefore from an educational viewpoint, it has to be considered that in mathematics as in physics, a more computational or instrumental role and a more relational or conceptual role can be distinguished (see Fig. 1.2). The instrumental role in mathematics implies that algorithms or calculational rules are used without thinking much about the background or reasons (Skemp 1976). Similarly if in physics a numerical procedure is applied without much thinking about the physical background or concepts behind, e.g. simply calculating the value of a formula, we can call this the computational aspect of physics. As opposed to superficially doing mathematics, Skemp defined "relational understanding" by which he means understanding the mathematical concepts instead of just applying rules. Hewitt (1983) on the physics side called for doing physics conceptually, meaning to focus on the physical concepts behind, e.g. everyday processes before trying to use mathematical elements. Concerning the interplay of mathematics and physics, we comprise superficial procedures in the "technical role of mathematics in physics" and a deeper entanglement we call the "structural role of mathematics in physics" (see Fig. 1.2).

The distinction between the technical and structural role in the interplay should serve to catch the most relevant aspects of insightful handling the mathematical elements in physics (Pietrocola 2008). According to Pietrocola, the technical role comprises aspects or activities mainly related to numerical procedures: calculating, using algorithms or drawing function graphs. This technical role dominates the perception of students as well as of many teachers or researchers. On the other hand, Pietrocola stresses the importance of the structural role of mathematics. He characterizes it by the following descriptions:

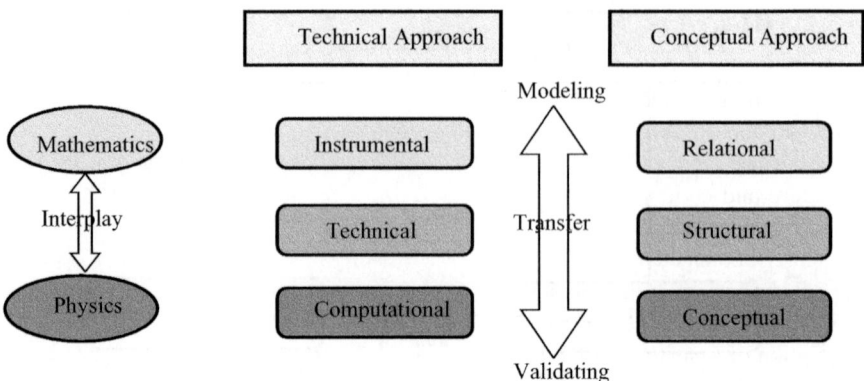

Fig. 1.2 Technical and structural aspects. The figure should be read as follows: Concentrating on the horizontal line labelled "Mathematics", the instrumental and relational aspects of mathematics are indicated, similarly in the line labelled "Physics" the corresponding computational and conceptual aspects. In the middle line labelled "Interplay" according to Pietrocola (2008), its technical and structural aspects are named. The vertical arrows indicate that there is a transfer between both domains: a mathematical model arises from the physics section and has to be evaluated and validated there. It is observed that in every constellation, the focus could lie on algorithmic aspects or on deeper understanding

- Physics inherits the formal operations and definitions of mathematical objects if these are used (use of vectors, derivatives, etc.) This point is related to the aspect of mathematics as a language.
- Mathematics orders the physical phenomena according to underlying patterns (e.g. analogies)
- Mathematics orders physical thought by the physical (concrete) meanings of its operations (limiting cases, functions, etc.)

The role of mathematics as a structural means hence builds the skeleton of physical theories and provides valuable general theorems allowing to proceed into the unknown. A famous example is the Noether theorems. Generally it can be said that the structuring role of mathematics becomes more and more important the more advanced the physical theory is, as is obvious, e.g. in the physical theories of the twentieth century. In these theories, often mathematics is needed as a guidance even for conceptual explanations or reasoning.

It has to be stressed that the transition from the technical to the structural role is along a continuum and there is no sharp separating line. Let us consider the concept of function as an example: functions can serve as a tool, e.g. evaluating them by inserting numbers, but by their properties, they also may clarify deeper relations between physical quantities or even allow for deduction of new insights. This example also shows that the technical role is a necessary part of doing physics but has to be informed by the structural aspects. Nevertheless the distinction of a technical and a structural role of mathematics could be helpful as a means to focus on the meaningful use of mathematical structures and elements in physics education.

1.3.2 Modelling Mathematization in Physics

A theory-based model of the interplay mathematics and physics from the perspective of understanding physics theories was developed by Greca and Moreira (2002). The starting point is the scientific theory requiring on the one hand the semantic structure and content of physics and on the other hand the syntactic structure of a mathematical model, including, e.g. equations or the formalism. These both are intimately connected with the mathematical model embedded into the physical model. Important is the relation of the physics model to the physical phenomena by itself requiring idealizations. The process of understanding also includes the formation of a mental model that connects all the parts: the physical phenomenon, the physical model and the mathematical model. They argue that this overall relation fits well from the viewpoint of well-established physical theories and physics taught at school but admit that this picture might not apply to physical theories of the twentieth century where physics and mathematics are much stronger intertwined with each other.

A different model but compatible to the Greca-Moreira model was developed by Hansson et al. (2015). It has in its centre the perspective of teaching at school with focus on classroom discourse. Specific in their model is the explicit reference to "reality" in different stages. "Reality" here means the appearances students experience with concrete objects or phenomena in their everyday life or in a learning environment at school as, e.g. experiments. The perspective of reality is related to a theoretical (physical) model and to a mathematical model and the relations between these three perspectives (see contribution of *Hansson et al. ("A Theoretical Framework for Ternary Analysis of Textbooks and the Teaching of Physics")* in this book). They also distinguish, as in Fig. 1.2, the technical or instrumental vs the structural or relational role of mathematical approaches. Their results hint to the importance of the role of the teacher and his or her use of textbooks in order to shape the interaction in the classroom, showing that the pedagogical content knowledge of teachers is important (for description of this aspect, see Sect. 1.5.3.1).

1.3.3 Models for Learning the Interplay of Mathematics and Physics

Getting to more concrete cases, there are models for the processes occurring in teaching and learning, more precisely in problem solving and mathematical-physical modelling. Those models often intend to capture strategies actually used by students. They could also serve for describing and analysing teaching and learning processes with respect to strategies used and difficulties experienced.

1.3.3.1   Modelling Physical-Mathematical Modelling

The key difficulty students face in problem solving or more general, in the process
of physical-mathematical modelling, is the transfer between the concrete physical
phenomena and the abstract mathematical world, blending the meaning of formal
elements in both worlds. Since the famous book by Polya (2014), often strategies
for problem solving have been proposed. However, mostly, only general rules
are given. In order to analyse the thinking and the strategies actually used by
students, more detailed models are necessary. As is natural, different research
contexts have led to different models of the interplay mathematics and physics
in students' problem solving. All these different models struggle with the diverse
possibilities of the modelling procedure – ideal modelling strategies vs the real
strategies, difficulties and ideas of students concerning the mathematical domain
as well as the physical domain. Even if the different models often were developed
and represented independently from each other depending on the research framings,
a close inspection shows a remarkable agreement between them.

In mathematics education, the mathematical modelling of a situation of everyday
life plays an important role. Accordingly there is a detailed model, the so-called
modelling cycle by Blum and Leiß (Blum and Borromeo 2009, for a discussion,
see also Phillips 2016). This model indicates the complexity of the process of
mathematical modelling. It defines several steps suggesting an ideal sequence in
the modelling process, starting from a real situation which is first transformed
into a model of the situation and then simplified or structured. It follows a step,
not described in detail, leading to the mathematical model in the mathematical
world which gives mathematical results. These then are interpreted and validated
in terms of the real situation. Empirical evidence shows that pupils do not follow
exactly this cycle but go back and forth in very different and individual paths (see,
e.g. Blum and Borromeo 2009; Borromeo 2006). However, knowing this model
seems to help school students in organizing their work. Analysis of this model and
comparison with the difficulties experienced by physics students in solving physics
problems with mathematical methods hints that the transfer between the simplified
situation model and the mathematical model is the most critical part of problem
solving. There seems to be a "gap" between the "real situation/problem" in the
physical world[2] and the mathematical model in the mathematical world (see, e.g.
Aufschnaiter et al. 2000; Brahmia 2014; Monk 1994). To make this gap visible
and find ways for analysing the modelling process in more detail, modified models
were developed. The two selected models (see Fig. 1.3) agree insofar as they fill
the gap between mathematics and physics and expand on the processes "between
the worlds" that were not addressed in the model of Blum and Borromeo (2009).
Both models are very similar in describing the ideal problem solving process even
if the drawing looks very different. Indeed the description and arrows from the
model by Czocher in Fig. 1.3a can be mapped one to one to the model of Uhden-

---

[2]"Real" again means objects or processes from everyday life.

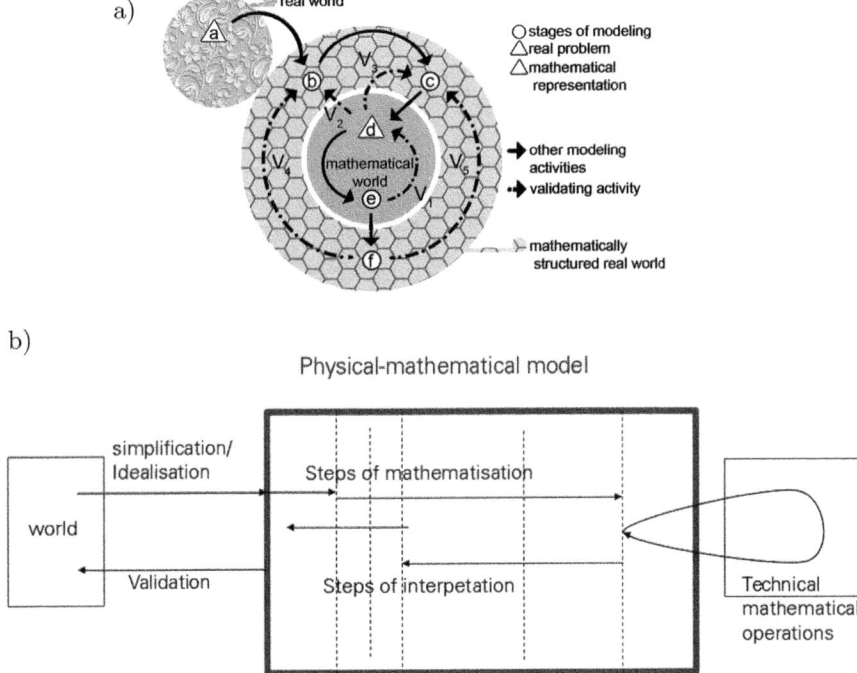

Fig. 1.3 Two graphically different representations of the mathematization process containing the same steps and procedures: (**a**) Model by Czocher as presented by Brahmia (2014), (**b**) Model based on Uhden et al. (2012). Both models strongly focus on the processes of mathematization and interpretation and validation

Karam in Fig. 1.3b. The arrows of Fig. 1.3a, signed by "b", correspond to the arrows in the central rectangle in Fig. 1.3b providing steps of increasing mathematization and interpretation. The detailed description of these steps gives hints for shaping instruction. On the other hand, the models can be used for detailed analysis of the transfer and translation processes and specific difficulties therein (Uhden et al. 2012; Uhden 2016) or for focus on the process as a whole and the interrelation of steps (Czocher 2018). Both models imply that students have to master the spectrum of understanding from procedural to conceptual or structural aspects, syntactics as well as semantics, both in mathematics and in physics.

1.3.3.2 Modelling Problem Solving

Whereas these models describe the interplay mathematics and physics from a theoretical viewpoint in order to analyse learning processes, Brahmia (2014) develops a pragmatic model derived from observations on the understanding of students (see Fig. 1.4).

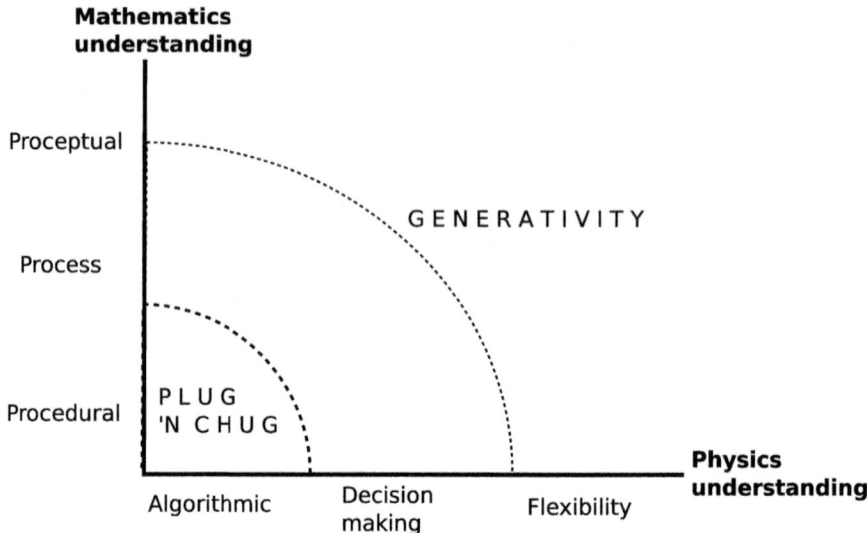

Fig. 1.4 Model of Brahmia, slightly adapted from Brahmia (2014)

This model suggests that mathematical and physical understanding are related to each other. It contains technical aspects, i.e. algorithms and mathematical procedures, as well as structural aspects: flexibility in applying physical concepts and proceptual[3] abilities in mathematics. Insofar it sheds light from a learning perspective on the different aspects of the interplay as described in Fig. 1.2. The desired result of instruction consists in mastering the higher levels of both sides and thus reaching "generativity", implying to be able to follow the emergence of the mathematical description from a physical phenomenon and the flexible use of mathematical elements and physical concepts. The importance of practical competences for a conceptual understanding may not be underestimated.

On the whole there seems to be a great consensus that should result in a unified representation of the intended mathematization process. Such a model depiction, leading the attention onto the processes and not the states during modelling of problem solving, is given in Fig. 1.5. Here the processes are depicted not as arrows but as rectangles in order to hint that there are fine-grained structures inside, leaving space to analyse the details of the processes. That this model indeed describes the process of physics problem solving of high school students was empirically validated (Trump 2015).

It is clearly seen in the model in Fig. 1.5 that the contributions of physics and mathematics overlap in a significant part of the process of problem solving. Therefore we prefer speaking about the interplay of mathematics and physics

---

[3]Brahmia uses the term of "proceptual" indicating that in mathematics the con**ceptual** understanding has to be related to **pro**cedural abilities.

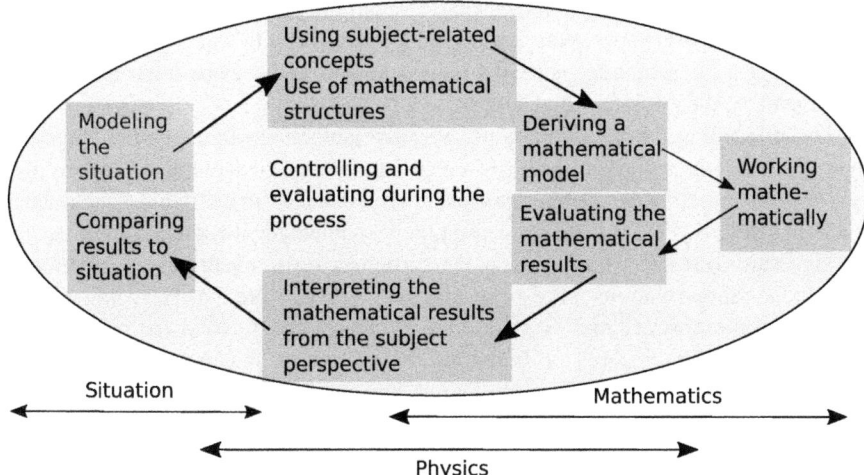

Fig. 1.5 Proposed unified model, suitable for scientific-mathematical modelling. It can also serve for displaying the procedures during problem solving (Trump and Borowski 2012; Müller et al. 2016, own graphical interpretation). In this model there are included the procedural, conceptual, structural and relational aspects already addressed in Fig. 1.2 and the proceptual aspect proposed by Brahmia (Fig. 1.4)

rather than about the application of mathematics in physics. It is the task of the teachers to make this interplay visible to the students, together with the relevance of understanding physics concepts. In order to achieve this goal successfully, the teacher needs relevant pedagogical content knowledge (see Sect. 1.5.3.1).

1.4 Mathematics as Language: Representations

In this book we see mathematical representations in physics as an essential aspect of mathematics as a language of physics. To put the typical representational forms into a bigger picture, a broader framework is unfolded in which representations in physics can be treated *(Kuske-Janßen and Geyer: "Mathematical Representations in Physics Lessons")*. From different resources, they derive a classification of representations most relevant for the use in physics education. Their approach serves as a joint basis for discussing the role of different representations, especially the switching between them. They also discuss which students' competences are required and which difficulties and advantages for students are connected with the use of multiple representations.

Here we describe mathematical representations in the framework set in the sections above. They make the interplay with signs and symbols visible and communicable. The power of symbolic representations for communication can easily be recognized in reading a physical paper in an unknown language: The mathematical symbols follow a fixed syntax satisfying general rules and conventions and represent relations between physical quantities. Partly independent of the verbal formulations, the semantics is telling how the symbols are to be interpreted in the framework of the physical knowledge.

For describing physical relations, various mathematical elements are used: numbers with units, diagrams such as line graphs, geometrical elements such as rays and arrows, functions and equations and more advanced mathematical techniques such as calculus. Perhaps the most prominent mathematical representations occur as algebraic forms (by which we mean formulas, resp., equations or functions). Graphical representations play a special role as they are partly iconic and at the same time abstract and symbolic. As graphs often are used for representing experimental data, formula or functions for describing the relations, they seem to play a bridging role between the iconic and the algebraic representations. All these representations, e.g. a line graph or an algebraic expression of a function, contain different information and highlight specific aspects of a physical relation (see, e.g. Larkin and Simon 1987). In this section we choose the most important representational forms, namely, formula and graphs for a deeper consideration. A central aspect is again the distinction of the technical and the structural role of mathematics in physics.

1.4.1 Graphical Representations

Among the graphical representations line graphs play a particularly prominent role. Such graphs occur, e.g. in evaluating experiments, where the tabulated data are visualized. Herewith they serve as an intermediate step between an experiment or phenomenon and the algebraic formulation of a physical law. During this transfer, some information (perhaps the precision of the measured value) is lost, but other information is gained, e.g. the underlying pattern of the data becomes visible by drawing a regression curve and thus describing the relation between the corresponding physical quantities. Such a visualized relation can be memorized more easily than an abstract formula. Therefore graphs may have an advantage over numbers or formula concerning learning. They could even contribute to reducing cognitive load and promote physical thinking.

However, some students show difficulties with graphs. For example, reading and making diagrams is not easy for school students. They have to systematically learn about sketching and reading graphs (Mevarech and Kramarsky 1997). This is underlined by the extensive research on graphs and their interpretation, mostly with college students in the context of kinematics (see, e.g. McDermott et al. (1987) and many others). Even if there is less research with school students, some fundamental papers give insight into the possible capabilities of young children (Disessa and Sherin 2000; Aberg-Bengtsson and Ottosson 2006; Leinhardt et al. 1990 or Wavering 1985 among many others). From these findings, it can be deduced that the competences of pupils have to be developed systematically. In this context it seems interesting that there is no recent model of the competences needed for handling graphs from the perspective of physics. A competence model has been developed for biology describing mostly technical aspects such as choosing the axes of the diagram, insertion of values or reading them off. The included structural aspects mostly refer to identification and interpretation of the graphs (Lachmayer et al. 2007). In addition there are extended studies from mathematics education (see, e.g. Kramarski 2004; Friel et al. 2001; Roth and McGinn 1997; Nitsch et al. 2014 and references therein). The question arises if the results from mathematics education can be transferred to physics. It might be conjectured that physics creates additional difficulties beyond the purely mathematical problems. Recently Planinic et al. (2013) have extensively studied the role of context if graphs are being used: graphs in mathematics, in physics and in other contexts. This is expanded on in the contribution of *Planinic et al. ("Student understanding of graphs in physics and mathematics")*. Their results hint that the technical aspects alone are not sufficient. They seem to imply that the interpretation and working with graphs need structural insights and that the transfer from the mathematics to physical concepts or vice versa requires corresponding experience and abilities.

The above-mentioned results provide arguments that graphing can take several roles in communicating and learning science and that students interpret graphs depending on the framing. Hence the isolated view on a fixed expected cognition might stress difficulties, whereas handling graphs is more a "practice" (comparable with language learning) being easier and more flexible the more often it is used (Roth and McGinn 1997). This stance is fruitful for the use of digital media where different graphical representations are easily be realized. With digital media, the physical process can be directly combined with graphical representations in real-time graphs. So students can use this direct connection for their understanding and conceptual development. This aspect is described in a study by *Alberto Stefanel ("Graph in physics education: from representation to conceptual understanding")*

1.4.2 Algebraic Representations

In the perception of the public as well as of school students, algebraic representations, named as formula (also equations or functions), serve as a symbol for physics: many reports on science contain formula, and pictures of science or scientists often show a blackboard full of formula (if they do not show an experiment). Correspondingly "formula" is the first association of students if they think about physics (Krey 2012).

In algebraic representations, the physical information is condensed with help of signs, following fixed conventions. These conventions allow for simplified communications among those who know the signs, the meaning of the signs and the meaning of the represented physical quantities. However, the use of algebraic representations can be an insurmountable obstacle for those who are not sufficiently acquainted with them (for the historical perspective, see Gingras 2001).

In physics the use of the terms formula, equation or function is generally not well defined. The transition from one to the other strongly depends on the situation or the problem at hand (for a deeper analysis, see *Heck and van Buuren "Students' understanding of algebraic concepts" in this book*). If a formula is being interpreted in a more general sense as an equation or as the definition of a function, then a full range of new possibilities arises, e.g. use of derivative or integral. The results of corresponding physical-mathematical reasoning can imply new insights into physical laws. As Sherin (2001) pointed out, also the understanding of the physical meaning of the mathematical operations themselves is relevant. Such theoretical implications make the algebraic representation a powerful tool of physics and show how intimately mathematical structures and techniques are intertwined with physics.

1.4.2.1   Formula or Equation

What is a formula? We could say that a formula is the amalgam of an algebraic term with a physical meaning: the letters carry a meaning or even create a new quantity, e.g. the density that is built from mass and volume. Formula serve for calculating concrete values and making precise quantitative predictions to be evaluated and validated. So the formula invites to technical application by just calculation, but the given characterization includes invariably the structural role, i.e. the transfer between the algebraic symbols and their physical meaning. In addition Hewitt (1983) argued that "formula are guides to thinking" and made strong the need of teaching concepts first but also to work with and interpret each formula.

1.4.2.2   Function

The concept of function is seen differently in mathematics and physics. In mathematics the term "function" is defined abstractly as a relation between two sets and is a central concept in mathematics. In physics, this definition of a function as a relation is not the prevalent feature but the functional thinking, recognizing and exploiting the covariance between physical quantities. Herewith line graphs play an important role by providing an overview of the function as a whole (Leinhardt et al. 1990). The intricacies of these notions are discussed in detail by *van Buuren and Heck ("Learning to use formulas and variables for constructing computer models in lower secondary physics education")*.

If a formula is thought of as a function, it becomes important to distinguish the dependent and independent variables and to identify constant parameters. Once the functional dependencies are identified, the mathematical techniques, e.g. differentiation or integration, can be used for analysing and exploring the scope of the function and derive additional insights into physical phenomena.

1.4.3 Language as Verbal Representation in Mathematization

Often verbal formulations are not treated as a separate representation in their own right. However, with language we describe physical situations and the relation between physical quantities in a qualitative way. Comparable to the transition from everyday language to special language where the terms and concepts have to be carefully built, the meaning of mathematical representations has to be imparted by means of language. It can be observed that many teachers tend to use very special figures of speech to describe, e.g. a proportional relation. In this process everyday understanding and mathematical meaning might interfere (Pospiech et al. 2012).

Up to now there is no model how meaning can be systematically assigned to a formula with help of (everyday) language and how a situation or verbal description can be systematically translated into a formula. Janßen and Pospiech (2015) developed such a model that distinguishes several steps bridging the gap between the formula itself and a verbal description of a situation in which the formula might play a role. It is currently being validated by analysing classroom discourse and is described in this book in chapter *Kuske-Janßen and Geyer ("Mathematical Representations in Physics Lessons")*.

1.4.4 Iconic Representations in Mathematization

Digital media may substantially enhance the interrelation of graphical and algebraic representation hence enabling high school students to enter into modelling and discussing the meaning of formula. This is shown with the example of ALGODOO by *Euler and Gregorcic: "Algodoo: Creatively Linking Mathematics and Physics"*.

A far broader understanding about graphical representation is described by *Lehavi et al. ("Taking the Phys-Math interplay from research into practice")*. Their invention of "Visual Mathematics" was developed and successfully evaluated in the context of Newtonian mechanics. Its refined and systematically introduced consistent graphical representation showed that students can learn a conceptual understanding in lower secondary school and even draw on it during a more formal and algebraic representation in high school. This approach has been extended to the topic of energy, which still has to be tested and evaluated.

1.4.5 Interrelating Representations

As every representation has its specific properties and information content, the whole picture of physics (or a physical situation) emerges through the use of several representations, complementing each other. Special attention is needed for the change between representations: each representation carries its own information, be it in an explicit form, e.g. numbers in a table, or be it implicitly encoded, e.g. the shape of a line graph. Learners have to relate the different representations to each other. Generally, Ainsworth (2008) discusses the benefits of multiple representations, the changing between them and also the problems these might have for inexperienced learners. The main problem seems to be that the use of several representations requires the learners to know all these and to relate them, causing cognitive load.

In the context of mathematics education, especially handling functions, models have been developed to describe problems of students in doing the required translation, e.g. by Nitsch et al. (2014). They applied their model for describing the competences of 15- to 16-year-old students in switching between different representations in mathematics lessons. They discovered that there are more simple transfers and others are more complicated but that each translation required specific competences. They could not identify a most basic translation between any of those representation types. Adu-Gyamfi et al. (2012) discuss a model for the translation between different mathematical representations. In a study with college students, they could identify three types of errors: interpretation, implementation and preservations error. They found that most implementation errors in translation occurred if an equation was involved. For the transfer of these models from mathematics education to physics education, see the contribution of *Kuske-Janßen and Geyer "Mathematical Representations in Physics Lessons"*.

1.5   Empirical Research

The topic "mathematics in physics education" has recently been rediscovered as a topic of physics education research. Most existing research focusses on problem solving on college or university level, but there is still relatively few research on lower secondary or high school level; however, this is changing by now.

On all stages of the educational career, there are many statements saying that students do not know mathematics sufficiently well in order to be successful in doing physics. However, these statements are too short-sighted: "It is undoubtedly the case that mathematics does make physics difficult for students. But any diagnosis that stops there is flawed" (Monk 1994). So many research groups tried to fill in this missing knowledge. On the basis of the theoretical analysis of the interplay between mathematics and physics in Sects. 1.2 and 1.3, the generally observed difficulties of students in coping with this interplay lead to the question after the deeper nature of their difficulties and the corresponding causes. Hence the goals of research may cover different aspects thereof:

• Research describes in detail the difficulties, strategies and abilities of students during problem solving and modelling or their views on certain aspects of this interplay.

- Research analyses the findings on the basis of models as presented in Sect. 1.3. This may result in the uncovering of patterns, e.g. the "epistemic games" by Tuminaro and Redish (2007).
- Research puts the observed or expected behaviour into a wider (psychologically or science theoretically based) framework in order to find deeper lying causes for certain types of difficulties, strategies and abilities.
- Research provides the basis for interventions on the basis of evidence, plans the interventions and evaluates them.

Most research approaches are aware of the distinction between the mathematics part, in itself divided into a more instrumental and a more relational understanding (Skemp 1976), and the physics part, also with more technical and more structural aspects (see Sect. 1.3.1 or Fig. 1.2). However, in order to find underlying mechanisms, also frameworks from psychology are needed and used.

1.5.1 Frameworks of Research Approaches

It is one thing to find, describe and analyse the difficulties, strategies and abilities of students. But for understanding the why of these observations and for developing suitable learning environments, it is necessary to know more about the underlying psychological reasons and patterns. An attempt concerning physics learning in general has been made, e.g. by di Sessa by applying "knowledge analysis" (diSessa 1993). Sherin embedded his approach in this framework, by stating that symbolic forms are kind of intermediate description being sufficiently small to capture the thinking of students and sufficiently big to give meaningful results (Sherin 2001).

Also in the aftermath of the identification of "epistemic games"[4], there were efforts to understand how these strategies emerge (Tuminaro and Redish 2007; Bing and Redish 2009). The influence of "epistemological framing" on the process of problem solving for upper level undergraduate students was found together with the observation that students might know the mathematics or physics they need for solving the problem at hand but get "on the wrong track". Then they do not have the variability of changing their strategy; they get stuck in their current "frame" which could be calculation, physical mapping, invoking authority or mathematical consistency. Hence it seems reasonable to infer that students' reasoning depends on the "framing" (What is expected from me?), leading them to use "epistemic resources" (Which elements of knowledge can I bring into play?) and, if challenged, to give "warrants" ("How do I support my choices?"). This framework is no fixed theory but well suited to model the emerging patterns in problem solving. With a focus on the potential abilities of students, it is used by *Eichenlaub and*

---

[4]The "epistemic games" range from "plug and chug" to "mapping mathematics to physics" or vice versa, also including incomplete strategies or graphical strategies. Tuminaro does not say that there only those six strategies exist but that those have been observed in the sample.

Redish ("Blending physical knowledge with mathematical form in physics problem solving") to take a positive approach and to describe competences of students.

Another framework is chosen by Greca and Ataide in their contribution *"The influence of epistemic views about the relationship between physics and mathematics in understanding Physics".* They use the so-called schemes framework derived from the idea of "conceptual fields" developed orginally by Vergnaud. His framework starts from procedural knowledge relying on experiences with many diverse situations from which "schemes" are developing. "Schemes" describe how the behaviour of a person is universally organized in certain situations. An example given by Ataide and Greca concerns the treatment of a motion with friction: the necessary forces, the definition and rules for velocity or acceleration and so on together build a scheme (which has not necessarily to be physically correct but guides the thinking and the action of the person). The learning (and understanding) happens by mental representations and increasingly explicit formulation of these schemes. For instance, students will have to recognize if one scheme can be applied to different situations or if they have to use different schemes. Some schemes can be very stable and hence difficult to change and therefore cause difficulties, e.g. in problem solving.

In this sense there are connections between the frameworks of "epistemic resources" and the "schemes framework". Both allude to the mental activation process of the students which might strongly depend on the context. Applied to the interplay of mathematics and physics, this activation process often is not realized in a favourable way as, e.g. an irrelevant scheme is activated or the students focus on technical aspects. In order to understand better their strategies in problem solving and to analyse in more detail their thinking also the "actor-oriented perspective", a research perspective from mathematics education (Lobato 2012), could enrich the research process. This perspective takes into account the previous experiences and interpretations made by students during the teaching and learning, mostly in classroom.

All these frameworks have in common that they consider cognitive as well as individual and situational or contextual aspects as relevant for learning and for transfer into new contents.

1.5.2   *Research Concerning Students*

Here we will describe research studies on the cognitive abilities, strategies and difficulties of students at high school or junior high school/middle school but also on their epistemic views and attitudes towards the mathematization in physics lessons.

Only relatively few studies ask students directly about their views and attitudes, while most studies try to find patterns among the problem solving strategies and analyse critical points in order to interpret these in the light of the underlying framework. This can be justified as research shows that cognition and affective or epistemic views are somehow interrelated.

1.5.2.1 Students' Views

From studies among college and university students, it is known that many students have an instrumental view of the role of mathematics in physics, more focusing on the technical role than the structural role. One of the rare studies with high school students has been performed by Krey and Mikelskis (2010) and Krey (2012). He analysed in detail the views of students at school and compared them to the views of students at university (physics majors and physics minors). The view of students on physics is characterized by a strong role of mathematics, mostly including formula as the dominant element. It could be shown that the attitude of students towards mathematics is not as negative as it is often assumed. Especially they recognize different roles of mathematics: as communicative tool and as computational tool. The views seem to develop during their educational career towards a more and more adequate view, also without explicitly teaching about the role of mathematics.

A study at junior high school with students of grade 8 (14 years old) hints that some of the students can distinguish the more technical and the structural role of formula already at this early stage. However, their interpretation of a given formula often remains on the technical level. Generally the students see the graphical representations, mostly line graphs, as a bridge between the physical situation and the algebraic representation (Pospiech 2013; Pospiech and Oese 2013). In interviews students stated that they appreciate graphs because they visualize relations between physical quantities and that they can remember them better than by a formula (Pospiech and Oese 2014). Especially weaker students prefer graphs and like them more than strong students. As we focus on students of (junior) high school, we cannot expect a refined viewpoint on the interplay math-phys because those students have relatively few experiences, but on the whole, their views are surprisingly balanced.

1.5.2.2 Students' Problem Solving

One step towards deeper analysis of problem solving is indicated in Sect. 1.3.3, resp., Sect. 1.3.3.2, where we placed the modelling and problem solving procedure into the context of the technical and the structural role of mathematics. There are several studies from mathematics education as well as from physics education underlining the need of taking into account conceptual understanding on both sides (see also Fig. 1.2). Once one is aware that the difficulties of students can lie on the technical as well as the structural side, both aspects and their mutual interplay have to be analysed in detail where the models described in Sect. 1.3.3 could serve as a basis.

Studies with students identify strategies as well as corresponding problems in different age groups, starting from secondary school (e.g. Trump and Borowski 2014; Uhden and Pospiech 2009; Uhden 2016) over college level (Bing and Redish 2007; Tuminaro and Redish 2007; Sherin 2001; Kuo et al. 2013, and many others) up to university (Britton et al. 2005; Kohl and Finkelstein 2008; Brahmia 2014 and many others). The general feature was that most students, not only at secondary school, focus on technical aspects. Among these many studies here I only mention a few because our focus is on the comparatively few studies with school students. A look into the details reveals interesting insights (Brahmia 2014; Byun and Lee 2014; Ivanjek et al. 2016). They show clearly that not so much deficiencies in the technical role of mathematics but missing awareness of its structural role is the cause for the apparent difficulties. Therefore, the conceptual understanding of physics and the structural function of mathematics have to be central in teaching and learning including the ability to explain physical phenomena and to apply or develop adequate physical models to a given situation. This is confirmed by a study of how students use equations in order to solve problems (Kuo et al. 2013). The focus of the study was on the distinction of a more algorithmic use or a blending of mathematical and physical considerations. For this purpose students from third semester were interviewed with concrete prompts for explanations of equations or solving a problem in an everyday context. As a result, it was found that blending techniques also can lead to success giving a hint for appropriate instruction of students. Similar strategies or difficulties have been observed (Meli et al. 2016; Niss 2017).

The transfer between mathematics and science, especially physics, cannot be described as single directed (Roorda et al. 2015; Marrongelle 2004). They analyse empirically the different possibilities students develop in making meaning of mathematical elements. Students might use elements from mathematics as well as from physics, depending on their individual strengths and learning progress. Physical events can help students to interpret mathematical concepts, e.g. in graphs (Marrongelle 2004). It was also observed that the students apply some techniques not directly after learning but only after some time of getting used to them and exercising (Roorda et al. 2015). In no case it is easy to transfer mathematical knowledge into the physics domain (see, e.g. Planinic et al. 2013; Ivanjek et al. 2016; Redish 2017). In addition the requirements during assessment might influence teaching and the strategies of students (Johansson 2016). Other studies concern the question if the difficulties of students are caused purely by lacking mathematical abilities or by applying mathematics in a context, e.g. in the interpreting and handling of graphs (Planinic et al. 2013; Ivanjek et al. 2016).

There are very few studies in lower secondary school. However, some examples are interesting. In a study with students in secondary school of age 15–16 years, special tasks were designed in order to evoke strategies not learned explicitly at school and to avoid standard routines (Uhden 2016; Uhden and Pospiech 2009). The students solved the tasks in pairs in order to have to discuss about possible solutions and the solution path. The analysis revealed different areas of difficulties. Most were related to the structural transfer between the physical situation and the mathematical formulation. Some difficulties clearly were due to lacking conceptual basis: missing understanding of physical concepts caused problems, especially if met by weaknesses on the mathematical side. On the other hand, mathematical weakness sometimes seemed to inhibit the correct implementation of physical conceptual considerations. In some instances, it was nearly impossible to tell if the mathematical or the physical difficulties were the reason for the inability to solve the problem at hand. There were also hints that students got better results if they tried to align the physics and the mathematics aspect from a more structural perspective.

1.5.2.3 Relation of Views and Problem Solving

There is evidence that the views of students on the role of mathematics in physics influences their strategies of problem solving (Al-Omari and Miqdadi 2014; Malone 2008; Mason and Singh 2016). Connected to the ways in which students (and teachers, Siswono et al. 2017; Turşucu et al. 2017) think about the transfer between mathematics and physics is their approach to problem solving. Expertlike beliefs are correlated with problem solving strategies of university students (Bodin and Winberg 2012; Malone 2008). Many studies show that the interrelation of views and strategies used is quite intricate and hint that several types of students can be characterized (Ataide and Greca 2013). Also different types of problems require different approaches to problem solving (Jensen et al. 2017). The deeper relation between epistemic views and problem solving approaches is being discussed in *Ataide/Greca ("The influence of epistemic views about the relationship between physics and mathematics in understanding Physics").*

As a summary, it seems crucial that the students acknowledge that the application of mathematics to physics problems has to be more deeply rooted than just to apply formulae to some physical problems by receipt. For this the students need a strong conceptual understanding of physics as well as of the meaning related to mathematical operations (as, e.g. described in Sherin (2001) and analysed by Brahmia (2014), see also Fig. 1.4). These findings still need confirmation with respect to completeness of findings. So additional studies will be necessary.

1.5.2.4 Use of Representations

Another broadly studied aspect refers to the role and use of representations, be it that the problem is posed with a specific representation or the students use different representations for solving the problem (e.g. De Cock 2012; Kohl and Finkelstein 2006, 2008; Meltzer 2005; Ibrahim and Rebello 2012). On the whole it is very difficult to draw general conclusions. However, it seems that the problem solving strategies strongly rely on details in representation and context of the problem supporting the framing aspect discussed in Sect. 1.5.1. So this might be taken as a hint that the framing of the given problem solving situation indeed strongly influences the invoked strategies. Students seem to use multiple representations quite well but depending on their experience often more in an exploratory than in a goal-oriented way. Especially they show difficulties in connecting visual representation with algebraic representations. However, some early studies showed that even young students are able to invent graphs in suitably designed learning environments (Beichner 1994; Hammer et al. 1991). Often it is possible to foster the competences of reading and making graphs with help of the interactive opportunities inherent to the use of computers or apps (Hale 2000; van den Berg et al. 2010). In the light of the possibilities of modern digital media, the competences of secondary and high school students are studied in detail in the contribution of *Stefanel ("Graph in physics education: from representation to conceptual understanding")*. He focusses on the activities and the reasoning of students in drawing and making sense of graphs in two different examples.

1.5.3 Research Concerning Teachers and Teacher Students

As was described in the preceding sections, the teacher plays an important role in imparting an appropriate view of physics and its method. Up to now, only little is known about the knowledge and views of teachers in the important area of shaping the interplay of mathematics and physics. The definition of teachers PCK as done in Sect. 1.5.3.1 is the first step towards a systematic evaluation. However, one has to be careful to infer from a good PCK that the teachers have corresponding success in their teaching as, e.g. is measured in students' learning success (see, e.g. Cauet et al. 2015; Kirschner et al. 2016). Nevertheless the construct of PCK is fruitful for domain-specific characterization of teachers' views and their teaching strategies.

1.5.3.1 Pedagogical Content Knowledge of the Interplay

Generally, for a high quality of teaching, the knowledge and competences of teachers are very relevant, especially their content knowledge and their pedagogical content knowledge (PCK) (for general aspects (Shulman 1987), concerning mathematics (Krauss et al. 2008)). In this section we want to focus on the PCK as this part has proven to be directly related to the learning success of students (Riese 2010; Baumert and Kunter 2013).

In the aftermath of the work of Shulman, different models of pedagogical content knowledge have been developed (for an overview, see Gramzow et al. 2013). A modified model in the context of chemistry education on the basis of case studies stressing the importance of the teachers' experiences in combination with their reflection was given by Park and Oliver (2008). Out of these, the model proposed by Magnusson et al. (1999, Fig. 6a), also adapted by Etkina (2010), seems to be a good choice in that it focusses on joint aspects of most models. The model from Fig. 1.6 is explained in detail in the contribution *"Role of Teachers as Facilitators of the Interplay Physics and Mathematics" by Pospiech et al.*. The detailed models described in Sect. 1.3 serve for deducing categories describing teachers' views and knowledge, e.g. with respect to technical/structural role or teaching strategies for facilitating the mathematization process for students in more detail.

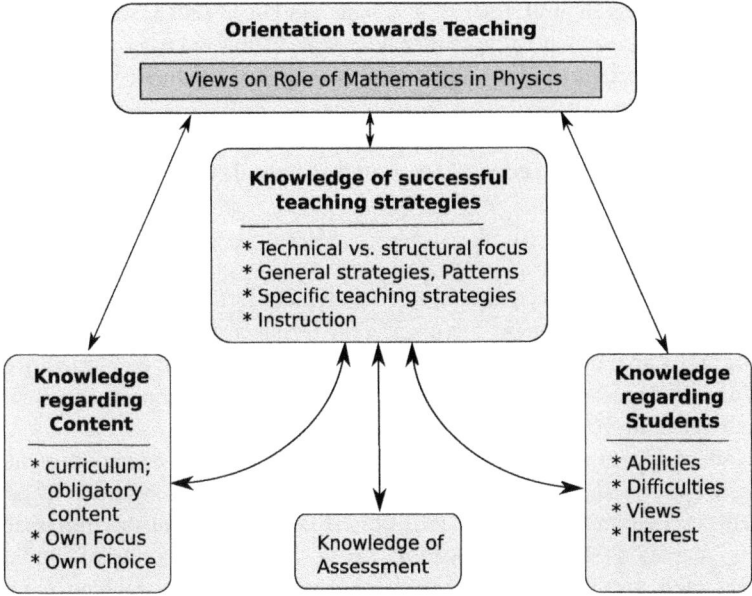

Fig. 1.6 The PCK model for the interplay mathematics physics. (Adapted from Lehavi et al. 2017)

1.5.3.2 Epistemic Views of Teachers and Teaching Problem Solving

Underlying the teachers' views on the interplay and their choice of teaching patterns are epistemic views. In studying this relation, it was found that traditionally oriented teachers tend to view the role of mathematics as instrumental, whereas more conceptually oriented teachers viewed mathematics as the language of physics and as suitable to derive models or new insights into physics (Mulhall and Gunstone 2007).

As an example the mathematical modelling of prospective teachers in a kinematic unit was studied (Carrejo and Marshall 2007). This study makes clear that the conscious "self-made" connection between the physical phenomenon and the mathematical model offers great learning opportunities for the teacher students with respect to their view on the interplay. In another case study, it was shown that the conceptions and actual teaching practices of teachers might diverge (Freitas et al. 2004). In a quantitative study, it was found that prospective teachers prefer a constructivist stance on mathematical-physical modelling. However, it could not be tested if this is reflected in their actual teaching strategies (Fazio and Spagnolo 2008).

According to a study with 34 teacher students, "a strong relationship between students' problem solving strategy, and their epistemological perception on the role mathematics plays in physics, learning and understanding physics, and solving problems in physics" exists (Al-Omari and Miqdadi 2014). Similar studies were conducted by Başkan et al. (2010) or Ataide and Greca (2013). However, it is not clear how teacher students reflect their views on the interplay of mathematics and physics in a concrete example.

1.5.3.3   Awareness of Students' Ideas and Proposals for Teaching Strategies

In order to shape the teaching-learning process, teachers have to be aware of the students' ideas. Only if they know the typical difficulties, they can think of appropriate teaching strategies. In a study the teachers stated that they achieve the best results by abandoning the "number crunching" and instead emphasizing careful explaining and reasoning as well of the mathematical operations as of the physical processes (Khalili 2016).

In order to learn about the teachers' views and attitudes towards the interplay of mathematics and physics, an interview study has been conducted in Israel (8 teachers) and Germany (14 teachers). This study showed that the introduced and slightly adapted PCK model from Fig. 1.6 fits to the statements of the teachers and that their views often are very much shaped by their teaching experience (see contribution *"Role of Teachers as Facilitators of the Interplay Physics and Mathematics" by Pospiech et al.*).

Besides foundational work on strategies and problems of students, several researchers at the same time also developed strategies for improving instruction towards a more sense-making use of mathematics. The overarching goal is to overcome the discrepancy between the apparent possibilities of learning and understanding physics by interpreting and analysing formula and the often predominant rote plug'n chug in solving physics problems.

Interrelated Treatment of Mathematics and Physics

In the view of the complex interplay, meanwhile many attempts have been made to let mathematics and physics understanding support each other, e.g. by integrating mathematics and physics on school level (see, e.g. Davison et al. 1995) and on the university level by Dunn and Barbanel (2000) or by invoking technology (Niess 2005; Vogel et al. 2007; Boujaoude and Jurdak 2010) using MBL teaching for mathematization at the example of kinematics.

Basson (2002) remarks that some textbooks (used in South Africa) use quite a lot of concepts not structured very clearly for the students. In order to establish a unifying view, he suggests to relate mathematics (concept of function) explicitly to physical concepts in order that those both school subjects benefit from each other and the other way around (Michelsen 2006). This would take into account that the transfer is not one way but bidirectional. The goal is that the interplay of mathematics and physics is used for better learning results.

In addition several proposals have been made to teach mathematics and physics in an interdisciplinary way (Michelsen 2015). In this book an example from teacher education is described by *Mäntylä/ Poranen ("Combining physics and mathematics learning: Discovering the latitude in pre-service subject teacher education").*

Meaning Making of Formula

Bagno et al. (2008) focus on the process of making meaning to a given formula, stressing the relevance of discussing the corresponding concepts. One example is the uniformly accelerated motion. They propose a step-by-step interpretation involving single work, discussion with a partner and then a whole class discussion. This approach led to fruitful exchange among teachers during professionalization and also among students in class.

Invention Tasks

One could also use unusual tasks as invention tasks where the students should develop their own "physical" quantity (Brahmia et al. 2016). Brahmia showed that the students develop a better understanding of the mathematization procedure even allowing to proceed quicker in the aftermath. A similar approach was tried by Uhden (2016) with students in lower grades at school developing a quantification to the concepts of density.

Also Uhden and Pospiech (2013) propose the variation of standard textbook tasks in order to let students activate their mathematical-physical reasoning. In the same direction, Dufresne et al. (1997) propose flexible problem solving with representations.

1.6 Outlook

From many studies, we can infer that the use of mathematics should be an integral part of teaching physics, as well at school as in teacher preparation. Generally, it might be useful to introduce students right from the beginning in the interplay of physical phenomena of objects, a physical model and a mathematical model. There is evidence such that some rules can be formulated:

- students need support in first thinking about the physics concepts before they go to a mathematical formulation
- teachers need encouragement to give a variety of different tasks which require flexible use of mathematics as well as physics concepts.

In view of the results, the next step would be to develop materials, tasks and problems thus implementing the favourable strategies. Hence the next research steps would be the evaluation of such teaching-learning sequences, perhaps in a design-based research approach.

References

Aberg-Bengtsson, L., & Ottosson, T. (2006). What lies behind graphicacy? Relating students' results on a test of graphically represented quantitative information to formal academic achievement. *Journal of Research in Science Teaching, 43*(1), 43–62.

Adu-Gyamfi, K., Stiff, L. V., & Bossé, M. J. (2012). Lost in translation: Examining translation errors associated with mathematical representations. *School Science and Mathematics, 112*(3), 159–170.

Ainsworth, S. (2008). The educational value of multiple-representations when learning complex scientific concepts. In *Visualization: Theory and practice in science education* (pp. 191–208). Springer.

Al-Omari, W., & Miqdadi, R. (2014). The epistemological perceptions of the relationship between physics and mathematics and its effect on problem-solving among pre-service teachers at Yarmouk University in Jordan. *International Education Studies, 7*(5), 39–48.

Ataide, A. R. P. d., & Greca, I. M. (2013). Epistemic views of the relationship between physics and mathematics: Its influence on the approach of undergraduate students to problem solving. *Science & Education, 22*, 1405–1421.

Aufschnaiter, S. v., Aufschnaiter, C. v., & Schoster, A. (2000). Zur Dynamik von Bedeutungsentwicklungen unterschiedlicher Schüler (innen) bei der Bearbeitung derselben Physikaufgaben. *Zeitschrift für Didaktik der Naturwissenschaften, 6*, 37–57.

Başkan, Z., Alev, N., & Karal, I. S. (2010). Physics and mathematics teachers' ideas about topics that could be related or integrated. *Procedia – Social and Behavioral Sciences, 2*(2), 1558–1562.

Bagno, E., Berger, H., & Eylon, B.-S. (2008). Meeting the challenge of students' understanding of formulae in high-school physics: A learning tool. *Physics Education, 43*(1), 75.

Basson, I. (2002). Physics and mathematics as interrelated fields of thought development using acceleration as an example. *International Journal of Mathematical Education in Science and Technology, 33*(5), 679–690.

Baumert, J., & Kunter, M. (2013). The COACTIV model of teachers' professional competence. In M. Kunter, J. Baumert, W. Blum, U. Klusmann, S. Krauss & M. Neubrand (Eds.), *Cognitive activation in the mathematics classroom and professional competence of teachers* (pp. 25–48). Boston: Springer US.

Beichner, R. J. (1994). Testing student interpretation of kinematics graphs. *American journal of Physics, 62*(8), 750–762.

Bing, T. J., & Redish, E. F. (2007). The cognitive blending of mathematics and physics knowledge. In *2006 Physics Education Research Conference* (Vol. 883, pp. 26–29).

Bing, T., & Redish, E. (2009). Analyzing problem solving using math in physics: Epistemological framing via warrants. *Physical Review Special Topics – Physics Education Research, 5*(2), 020108.

Blum, W., & Borromeo, F. R. (2009). Mathematical modelling: Can it be taught and learnt? *Journal of mathematical modelling and application, 1*(1), 45–58.

Bodin, M., & Winberg, M. (2012). Role of beliefs and emotions in numerical problem solving in university physics education. *Physical Review Special Topics – Physics Education Research, 8*(1), 010108.

Borromeo, F. R. (2006). Theoretical and empirical differentiations of phases in the modelling process. *ZDM, 38*(2), 86–95.

Boujaoude, S. B., & Jurdak, M. E. (2010). Integrating physics and math through microcomputer-based laboratories (MBL): Effects on discourse type, quality, and mathematization. (May 2008), 1019–1047.

Brahmia, S. M. (2014). *Mathematization in introductory physics*. PhD Thesis, Rutgers University-Graduate School, New Brunswick.

Brahmia, S., Boudreaux, A., & Kanim, S. E. (2016). Developing mathematization with physics invention tasks. arXiv preprint arXiv:1602.02033.

Britton, S., New, P. B., Sharma, M. D., & Yardley, D. (2005). A case study of the transfer of mathematics skills by university students. *International Journal of Mathematical Education in Science and Technology, 36*(1), 1–13.

Brush, S. G. (2015). Mathematics as an instigator of scientific revolutions. *Science & Education, 24*(5–6), 495–513.

Byun, T., & Lee, G. (2014). Why students still can't solve physics problems after solving over 2000 problems. *American Journal of Physics, 82*(9), 906–913.

Carrejo, D. J., & Marshall, J. (2007). What is mathematical modelling? Exploring prospective teachers' use of experiments to connect mathematics to the study of motion. *Mathematics Education Research Journal, 19*(1), 45–76.

Cauet, E., Liepertz, S., Borowski, A., & Fischer, H. E. (2015). Does it matter what we measure?: Domain-specific professional knowledge of physics teachers. *Schweizerische Zeitschrift für Bildungswissenschaften, 37*(3), 462–479.

Czocher, J. A. (2018). How does validating activity contribute to the modeling process?. *Educational Studies in Mathematics, 99*(2), 137–159.

Davison, D. M., Miller, K. W., & Metheny, D. L. (1995). What does integration of science and mathematics really mean? *School Science and Mathematics, 95*(5), 226–230.

de Berg, K. C. (1992). Mathematics in science: The role of the history of science in communicating the significance of mathematical formalism in science. *Science & Education, 1*(1), 77–87.

De Berg, K. C. (1995). Revisiting the pressure-volume law in history-what can it teach us about the emergence of mathematical relationships in science? *Science & Education, 4*(1), 47–64.

De Cock, M. (2012). Representation use and strategy choice in physics problem solving. *Physical Review Special Topics – Physics Education Research, 8*(2).

Dirac, P. A. (1939). The relation between mathematics and physics. In *Proceedings of the Royal Society,* Edinburgh (Vol. 59, p. 122).

diSessa, A. A. (1993). Toward an epistemology of physics. *Cognition and Instruction, 10*(2–3), 105–225.

Disessa, A. A., & Sherin, B. L. (2000). Meta-representation: An introduction. *The Journal of Mathematical Behavior, 19*(4), 385–398.

Dufresne, R. J., Gerace, W. J., & Leonard, W. J. (1997). Solving physics problems with multiple representations. *Physics Teacher, 35*, 270–275.

Dunn, J. W., & Barbanel, J. (2000). One model for an integrated math/physics course focusing on electricity and magnetism and related calculus topics. *American Journal of Physics, 68*(8), 749–757.

Einstein, A. (1921). Geometrie und Erfahrung.

Etkina, E. (2010). Pedagogical content knowledge and preparation of high school physics teachers. *Physical Review Special Topics – Physics Education Research, 6*(2).

Fazio, C., & Spagnolo, F. (2008). Conceptions on modelling processes in Italian high-school prospective mathematics and physics teachers. *South African Journal of Education, 28*(4), 469–487.

Freitas, I. M., Jiménez, R., & Mellado, V. (2004). Solving physics problems: The conceptions and practice of an experienced teacher and an inexperienced teacher. *Research in Science Education, 34*(1), 113–133.

Friel, S. N., Curcio, F. R., & Bright, G. W. (2001). Making sense of graphs: Critical factors influencing comprehension and instructional implications. *Journal for Research in Mathematics Education, 32*(2), 124–158.

Gingras, Y. (2001). What did mathematics do to physics? *History of science, 39*, 383–416.

Gramzow, Y., Riese, J., & Reinhold, P. (2013). Modellierung fachdidaktischen Wissens angehender Physiklehrkräfte- Modelling Prospective Teachers' knowledge of Physics Education. *ZfDN (Zeitschrift für Didaktik der Naturwissenschaften), 19*, 7–30.

Greca, I. M., & Moreira, M. A. (2002). Mental, physical, and mathematical models in the teaching and learning of physics. *Science Education, 86*(1), 106–121.

Hale, P. (2000). Kinematics and graphs: Students' difficulties and CBLs. *Mathematics Teacher, 93*(5), 414–417.

Hammer, D., Sherin, B., Kolpakowski, T., and others (1991). Inventing graphing: Meta-representational expertise in children. *Journal of Mathematical Behavior, 10*(2), 117–160.

Hansson, L., Hansson, O., Juter, K., & Redfors, A. (2015). Reality–theoretical models–mathematics: A ternary perspective on physics lessons in upper-secondary school. *Science & Education, 24*(5–6), 615–644.

Hestenes, D. (1986). A unified language for mathematics and physics. In J. S. R. Chisholm, A. K. Common (Eds.), *Clifford algebras and their applications in mathematical physics* (pp. 1–23). Dordrecht: Springer.

Hewitt, P. G. (1983). Millikan lecture 1982: The missing essential—A conceptual understanding of physics. *American Journal of Physics, 51*(4), 305–311.

Hewitt, P. G. (2011). Equations as Guides to thinking and problem solving. *The Physics Teacher, 49*(5), 264.

Hu, D., & Rebello, N. S. (2013). Using conceptual blending to describe how students use mathematical integrals in physics. *Physical Review Special Topics – Physics Education Research, 9*(2).

Hund, F. (1975). *Geschichte der Quantentheorie*. BI Wissenschaftsverlag.

Ibrahim, B., & Rebello, N. S. (2012). Representational task formats and problem solving strategies in kinematics and work. *Physical Review Special Topics-Physics Education Research, 8*(1), 010126.

Ivanjek, L., Susac, A., Planinic, M., Andrasevic, A., & Milin-Sipus, Z. (2016). Student reasoning about graphs in different contexts. *Physical Review Physics Education Research, 12*(1).

Janßen, W., & Pospiech, G. (2015). Versprachlichung von Formeln und physikalisches Formelverständnis. In S. Bernholt (Ed.), *Heterogenität und Diversität – Vielfalt der Voraussetzungen im naturwissenschaftlichen Unterricht*. Volume 35 of Gesellschaft für Didaktik der Chemie und Physik#Bd.#35; Jahrestagung der GDCP. 2014 (pp. 636–638). Kiel: IPN.

Jensen, J. H., Niss, M., & Jankvist, U. T. (2017). Problem solving in the borderland between mathematics and physics. *International Journal of Mathematical Education in Science and Technology, 48*(1), 1–15.

Johansson, H. (2016). Mathematical reasoning requirements in Swedish national physics tests. *International Journal of Science and Mathematics Education, 14*(6), 1133–1152.

Khalili, P. (2016). *Mathematical needs in the physics classroom*. PhD thesis, Education: Faculty of Education.

Kirschner, S., Borowski, A., Fischer, H. E., Gess-Newsome, J., & von Aufschnaiter, C. (2016). Developing and evaluating a paper-and-pencil test to assess components of physics teachers' pedagogical content knowledge. *International Journal of Science Education, 38*(8), 1343–1372.

Kjeldsen, T. H., & Lützen, J. (2015). Interactions between mathematics and physics: The history of the concept of function—Teaching with and about nature of mathematics. *Science & Education, 24*(5–6), 543–559.

Kneubil, F. B., & Robilotta, M. R. (2015). Physics teaching: Mathematics as an epistemological tool. *Science & Education, 24*(5–6), 645–660.

Kohl, P., & Finkelstein, N. (2006). Effects of representation on students solving physics problems: A fine-grained characterization. *Physical Review Special Topics – Physics Education Research, 2*(1).

Kohl, P., & Finkelstein, N. (2008). Patterns of multiple representation use by experts and novices during physics problem solving. *Physical Review Special Topics – Physics Education Research, 4*(1).

Kramarski, B. (2004). Making sense of graphs: Does metacognitive instruction make a difference on students' mathematical conceptions and alternative conceptions? *Learning and Instruction, 14*(6), 593–619.

Krauss, S., Baumert, J., & Blum, W. (2008). Secondary mathematics teachers' pedagogical content knowledge and content knowledge: Validation of the COACTIV constructs. *ZDM, 40*(5), 873–892.

Krey, O. (2012). *Zur Rolle der Mathematik in der Physik: wissenschaftstheoretische Aspekte und Vorstellungen Physiklernender.* Berlin: Logos.

Krey, O., & Mikelskis, H. F. (2010). The role of mathematics – The students point of view. In M. Tasar & G. Cakmakci (Eds.), *Contemporary science education research.* Volume book 4: Learning and assessment. Istanbul: Academia.

Kuo, E., Hull, M. M., Gupta, A., & Elby, A. (2013). How students blend conceptual and formal mathematical reasoning in solving physics problems. *Science Education, 97*(1), 32–57.

Lachmayer, S., Nerdel, C., & Prechtl, H. (2007). Modellierung kognitiver Fähigkeiten beim Umgang mit Diagrammen im naturwissenschaftlichen Unterricht (Modelling of cognitive abilities regarding the handling of graphs in science education). *Zeitschrift für Didaktik der Naturwissenschaften, 13*, 161–180.

Larkin, J. H., & Simon, H. A. (1987). Why a diagram is (sometimes) worth ten thousand words. *Cognitive Science, 11*(1), 65–100.

Lehavi, Y., Bagno, E., Eylon, B.-S., Mualem, R., Pospiech, G., Böhm, U., Krey, O., & Karam, R. (2017). Classroom evidence of teachers' PCK of the interplay of physics and mathematics. In *Key competences in physics teaching and learning,* (pp. 95–104). Springer.

Leinhardt, G., Zaslavsky, O., & Stein, M. K. (1990). Functions, graphs, and graphing: Tasks, learning, and teaching. *Review of Educational Research, 60*(1), 1–64.

Lobato, J. (2012). The actor-oriented transfer perspective and its contributions to educational research and practice. *Educational Psychologist, 47*(3), 232–247.

Magnusson, S., Krajcik, J., & Borko, H. (1999). Nature, sources, and development of pedagogical content knowledge for science teaching. In *Examining pedagogical content knowledge* (pp. 95–132). Heidelberg: Springer.

Malone, K. (2008). Correlations among knowledge structures, force concept inventory, and problem-solving behaviors. *Physical Review Special Topics – Physics Education Research, 4*(2), 020107.

Marrongelle, K. A. (2004). How students use physics to reason about calculus tasks. *School Science and Mathematics, 104*(6), 258–272.

Mason, A. J., & Singh, C. (2016). Surveying college introductory physics students' attitudes and approaches to problem solving. *European Journal of Physics, 37*(5), 055704.

McDermott, L. C., Rosenquist, M. L., & Van Zee, E. H. (1987). Student difficulties in connecting graphs and physics: Examples from kinematics. *American Journal of Physics, 55*(6), 503–513.

Meli, K., Zacharos, K., & Koliopoulos, D. (2016). The integration of mathematics in physics problem solving: A case study of Greek upper secondary school students. *Canadian Journal of Science, Mathematics and Technology Education, 16*(1), 48–63.

Meltzer, D. E. (2002). The relationship between mathematics preparation and conceptual learning gains in physics: A possible "hidden variable" in diagnostic pretest scores. *American Journal of Physics, 70*(12), 1259.

Meltzer, D. E. (2005). Relation between students' problem-solving performance and representational format. *American Journal of Physics, 73*(5), 463.

Mevarech, Z. R., & Kramarsky, B. (1997). From verbal descriptions to graphic representations: Stability and change in students' alternative conceptions. *Educational Studies in Mathematics, 32*(3), 229–263.

Michelsen, C. (2006). Functions: A modelling tool in mathematics and science. *ZDM, 38*(3), 269–280.

Michelsen, C. (2015). Mathematical modeling is also physics—Interdisciplinary teaching between mathematics and physics in Danish upper secondary education. *Physics Education, 50*(4), 489.

Monk, M. (1994). Mathematics in physics education: A case of more haste less speed. *Physics Education, 29,* 209–211.

Mulhall, P., & Gunstone, R. (2007). Views about physics held by physics teachers with differing approaches to teaching physics. *Research in Science Education, 38*(4), 435–462.

Müller, J., Fischer, H. E., Borowski, A., & Lorke, A. (2016). Physikalisch-Mathematische Modellierung und Studienerfolg. In *Implementation fachdidaktischer Innovation im Spiegel von Forschung und Praxis,* Zürich (pp. 75–78).

Nguyen, D.-H., & Rebello, N. S. (2011). Students' understanding and application of the area under the curve concept in physics problems. *Physical Review Special Topics – Physics Education Research, 7*(1), 010112.

Niess, M. (2005). Preparing teachers to teach science and mathematics with technology: Developing a technology pedagogical content knowledge. *Teaching and Teacher Education, 21*(5), 509–523.

Niss, M. (2017). Obstacles related to structuring for mathematization encountered by students when solving physics problems. *International Journal of Science and Mathematics Education, 15*(8), 1441–1462.

Nitsch, R., Fredebohm, A., & Bruder, R. (2014). Competencies in working with functions in secondary mathematics education – empirical examination of a competence structure. *Mathematics Education* (1985)

Park, S., & Oliver, J. S. (2008). Revisiting the conceptualisation of pedagogical content knowledge (PCK): PCK as a conceptual tool to understand teachers as professionals. *Research in Science Education, 38*(3), 261–284.

Pask, C. (2003). Mathematics and the science of analogies. *American Journal of Physics, 71*(6), 526–534.

Phillips, C. G. (2016). An improved representation of mathematical modelling for teaching, learning and research. *International Journal of Innovation in Science and Mathematics Education (formerly CAL-laborate International), 23*(4).

Pietrocola, M. (2008). Mathematics as structural language of physical thought. In M. Vicentini & E. Sassi (Eds.), *Connecting research in physics education with teacher education.* Volume 2 of ICPE – book. International Commission on Physics Education.

Planinic, M., Ivanjek, L., Susac, A., & Milin-Sipus, Z. (2013). Comparison of university students' understanding of graphs in different contexts. *Physical Review Special Topics – Physics Education Research, 9*(2).

Polya, G. (2014). *How to solve it: A new aspect of mathematical method.* Princeton: Princeton University Press.

Pospiech, G. (2013). Mathematisierung aus Sicht von Schülern der Sekundarstufe I. In S. Bernholt (Ed.), *Inquiry-based Learning – Forschendes Lernen* (pp. 326–328). Kiel: IPN.

Pospiech, G., & Oese, E. (2013). Wahrnehmung der Mathematisierung im Physikunterricht der Sekundarstufe 1. *PhyDid B, Didaktik der Physik, Beiträge zur DPG-Frühjahrstagung.*

Pospiech, G., & Oese, E. (2014). Use of mathematical elements in physics – Grade 8. In *Active learning – in a changing world of new technologies* (pp. 199–206). Prag: Charles University in Prague, MATFYZPRESS Publisher.

Pospiech, G., Böhm, U., & Geyer, M. (2012). Making meaning of graphical representations in beginners' physics lessons. *Discourse and Argumentation in Science Education,* 65.

Quale, A. (2010). On the role of mathematics in physics: A constructivist epistemic perspective. *Science Education, 20*(7–8), 609–624.

Redish, E. F. (2005). Problem solving and the use of math in physics courses.

Redish, E. F. (2017). Analysing the competency of mathematical modelling in physics. In *Key competences in physics teaching and learning* (pp. 25–40). Springer.

Redish, E. F., & Kuo, E. (2015). Language of physics, language of math: disciplinary culture and dynamic epistemology. *Science & Education, 24*(5–6), 561–590.

Richland, L. E., Stigler, J. W., & Holyoak, K. J. (2012). Teaching the conceptual structure of mathematics. *Educational Psychologist, 47*(3), 189–203.

Riese, J. (2010). Empirische Erkenntnisse zur Wirksamkeit der universitären Lehrerbildung – Indizien für notwendige Veränderungen der fachlichen Ausbildung von Physiklehrkräften. *PhyDid A-Physik und Didakt. Schule und Hochschule, 9*(1), 25–33.

Roorda, G., Vos, P., & Goedhart, M. J. (2015). An actor-oriented transfer perspective on high school students' development of the use of procedures to solve problems on rate of change. *International Journal of Science and Mathematics Education, 13*(4), 863–889.

Roth, W.-M., & McGinn, M. K. (1997). Graphing: Cognitive ability or practice? *Science Education, 81*(1), 91–106.

Sherin, B. L. (2001). How students understand physics equations. *Cognition and Instruction, 19*(4), 479–541.

Shulman, L. S. (1987). Knowledge and teaching: Foundations of the new reform. *Harvard Educational Review, 57*(1), 1–23.

Siswono, T. Y. E., Kohar, A. W., & Hartono, S. (2017). Secondary teachers' mathematics-related beliefs and knowledge about mathematical problem-solving. *Journal of Physics: Conference Series, 812*, 012046.

Skemp, R. R. (1976). Relational understanding and instrumental understanding 1 (pp. 20–26).

Trump, S. S. (2015). *Mathematik in der Physik der Sekundarstufe II!?* Doctoral thesis, Universität Potsdam, Potsdam.

Trump, S. S., & Borowski, A. (2012). Mathematikkompetenz beim Lösen von Physikaufgaben. *PhyDid B-Didaktik der Physik-Beiträge zur DPG-Frühjahrstagung.*

Trump, S., & Borowski, A. (2014). Die Anwendung von Mathematik in Physik. In S. Bernholt (Ed.), *Naturwissenschaftliche Bild. zwischen Sci. und Fachunterricht. Gesellschaft für Didakt. der Chemie und Phys. Jahrestagung München 2013* (pp. 288–290). Kiel: IPN.

Tuminaro, J., & Redish, E. (2007). Elements of a cognitive model of physics problem solving: Epistemic games. *Physical Review Special Topics – Physics Education Research, 3*(2), 020101.

Turşucu, S., Spandaw, J., Flipse, S., & de Vries, M. J. (2017). Teachers' beliefs about improving transfer of algebraic skills from mathematics into physics in senior pre-university education. *International Journal of Science Education, 39*(5), 587–604.

Uhden, O. (2016). Verst{ä}ndnisprobleme von Schülerinnen und Schülern beim Verbinden von Physik und MathematikStudent's problems of understanding concerning the translation between physics and mathematics. *Zeitschrift für Didaktik der Naturwissenschaften, 22*(1), 13–24.

Uhden, O., & Pospiech, G. (2009). Translating between mathematics and physics: Analysis of student's difficulties. In *GIREP-EPEC Conference Frontiers of Physics Education* (pp. 26–31).

Uhden, O., & Pospiech, G. (2013). Die physikalische Bedeutung der mathematischen Beschreibung – Anregungen und Aufgaben für einen neuen Umgang mit der Mathematik. *Praxis der Naturwissenschaften – Physik in der Schule, 62*(2), 13–18.

Uhden, O., Karam, R., Pietrocola, M., & Pospiech, G. (2012). Modelling mathematical reasoning in physics education. *Science & Education, 21*(4), 485–506.

van den Berg, E., Schweickert, F., & Manneveld, G. (2010). Learning graphs and learning science with sensors in learning corners in fifth and sixth grade. In M. F. Tasar & G. Cakmakci (Eds.), *Contemporary science education research: Teaching* (Vol. 1, pp. 383–394). Ankara: Pegem akademi.

Vogel, M., Girwidz, R., & Engel, J. (2007). Supplantation of mental operations on graphs. *Computers & Education, 49*(4), 1287–1298.

Wavering, M. J. (1985). The logical reasoning necessary to make line graphs. French Lick, Indiana. National Association of Science Teachers.

Wigner, E. (1960). The unreasonable effectiveness of mathematics in the natural sciences. *Communications on Pure and Applied Mathematics, 13*(1), 1–14.

Part I
Perspectives on Mathematics in Physics Education

Chapter 2
The "Math as Prerequisite" Illusion: Historical Considerations and Implications for Physics Teaching

Ricardo Karam, Olaf Uhden, and Dietmar Höttecke

2.1 Introduction

Mathematics is commonly seen as a prerequisite for the learning of physics, and the lack of basic mathematical skills is often regarded as one of the main reasons for students' failure in physics courses. This widespread view is found, for instance, in the way typical physics curricula at university level are structured, where disciplines like calculus, linear algebra, or mathematical methods for physicists are supposed to provide the tools that students will need in their subsequent physics courses.

This attitude is grounded in an implicit epistemological conviction about the relationship between mathematics and physics, namely, that mathematicians first develop elegant and abstract structures and then physicists make use of them in their theories. This image is, of course, historically inaccurate, and there are plenty of cases in the history of mathematics showing that several theories (e.g., calculus, trigonometry, tensor analysis to name a few) were originally related to the search for solutions to physics problems (Kline 1959). An evidence of this close relationship is the fact that some of the most influential scholars of the eighteenth

This is an extended version of a paper published in Karam, R., Uhden, O. & Höttecke, D. (2016). Das habt ihr schon im Matheunterricht gelernt! Stimmt das wirklich? Ein Vergleich zwischen dem Umgang mit mathematischen Konzepten in der Mathematik und in der Physik. *Unterricht Physik* 153/154, S. 22–27.

R. Karam (✉)
Department of Science Education, University of Copenhagen, København, Denmark
e-mail: ricardo.karam@ind.ku.dk

O. Uhden · D. Höttecke
Faculty of Education, Physics Education, University of Hamburg, Hamburg, Germany

© Springer Nature Switzerland AG 2019
G. Pospiech et al. (eds.), *Mathematics in Physics Education*,
https://doi.org/10.1007/978-3-030-04627-9_2

and nineteenth centuries (e.g., Newton, d'Alembert, Euler, Lagrange, Fourier) made sound contributions to both mathematics and physics.

Despite this mutual interplay, it is imprecise and even unfair to say that mathematics and physics are just different manifestations of the same activity. As Feynman (1985, p. 55) rightly pointed out, although one helps the other "physics is not mathematics and mathematics is not physics." Even in the cases where the same person made important contributions to both fields, it is still possible to find different parts of their work that can be classified as being either mathematical or physical.

This chapter focuses precisely on some important differences between the way physicists use mathematics to think about the world and what is usually recognized as mathematics by mathematicians. In this sense, we are aligned with other chapters in this book (e.g., Heck & Buuren and Planinic) as well as other authors in the literature such as Redish and Kuo (2015), who clearly state that "using math in physics has a different language, even a distinct semiotics, from pure mathematics." The explicit acknowledgment of these differences has important didactical implications, as will be shown with specific examples in the following sections. Our focus here is different from the contributions of Heck and Buuren and Planinic et al., since it aims at identifying aspects of the origin of these differences, by taking a closer look at their *historical genesis*.

2.2 Multiplication

Let us start with a quite simple mathematical operation, namely, multiplication. In mathematics lessons, we learn to interpret multiplication as some kind of sum, for instance, when we say "2×3," we mean two threes, i.e., "$3 + 3$." If we then calculate "3×2," we mean something else, i.e., three twos "$2 + 2 + 2$," which turns out to be also 6 and that is why we say that multiplication is a commutative operation.

Another plausible interpretation of the multiplication of two numbers we learn from mathematics lessons is related to areas of planar figures. Although it may seem that calculating an area is something very different, the idea of a sum is still deeply present. Consider, for instance, a rectangle with sides 2 m and 3 m (Fig. 2.1). To calculate its area is to add the total number of squares (1 m^2). And in order to count the number of squares that compose the rectangle on the left of Fig. 2.1, we can

Fig. 2.1 Visualizing $2 \times 3 = 3 \times 2 = 6$

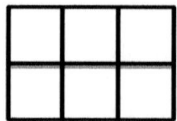
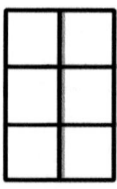

either say there are two rows of three squares or three columns of two squares. One important difference of this example is that now the result of the multiplication has a different unit ($2\,\text{m} \times 3\,\text{m} = 6\,\text{m}^2$). Reasoning with units is crucial in physics, and we will return to this point later.

With these images associated to multiplication in mind, what kind of meaning can be extracted from equations like $p = mv$, $U = RI$, or $F = ma$? Does it make sense to say that to calculate p is to add v m times or that U is the area of a rectangle with sides R and I?

These simple examples suffice to highlight the fact that in physics we assign a different meaning to multiplication. Usually, when physicists define a magnitude as the product of two others, they want to express a mutual dependence.[1] Consider, for instance, the physical quantity that expresses the capability of a force to rotate a given body, often called torque (τ). Disregarding vector considerations, we define torque as $\tau = Fd$, which means that τ depends *both* on the intensity of the force *and* on the distance from the point in which the force is applied to the rotation axis. It is also important to stress that the multiplication gives rise to a new/different quantity, whereas in mathematics the result is usually just another number.

This way of multiplying different magnitudes is more recent than one would think. It is definitely not found in the original works on statics from the classical antique, e.g., Archimedes' studies of equilibrium (Fig. 2.2).

Archimedes would express the relationship between the bodies' weights and their distances to the fulcrum in the equilibrium situation as follows:

Magnitudes are in equilibrium at distances reciprocally proportional to their weights

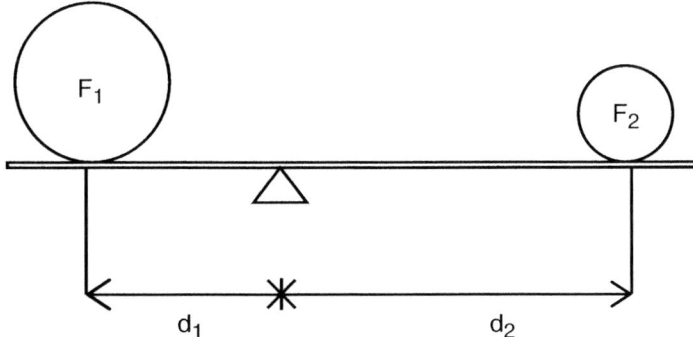

Fig. 2.2 Static equilibrium

---

[1]This is not always the case. Consider, for instance, the equation $v = \lambda f$, expressing the relation between a wave's velocity, frequency, and wave length. Phenomenologically, v depends on the characteristics of the medium, whereas f is the frequency of the source. Thus, it is usually not correct to say that v increases when f increases. Since v and f are determined by different causes, λ is the factor that is adjusted to maintain this equality. Overall, this exemplifies how physics makes a rather flexible use of mathematics.

Mathematically, this inverse [reciprocal] proportionality would be represented as

$$\frac{F_1}{F_2} = \frac{d_2}{d_1}$$

This is how the Greeks would represent relationships between quantities, namely, in the language of ratios. Thus, there is no concept of torque as the multiplication of force by distance (Fd) in Archimedes' work, since this was not conceptually conceivable by the Greeks (Bochner 1963). A multiplication between two lengths could be interpreted as an area, but it did not make sense to multiply two different magnitudes let alone to obtain a third one.

This begins to be conceptually possible in the seventeenth century. John Wallis (1616–1703), who also made significant contributions for the development of our modern algebraic notation, introduces this kind of reasoning in his treatise entitled *Mechanica, sive Tractatus de Motu* (Mechanics, or Tract on Motion, 1669–1671). When defining the quantity *Impedimenta* (I), somehow similar to torque, by the product of *Pondera* (P), equivalent to weight, by the arm (L), he writes the following (Fig. 2.3).

Here we see his clear emphasis on the commutative property of multiplication, something we would nowadays simply take for granted. Later in his treatise, Wallis expresses another quantity called *Moment* (M) by the product of the driving force (V) by its time of action (T) (Fig. 2.4).

Once again, we notice Wallis' careful and clear presentation that goes from specific to general. This is justified due to the novelty of both the reasoning and the notation.

Another example of the physical use of multiplication is Newton's (1642–1727) definition of quantity of motion (linear momentum) found in the very beginning of his celebrated *Principia*:

> The quantity of motion is the measure of the same, arising from the velocity and quantity of matter conjointly.

P.	2 P.	2 P.	3 P.	n P.	m P.
2 L.	L.	3 L.	2 L.	m L.	n L.
2PL. I ::	2PL. I.	6PL. I ::	6PL. I.	mnPL.	I :: mnPL. I.

Fig. 2.3 Wallis' definition of *Impedimenta*

V.		2V.		2V.		nV.	
T. M. ::		2T. 4M. ::		3T. 6M. ::		mT. nmM.	
VT.		4VT.		6VT.		nmVT.	

Fig. 2.4 Wallis' definition of *Moment*

The mutual dependence is emphasized by the word *conjointly*. Although not explicitly represented by an algebraic expression as Wallis, Newton does make clear that he means multiplication by stating that "in a body double in quantity, with equal velocity, the motion is double; with twice the velocity, it is quadruple." This need to express the relationship with words shows once more its innovative character.

Together with the notion of torque, the quantity of motion is one of the first appearances in history of a magnitude defined as the product of two others (Bochner 1963).[2] Of course, previous works have studied how multiple quantities are related to each other, but they did not express this relationship by means of multiplication of different quantities. In sum, the physical way of reasoning about multiplication as mutual dependence was developed in the seventeenth century and is essentially different from our first mathematics lessons. Being physics specific, one should assume that physics teachers are the ones responsible for introducing such way of thinking.

2.3 Division and Proportion

Similar differences between the typical reasoning in physics and mathematics can be found in the operation of division. In our mathematics lessons, we learn that division has something to do with sharing. If an adult wants to distribute ten bonbons equally among five children (10:5), it means that each child will receive two bonbons (10:5 = 2). This is already a very different operation when compared to multiplication, since it is not commutative (5:10 = 0.5). In the latter example, although the "sharing" idea is still valid since five bonbons to ten children is equal to 0.5 (half a bonbon) per child, dividing a smaller by a greater number poses serious learning difficulties for beginners.

In order to promote a better understanding of such cases in which the numerator is smaller than the denominator, another image intensively promoted in mathematics lessons are fractions. One learns to divide a whole (e.g., a circle) in equal parts given by the denominator and "take" or "color" the number of parts expressed in the numerator. In this sense, 3/4 means three parts of a whole that was divided into four equals parts (Fig. 2.5).

Once again, we are tempted to ask the question: With these images of division in mind, how are we supposed to make meaning from $\rho = m/V$, $R = U/I$, or $f = v/\lambda$? Can the physical way of reasoning about division be naturally transferred from mathematics lessons?

---

[2]The first definition of Newton's *Principia* is the "quantity of matter" which is "*arises from its density and bulk* [volume] *conjointly*." In Wallis' work we also find several other examples of such mutual dependence. Furthermore, Huygens' vis viva, represented by the scalar quantity mv^2, is also among the first examples of the essentially physical use of multiplication to represent a new quantity.

Fig. 2.5 Mathematical image
of 3/4

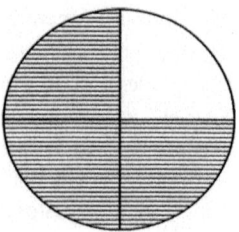

As previously mentioned, expressing a physical quantity as a quotient of two
other quantities was not possible for the Greeks, since a ratio had to be the division
between two values of the same magnitude. Thus, a quantity like U/I would not be
conceivable. If one wanted to express a direct proportionality between U and I, one
would write $U_1/U_2 = I_1/I_2$.

In physics, a division between two quantities is often related to the idea
of "per" unit of the quantity in the denominator. Some common examples are
velocity (meters *per* second), density (kilograms *per* cubic meter), and heat capacity
(calories *per* Kelvin). In some situations this can actually resemble a bit the
idea of sharing bonbons, although this is usually not the case. Furthermore, if
more than one quantity appears in the denominator, a situation rarely approached
in mathematics lessons, the conceptual understanding, is quite challenging, e.g.,
specific heat (calories *per* kilogram *per* Kelvin) and water flux (cubic meters *per*
second *per* square meter). It is worth noticing that a systematic reasoning about
units and dimensions came much later as we will show in the next section.

A common question studied in physics is how rapidly things change, and
therefore it is quite usual to express quantities with *time* in the denominator. Galileo
strongly emphasized the time dependence in his studies of motion, but he still
reasoned in terms of ratios (we do not find something like s/t in his work). Wallis
(1669), on the other hand, does express the quantity *celeritas* (speed) by the quotient
L/T, but similarly to his approach to describe multiplication, this is done in a very
didactic step-by-step way.

When treating the relation C = L/T, Wallis considers separately each case in
which one of the magnitudes is constant. For situations where time is constant,
he writes the following (Fig. 2.6): meaning that distance (*longitudo*, L) and speed
are directly proportional. For constant distance, the effect in the change of time is
expressed as in Fig. 2.7, and for the case of constant speed, it is expressed as in
Fig. 2.8.

A modern reader might not understand why Wallis bothers to describe each single
situation, since C = L/T implies all of this in one single formula. What the modern
reader must understand is that, similarly to the case of multiplication, this kind of
notation and reasoning appears only in the seventeenth century and is radically new
for the time. In fact, understanding some of the historical struggles can be helpful
to appreciate some difficulties faced by our students when trying to make sense of

$$L . C :: 2 L . 2 C :: n L . n C :$$

Fig. 2.6 Relation between speed (celeritas) and distance for constant time

$$\frac{L}{T} . C :: \frac{L}{2 T} . \frac{I}{2} C :: \frac{L}{n T} . \frac{I}{n} C .$$

Fig. 2.7 Relation between speed (celeritas) and time for constant distance

$$L . T :: 2 L . 2 T :: 3 L . 3 T :: n L . n T :$$

Fig. 2.8 Relation between time and distance for constant speed (celeritas)

the condensed representations we teach them today. For instance, when we write $s = vt$ and claim that it is just the same as $v = s/t$, Wallis' presentation could help us realize that this is far from obvious for newcomers. The main reason is that, although mathematically equivalent, the two representations may refer to two different physical situations.

This example illustrates another interesting difference found in the way proportionality relations are expressed in physics and mathematics lessons. If two quantities (A and B) are said to be directly proportional, a mathematics teacher would highlight that the quotient between these quantities remains the same and would express this by $A/B = k$ (where k is a constant). If A and B were inversely proportional, then the product between them is constant ($AB = k$).

Students in a physics lesson would learn something apparently different. When a physics teacher says that the quantity A is directly proportional to the quantity B, she usually writes $A \propto B$ or the following equation $A = kB$ (where k is constant). If A were inversely proportional to B, she would write $A \propto 1/B$ and $A = k/B$. This is probably because in physics one is often connecting the mathematical representation to an experimental setting, in which the independent variable is what one can vary in the experiment and the dependent is what is measured. The two approaches to proportionality are of course equivalent, but it is pedagogically relevant to notice that they are "psychologically" different. Thus, one should not expect this specifically physical kind of reasoning to be transferred automatically from mathematics lessons.

2.4 The Crucial Role of Units in Physics

The way physicists emphasize the importance of units is another crucial difference when compared to mathematicians. In math, the equation $a = b + c$ is usually not seen as problematic, whereas a physicist would immediately protest that this is only valid if a, b, and c have the same units. In fact, physics equations do not involve sums of quantities as often as products. When assuming that a particular quantity depends on two others – for instance, that linear momentum depends both on mass and velocity – physicists always express this as a multiplication ($p = mv$) and never as a sum ($p = m + v$). It is simply forbidden to add different physical quantities; in physics the rule "apples with apples and oranges with oranges" always applies. One of the reasons for using multiplication is that if a quantity depends mutually on two others, this means that when one of them is zero (e.g., $v = 0$), then the original quantity must also be zero ($p = 0$).

The differences regarding the ways physicists and mathematicians worry about units can be evidenced by looking at some common problems proposed in mathematics textbooks that are supposed to be related to the physical world. Consider one typical example below (Fig. 2.9).

For a physicist, the equation $y = -\frac{5}{v^2}x^2 + h$ is clearly wrong because the first term on the right side has unit s^2 and therefore cannot be added with h (in m). In physics lessons, on the other hand, one learns specific rules about how to consider the units in the calculations and always check if they match, as exemplified in the following example typically found in physics textbooks (Fig. 2.10).

As we can see, this methodical concern with units is essentially physical and will usually not be taught in mathematics lessons. Interestingly, a systematic approach to units is something quite recent. One of its first examples can be traced back to Fourier's (1768–1830) masterpiece *Analytical Theory of Heat* (Fourier 1822).

In this work, Fourier investigates the problem of finding the temperature at a given point in space and time ($T = f(r,t)$) of different surfaces and bodies under certain boundary conditions. The most important physical result of this work is a partial differential equation called the *heat equation*, and the mathematical one is

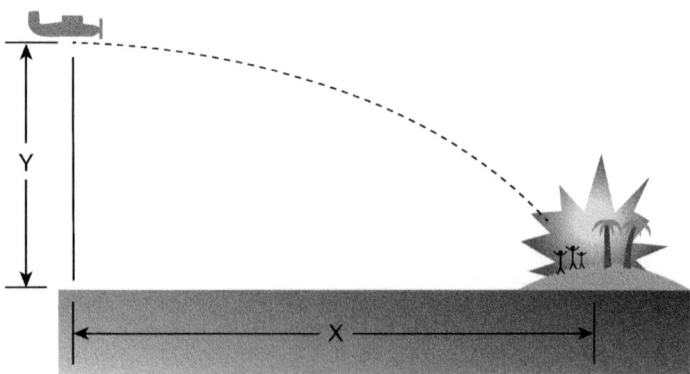

Fig. 2.9 Typical problem in mathematics textbooks: an airplane flying at a height h (in m) with a speed v (in m/s) delivers a package, which has approximately a parabolic trajectory described by the equation $y = -\frac{5}{v^2}x^2 + h$, where y is the package's height and x its horizontal distance from the point it was abandoned

Peter is standing with his motorbike at a stop light. After it turns green he accelerates uniformly. After four seconds he has reached the velocity of 72 kilometers per hour. What is his acceleration? What distance has he travelled?

Find: a, s   *Given: t = 4s, v = 72 kilometers/hour = 20 m/s*

Solution:

1. Step: Calculation of a 2. Step: Calculation of s

$$a = \frac{v}{t}$$ $$S = \frac{1}{2}\,at^2$$

$$a = \frac{20m}{4\ \text{s·s}} = 5\,\frac{m}{s^2}$$ $$s = \frac{1}{2} \cdot 5\,\frac{m}{s^2} \cdot 4^2 s^2 = 40m$$

Fig. 2.10 Typical problem in physics textbooks showing how to perform calculations with units

Table 2.1 Fourier (1822, p. 130)

Quantity or Constant.	Length.	Duration.	Temperature.
Exponent of dimension of x ...	1	0	0
" " t ...	0	1	0
" " v ...	0	0	1
The specific conducibility, K ...	−1	−1	−1
The surface conducibility, h ...	−2	−1	−1
The capacity for heat, c ...	−3	0	1

an innovative way to conceive functions as trigonometric series (Fourier series). In his physical considerations, Fourier separated the specific (internal) conductivity from the external (surface) conductivity of materials. Another important material-dependent quantity for his calculations was the specific heat capacity. Since the several equations in his work involved many of these quantities, Fourier felt the need to introduce a method to represent all of them in terms of three basic quantities *length, duration*, and *temperature*, as shown in the table (Table 2.1).

Fourier justifies the importance of this method as follows:

It must now be remarked that every undetermined magnitude or constant has one *dimension* proper to itself, and that the terms of one and the same equation could not be compared, if they had not the same *exponent of dimension*. We have introduced this consideration into the theory of heat, in order to make our definitions more exact, and to serve to verify the analysis; it is derived from primary notions on quantities; for which reason, in geometry and mechanics, it is the equivalent of the fundamental lemmas which the Greeks have left us without proof. (*ibid*, p. 128)

A specific concern with units was also motivated by studies of electricity and magnetism. The main reason is that many phenomena were manifested by forces, which were themselves expressed in mechanical units. This generated the question of defining the units to measure electric/magnetic charges (Coulomb's laws) as well as currents (Ampère's force law). Wilhelm Weber (1804–1891) took this project very seriously and designed extremely precise instruments to measure relations between electrostatic and electromagnetic units. These results turned out to be essential for the unification of electromagnetism and optics (see Assis 2003).

In the very first pages of his celebrated *Treatise of Electricity and Magnetism* (1873), Maxwell (1831–1879) highlights the importance of dimensional analysis and presents it in a very familiar way for the modern reader. A general rule of dimensional coherence for equations is stated as follows:

> A knowledge of the dimensions of units furnishes a test which ought to be applied to the equations resulting from any lengthened investigation. *The dimensions of every term of such an equation, with respect to each of the three fundamental units, must be the same. If not, the equation is absurd*, and contains some error, as its interpretation would be different according to the arbitrary system of units which we adopt. (Maxwell, 1873, p. 1, our emphasis)

Later on, he defines *mass*, *length*, and *time* as fundamental units and refers to other units as derived. The dimensions of velocity, acceleration, and density, for instance, are written as $[LT^{-1}]$, $[LT^{-2}]$, and $[ML^{-3}]$, respectively, pretty much how we would learn/teach today. In sum, these historical examples suffice to show that a systematic treatment of units is a characteristic trait of physics and appeared mostly in the nineteenth century in investigations about heat and electromagnetism. Once again, it is naïve to expect that pupils will learn this in their mathematics lessons.

2.5   Functions and Diagrams

The way functions are treated in mathematics and physics lessons is also rather different. In mathematics, functions are defined in the context of set theory as abstract assignments of elements from one set to another. According to a functional rule often referred to as f, elements of a domain set X are assigned to a codomain set Y by a correspondence, so that exactly one y of Y can be assigned to each x of X. The pairings of x and the associated y are represented by a Cartesian product of the elements of the sets X and Y.

The independent variable is usually x and can take all values from the domain set. Parameters which are common in general forms of functions, e.g., a, b, c in $f(x) = ax + b$ or $f(x) = ax^2 + bx + c$, are usually characterized by the choice of specific letters. In mathematics lessons, properties of the functions are discussed using the respective parameters a, b, and c. The letters have a high recognition value and are generally not modified to avoid cognitive overload. The consequence may be that students cannot deal with different representations of the same function, because they associate the concept of function with representations commonly used in mathematics lessons. If a function is not expressed by $f(x)$ and/or if the parameters are not a, b, and c, the function may not be identified as a linear or quadratic function.[3]

But how are functions treated in physics lessons? Well firstly the notation $f(x)$ is not used at all! That this does not mean that there are no mathematical functions in physics is far from obvious to students. One of the first contacts with functions in physics lessons is in kinematics, for instance, when equations of motion are represented as

$$s(t) = s_0 + vt \quad \text{or} \quad s(t) = s_0 + v_0 t + at^2/2.$$

Here at least the term $s(t)$ indicates that position is a function of time. Nevertheless the connection with the functions of mathematics lessons is still not trivial, and it is rarely the case that students spontaneously transfer their knowledge from one domain to the other.

In reality the situation is even worse. Formulas in physics are typically functions, but this is seldom explicitly expressed; one writes $U = RI$, $E = mgh$, $P = UI$, or $p = mv$. In fact, in physical formulas the very role of dependent and independent variables is often blurred, and the representation of a formula as a function depends on the physical situation and the measurement process. If, in the case of a constant resistance, the current intensity is measured as a function of the voltage, Ohm's law is actually a function of the form $I(U) = U/R$. However, there are other measurement processes that are better represented by functions of the form $U(R) = IR$. This shows that the representation of physical functions can be highly dependent on the design of the respective experiments. In general, Ohm's law is a function of more than one variable, $I(U, R) = U/R$, which usually has no correspondent in mathematics lessons at school level.

Significant differences[4] are also identified in the way graphs of functions are built in physics and mathematics lessons. Let us consider the simple example $f(x) = 2x + 3$ from math and compare it with $s(t) = 2t + 3$ from physics lessons. Although the function is the same, their representations are rather different (Fig. 2.11). For instance, the angle with the horizontal has a specific meaning in mathematics (its tangent is the slope), whereas it makes no sense to talk about an angle in physics since the scale is arbitrary, and the axes contain two different quantities. Moreover, the physical situation may impose some constraints; the graph on the right is only physically valid for $t \geq 0$.

---

[3]There are many other subtleties in the way functions are taught/learned at school (see Ellermeijer and Heck 2002 as well as the chapter from Heck and Buuren in this collection).

[4]See also Ellermeijer and Heck 2002, and the chapter from Planinic in this collection.

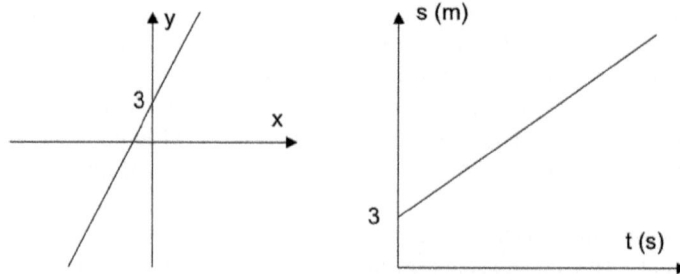

Fig. 2.11 Differences between graphs in mathematics and physics

Fig. 2.12 Oresme's
classification of motion.
(Clagett 1968)

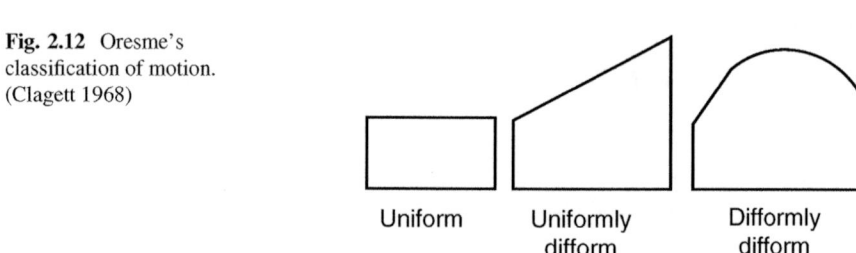

When we look at the historical development of the concept of function, it is interesting to notice how it has been greatly influenced by physics. Already in one of the first manifestations of graphs of functions, the French medieval scholar Nicole Oresme represented the change of velocity with time and distinguished between uniform, uniformly difform, and difformly difform motions. In Fig. 2.12, the baseline (*longitudo*) is the time, and the perpendiculars raised on the baseline (*latitudines*) represent the velocity from instant to instant in the motion.

A similar influence is found in Newton's celebrated *Method of Fluxions* (written in 1671, published in 1736), which is one of the first publications on differential and integral calculus. In the beginning of this work, Newton states that all problems solved can be divided into two categories. In his own words:

> [...] it may be observed, that all the difficulties [...] may be reduced to these two problems only, which I shall propose concerning a Space described by local Motion, any how accelerated or retarded.
>
> I. The Length of the Space described being continually (that is, at all Times) given; to find the Velocity of the Motion at any Time proposed.
>
> II. The Velocity of the Motion being continually given; to find the Length of the Space described at any Time proposed. (Newton 1736, p. 19)

For the modern reader, those are the well-known differentiation and integration operations. In mathematics lessons it is common to define the derivative in an abstract manner (limit of a function) and afterwards present the velocity as an application of this concept. A quick look at its original formulation gives a different picture and illustrates how velocity and derivative are deeply interrelated since the very genesis of calculus.[5]

Nevertheless, the abstract set-theoretic definition of function we teach today in math lessons is less than 100 years old and was driven by intrinsic foundational considerations.[6] Although it is pedagogically relevant to learn about the "physical origins" of the concept of function, it is equally important that students become acquainted with authentic (pure) mathematical reasoning in order to recognize the disciplinary specificities of these two intellectual enterprises.

2.6   Final Discussion and Didactical Implications

Physics teachers often complain that their students don't know enough math. When solving problems where some kind of mathematical operation is needed, it is not uncommon to hear them telling their students: "You should have learned this in your math courses!". The main goal of this chapter is to show that this assumption is flagrantly incorrect since physicists often use mathematics in a very different way compared to how these things are taught in math lessons.

These differences have historical roots as we have tried to show in the examples discussed. The list could go on and on (e.g., trigonometric functions, complex numbers, matrices, differentials, etc.), but the take-home message is the same: Physics has developed a specific way to reason mathematically about the world that serves its own purposes. Therefore it should be taught/learned in physics lessons. Some of these differences may have more sociological origins (it is just a matter of convention), but most of them are due to important epistemological differences that pertain to the nature of physics and mathematics.

What does this imply for the teaching of physics? Here are some general ideas:

– *Make differences explicit to students.*

An important methodological consequence of what has been presented is simply to address these differences explicitly with students. Instead of implicitly meaning "this is just like what you learn in math," physics teachers should prefer "this is different from what you learned in math because . . . ." Having the same teacher teaching both disciplines to the same class can be a useful strategy because then she can always refer to the other discipline and cannot just blame the math teacher. Even better is the idea of having fully integrated courses, taught conjointly

---

[5]The interested reader can find another episode that illustrates the great influence physics had in the conceptualization of function, namely, the so-called vibrating string controversy (Wheeler and Crummett 1987).

[6]See Kjeldsen and Lützen (2015) for an overview on the historical development of the concept of function.

by a mathematician and a physicist, where epistemological and methodological differences become inevitably apparent. An interesting experience is described in Dunn and Barbanel (2000), where vector calculus is taught together with electromagnetism and explicit discussions about substantial differences in context, notation, and philosophy are presented as pedagogically beneficial.

– *Express variables explicitly; compare with known functions from mathematics.*

A possibly fruitful approach for physics lessons is to express physics formulas in a more mathematically coherent way so that the distinction between dependent and independent variables is explicit. For instance, if the gravitational potential energy is expressed by $E(h) = m \cdot g \cdot h$, then it should be clear the association with a physical situation where one is varying the height of a body with (constant) mass m immersed in a (fixed) gravitational field g and estimating the potential energy. Now it even becomes possible to relate this formula with the general linear function from the mathematics lessons $f(x) = a \cdot x + b$. Then, the slope a can be associated with mg, x with h and the point where the graph crosses the y-axis interpreted as the potential energy at height zero (physically it could be assumed to have a non-zero value E_0). This back and forth from mathematical formalism to physical meaning is extremely beneficial in order to develop the competence of transfer/translation, which is essential for physics.

– *"Play" with conventions to aim at deep conceptual understanding.*

Several differences are due to conventions adopted by the communities of mathematicians and physicists and have no objective justification. A potentially meaningful teaching strategy is to "play" with these conventions and explore whether or not students have a deep conceptual understanding of the matter. Consider, for instance, the kinematic expression $s(t) = 2\,t + 3$ mentioned above. Instead of plotting its graph with t in the horizontal axis, one could consider the function $t(s) = s/2 - 3/2$, draw a *time* x *position* diagram, and discuss its meaning. Using the international system of units, in the first expression, one would identify the body's velocity as being 2 m/s, whereas in the second case, a new quantity whose unit is s/m could be thought. Instead of how many meters are traversed per second, the latter induces the question of how many seconds a body takes to traverse 1 meter, which is still physically meaningful. In general, highlighting the differences between mere conventions and objectively right or wrong statements is of major educational value, and the "playing with conventions" strategy can be quite useful.

– *Use original sources.*

Although the historical examples mentioned in this chapter were primarily thought to be discussed with pre- or in-service physics teachers, we are confident that original sources can also play an important role in the classroom. Already the fact that the notation is usually very different shows the dynamic nature of our conventions and the evolution of our representations. Sometimes, since the abstract mathematical formalism was not yet established, more accessible, concrete, and visual representations are found, which possess a high didactical potential (e.g.,

Galileo's or Newton's geometrical reasoning). Examples of early appearances of particular kinds of reasoning and notations, such as Wallis' definitions of torque and speed, can also highlight the nontriviality of topics students often struggle to learn.

By referring to the "math as prerequisite" illusion in the title, it may appear that we are against popular math "crash courses" or that we advocate that only physicists should teach math to physics students. This is not the case. Pure math courses are very important – among other reasons – because they convey a genuine mathematical approach and possibly make physics students even learn more about the nature of physics (by contrast with the nature of mathematics). The main argument is that even if such courses are necessary, they are certainly not sufficient because of the specificities of physics we have underlined. But although this may seem a general methodological recommendation, one needs to be aware of the complex and multifaceted nature of the relationship between physics and mathematics. Instead of looking for "always valid" assertions, one learns much more by studying concrete cases and drawing specific conclusions related to them.

References

Assis, A. K. T. (2003). On the first electromagnetic measurement of the velocity of light by Wilhelm Weber and Rudolf Kohlrausch. In F. Bevilacqua & E. A. Giannetto (Eds.), *Volta and the history of electricity* (pp. 267–286). Milano: Università degli Studi di Pavia and Editore Ulrico Hoepli.

Bochner, S. (1963). The significance of some basic mathematical conceptions for physics. *Isis, 54*(2), 179–205.

Clagett, M. (1968). *Nicole Oresme and the medieval geometry of qualities and motions*. Madison: The University of Wisconsin Press.

Dunn, J. W., & Barbanel, J. (2000). One model for an integrated math/physics course focusing on electricity and magnetism and related calculus topics. *American Journal of Physics, 68*(8), 749–757.

Ellermeijer, T., & Heck, A. (2002). Differences between the use of mathematical entities in mathematics and physics and consequences for an integrated learning environment. In M. Michelini & M. Cobal (Eds.), *Developing formal thinking in physics* (pp. 52–72). Udine: Forum, Editrice Universitaria Udinese.

Feynman, R. P. (1985). *The character of physical law*. Cambridge, MA: The MIT Press.

Fourier, J. (1822/1878). *The analytical theory of heat*. (Translated, with notes, by A. Freeman). Cambridge: University Press.

Kjeldsen, T. H., & Lützen, J. (2015). Interactions between mathematics and physics: The history of the concept of function – Teaching with and about nature of mathematics. *Science & Education, 24*(5), 543–559.

Kline, M. (1959). *Mathematics and the physical world*. London: John Murray (Publishers) Ltd.

Maxwell, J. C. (1873/1954). A treatise on electricity and magnetism. New York: Dover.

Newton, I. (1736). *The method of fluxions and infinite series: With its application to the geometry of curve-lines* (Translated from the Author's Latin Original Not Yet Made Public). https://books.google.dk/books?id=WyQOAAAAQAAJ&redir_esc=y

Newton, I., Cohen, I. B., & Whitman, A. M. (1999). *The principia: Mathematical principles of natural philosophy*. Berkeley: University of California Press.

Redish, E. F., & Kuo, E. (2015). Language of physics, language of math: Disciplinary culture and dynamic epistemology. *Science & Education, 24*(5), 561–590.
Wheeler, G. F., & Crummett, W. P. (1987). The vibrating string controversy. *American Journal of Physics, 55*(1), 33–37.

Chapter 3
Students' Understanding of Algebraic Concepts

André Heck and Onne van Buuren

3.1 Introduction

In mathematics and physics education, teachers and researchers working at all levels of education have identified many difficulties of their students with mathematics and physics at conceptual and procedural level. Because it is well accepted that mathematics plays an important role in physics, to such an extent that many call it the language of physics, it comes to no surprise that much research has been done to identify the students' problems with mathematical formalism in physics and to design tasks that help students get a better understanding of mathematics and its methods in the context of physics. Lack of algebraic expertise and lack of representational fluency of the students and their inability to recognize concepts and procedures of mathematics when dealing with problems in physics can be considered as causes of the difficulties, for which physics teachers easily tend to blame their students or the mathematics teachers that they did not do a good job before.

But there are more causes: one of them seems to be the unawareness or neglect of teachers of the many differences that exist between mathematics used in its own discipline and mathematics in physics. This holds both for the algebraic, graphical, and other representations used during mathematical work and for the language spoken and written alongside. The coin is two-sided: on the one side, there is the complex meaning of mathematical formalism in physics that many students

A. Heck (✉)
Korteweg-de Vries Institute for Mathematics, University of Amsterdam, Amsterdam, The Netherlands
e-mail: a.j.p.heck@uva.nl

O. van Buuren
Faculty of Behavioural and Movement Sciences, Vrije Universiteit Amsterdam, Amsterdam, The Netherlands
e-mail: o.p.m.van.buuren@vu.nl

© Springer Nature Switzerland AG 2019
G. Pospiech et al. (eds.), *Mathematics in Physics Education*,
https://doi.org/10.1007/978-3-030-04627-9_3

apparently not understand deeply enough (Uhden 2012; Uhden and Pospiech 2010, 2012), but on the other side, physics itself adds complexities. Mathematics and physics are closely intertwined as disciplines, but this seems often not sufficiently addressed in education. When not discussed in class, the differences and similarities in mathematical work may not be noticed by the students and lead to mistakes that they cannot really understand well, or these differences and complexities may confuse students and challenge them in an unproductive manner. Redish and Kuo (2015, p. 583) gave in their paper on the language of physics and mathematics for higher physics education the following take-away message:

> How mathematical formalism is used in the discipline of mathematics is fundamentally different from how mathematics is used in the discipline of physics—and this difference is often not obvious to students. For many of our students, it is important to explicitly help them learn to blend physical meaning with mathematical formalism.

We argue that this message already holds in lower secondary physics education, in particular when mathematical modelling comes into play in school physics, and that the students are not to blame for being confused and failing to transfer between mathematics and physics. We discuss the challenges that these students face in algebra when using and building mathematical models of physical systems. This chapter is organized as follows: in subsequent sections, we focus on the different ways of using variables, equations, formulas, and functions in mathematics and physics. We are of opinion that the differences cannot be ignored in physics education and that discussion can better start at the lower secondary level, because at that stage also the teaching and learning path from arithmetic to algebra starts in mathematics classes. Many of the data and examples of student challenges presented in this chapter are taken from earlier work of the first author (Ellermeijer and Heck 2002; Heck 2001) and from the doctoral and postdoctoral work of the second author on the development of a modelling learning path (Van Buuren 2014).

Two other chapters in this volume discuss the differences in mathematical formalism between mathematics and physics and how it is taught at secondary school level. But their aims differ from each other. Van Buuren and Heck (this volume) narrow the discussion to the use of variables and formulas for constructing computer models in lower secondary physics education. Karam, Uhden, and Höttecke (this volume) aim at identifying aspects of the origin of the differences by taking a closer look at their historical genesis. Altogether, the chapters give a comprehensive and helpful overview of the subject and offer plenty of food for thought.

3.2   The Meaning of Variable Is Variable

Tarski (1956, p. 1909) characterized the significance of the notion of variable as follows:

> Without exaggeration it can be said that the invention of variables constitutes a turning point in the history of mathematics; with these symbols man acquired a tool that prepared the way

for the tremendous development of the mathematical sciences and for the solidification of its logical foundations.

Skemp (1986, p. 213) claimed that

the idea of a set and that of a variable are two of the most basic in mathematics... The idea of a variable is in fact a key concept in algebra—although many elementary texts do not explain or even mention it.

Further on he wrote: "In mathematics, an unspecified element of a given set is called a variable." This last sentence already marks a significant difference between (1) the use of a variable in mathematics and logics as a placeholder for things from a particular replacement set and (2) the use of a variable in physics as a name for a measurable physical quantity with a unit, certain precision, and possibly a direction. In essence, the several meanings of variables (see, e.g., Kücheman 1981; Schoenfeld and Arcavi 1988), depending on the context in which they are used, and the lack of discussion about this in the classroom are what makes it hard for students to understand.

Results of research studies on the role of variables in students' algebraic thinking (Kücheman 1981; Malisani and Spagnoloi 2009; Trigueros and Ursini 2003) suggest that different conceptions of variable have different degrees of difficulty: the "letter as an unknown" is simpler to grasp than the "generalised number" and "variable in functional relationship." Several studies have linked students' difficulties with obstacles that have been met in history in the slow development of symbolic language. For example, Karam, Uhden, and Höttecke (this volume) show how the way physicists make use of some basic mathematical concepts such as multiplication, division, and functions is specific to physics by identifying their historical genesis and contrasting with the way these concepts are usually taught in math lessons. Malisani (2006) and Radford (1996) highlighted that the notion of variable as unknown and as a thing that varies has a totally different genesis and evolution. The notion of the unknown has its origin in the resolution of equations. The introduction by Viète (1540–1603) of letters and signs to represent quantities and operations is a landmark in the development of the language of symbolic algebra. The notion of a variable as a thing that varies is very ancient, but its origin in the history of mathematics is unclear. Its evolution is very slow and goes from relationships among numbers on Babylonian tablets, through curves described in kinematical terms and functional relationships among variable quantities alongside the development of calculus and analysis during the seventeenth and eighteenth centuries, to the set-theoretical definition of the concept of function in the twentieth century.

In secondary education, one often deals with the various purposes and meanings of variables by treating them as primitive terms that are best learned by practice. After all, trying to fit the notion of variable into a single conception would oversimplify it and actually distort the purposes of algebra. Basic idea of the pragmatic approach is that students will learn from the examples and the exercises and that they will gradually sense the meanings of variable. One obstacle in this learning process is however that mathematical meaning is often determined by context, rather than by formal rules and notation. For example, what does the symbolism $v(t - \theta)$ mean? Is it a function v with argument $t - \theta$, a generalized number $v \times (t - \theta)$, or an expression in which the symbol v represents the velocity of an object and the time t is transformed via a constant θ? And how does $v(t - \theta)$ compare with $a(x - y)$? For a secondary school student, it takes time and practice to get used to the fact that a variable actually gets meaning in mathematics and physics through its use (as indeterminate, as unknown, as parameter, etc.), through its domain of values, and through the context in which it is used. We quote a 16-year-old student in upper vocational physics, who struggled initially with the resemblance of Hooke's law for a spring with a mathematical equation of a straight line but acquired the structural insight while he was doing regression analysis in the computer learning environment Coach (Heck et al. 2009):

> Only after the practical work, in which I had to do a function fit, I realised that $F_{spring} = C \cdot u$ actually is $y = ax + b$. There you see the link with mathematics. This should have been explicitly mentioned in the textbook.

In the above example, there is a link between the two equations, but it cannot be ignored that they are different in several aspects: in the first equation, the symbols stand for the physics notion of force, stiffness, and displacement. The equation is used to make sense of the process of extension and compression of a spring. The second equation primarily activates mathematical knowledge about straight lines. What students need to learn is to transfer back and forth between these two perspectives, and they definitely need the help of both their mathematics and physics teacher herein.

3.3   Variables in Mathematics

In mathematics education, researchers and educators are constantly searching for ways to familiarize students with variables (see, e.g., Arcavi et al. 2017; Kieran 1992, 1997, 2007) and to make the transition from arithmetic to algebra, or generalized arithmetic as it is commonly called, easier for them. Besides algebra as generalized arithmetic, Usiskin (1988) identified three other conceptions of algebra, namely, algebra as a study of procedures for solving certain kinds of problems, as a study of structures, and as the study of relationships among quantities (including modelling and functions). He correlated each of these four approaches with the different relative importance given to various uses of variables: pattern generalizer, unknown, arbitrary object, and argument or parameter, respectively.

Ursini and Trigueros (2001) distinguished three main uses of variables in elementary algebra and described in detail different aspects that underlie a basic understanding of these three uses of variable. In their model, they distinguished variable as unknown (as in $2x + 3 = 5$), variable as general number (as in $2x + 3$, but also as parameter in $ax + b$), and variables in a functional relationship (as in $2x + 3 = y$). They found that secondary school students and teachers have problems to differentiate between the uses of variable as unknown and as general number, to accept an open expression as valid, and they feel compelled to find a specific result when an equal sign is present. In another study (Trigueros and Ursini 2003), these authors obtained similar results for first year undergraduates in economics, administration, and accounting at a private Mexican university.

The framework of Ursini and Trigueros resembles the distinction of the three uses of variable made by Freudenthal (1983): variable as a

- polyvalent name, i.e., a name for an object that can take a multitude of values (as in the task to solve the equation $x^2 = 2$)[1];
- placeholder, which denotes the places in an expression where the same object is meant (as in the equivalence $(a + b)^2 = a^2 + 2ab + b^2$ or in the function definition $f(x) = x^2$);
- variable object, i.e., a symbol for an object with varying value (as in the statement "2^{-n} for n from 1 upward" or when the object is a quantity with varying values).

This distinction can also be applied to the notion of parameter. Although this can be illustrated within a mathematical context, we do this with examples from physics. In the role of a placeholder, the parameter has one value at a time. For example, in the formula $T = 2\pi \sqrt{L/g}$, for the period of the mathematical pendulum of length L, the letter g stands for the acceleration of gravity. You may study the motion of the pendulum on earth or moon, but always it has one value only (although this value is different in different contexts). The given formula expresses the relationship between the period of the pendulum and the length of the pendulum. Then the letter L can play the role of a variable object (in a thought experiment taking pendulum to the moon and to Mars, L of course stays the same). Thirdly, as a polyvalent name, a parameter supports the writing of a general formula or a distinction of various cases. For example, the letters A (for amplitude), ω (angular frequency), and ϕ (phase) are used in the general formula $u = A \sin(\omega t + \phi)$ to describe harmonic motion mathematically. Another example is the role of initial height y_0, the initial velocity v_0, and the release angle α in the mathematical formula $y = y_0 + v_0 t \sin(\alpha) - \frac{1}{2}gt^2$ that gives the vertical position of an object thrown away. Typically, one keeps values of all parameters except one fixed and explores the formula for several values of the free parameter. In other words, the parameter in its generalizing role is used to distinguish several cases: the values of a parameter can vary, but they are temporarily thought of as fixed, and a change of the parameter value will not just change one quantity but a complete expression, function, or graph at hand. These examples also show that, in mathematics and physics, the type of a variable depends on the way it is used in the context. For example, the acceleration of gravity g can be seen in a

---

[1]Although the term "polyvalent name" suggests that the variable represents more than one value, this is not necessary. Often, one does not know in advance how many values are possible: for example, when asked to find the real roots of a third-degree polynomial with real coefficients, one does not easily whether the answers will consist of one or three solutions.

physics problem as a parameter, but also as a variable quantity, in case we study the dependence of g on the height h above the surface of a planet.

Students have to learn to distinguish and give meaning to letters as parameters and other variables in algebraic expressions and problems involving parameters. Complication factor is that this meaning is in mathematics and physics problems often context dependent. Ursini and Trigueros (2004) found that although students have difficulties in differentiating the role of a parameter and often use memorized facts to make sense of the role of a parameter, their difficulties decrease when they can attribute a referent from a familiar situation to it. In mathematics lessons, the referent of the equation $ax + by + c = 0$ can be that of the straight line in plane geometry, and the equation represents families of straight lines. In physics problems, the context situation provides in many cases the referent, and this is reflected in the choice of letters as abbreviations of full names for quantities involved. The formula $A = A_0 \cdot e^{-r \cdot t}$ not only allows the calculation of the radioactive amount A from the initial amount A_0, the decay rate r, and time t, but it also represents families of exponential decay functions.

3.4   The Process-Object Duality

Mathematics education research on the teaching and learning of algebra has led to several frameworks for understanding difficulties of students in the transition from arithmetic to algebra; we discuss some of them. They all take the viewpoint that at the beginning of learning algebra, the link with arithmetic is emphasized. At this early stage, an algebraic expression is still considered as an arithmetic operation upon numbers, and algebraic manipulation is more or less a change of the recipe to compute a result: $x + x$ and $2x$ are algebraically equivalent because the numerical results when x is replaced by a concrete numerical value are the same, ignoring that the expressions are different from computational point of view. But in order to become successful in algebra, a student must learn to see $2x$ not just as that process of doubling a value but also as a product with which one can continue to compute, say, for example, adding it to 4 to get $4 + 2x$. Also in arithmetic, this process-product duality already comes to the fore when one sees $\frac{4}{6}$ and $\frac{6}{9}$ as equivalent representative of the fraction $\frac{2}{3}$. How small this step may seem, for learners it is in fact a big step in their learning of algebra. It explains why students have difficulty in accepting $4 + 2x$ as an answer and why many still want to simplify it to $2 + x$ or $6x$, or even ignore the variable in the end.[2]

---

[2] Note that arithmetic computations like $4 + 2\frac{1}{2} = 6\frac{1}{2}$ and $4 + 2\frac{1}{3} = 6\frac{1}{3}$ may seduce a learner to generalize to $4 + 2x = 6x$. Similarly, an arithmetic computation like $3\frac{1}{2} - \frac{1}{2} = 3$ may explain the mistake $3x - x = 3$, and the calculation $2 + \frac{1}{2} = 2\frac{1}{2}$ may make a learner believe that $2 + x$ is equal to $2x$.

Sfard (1991) generalized the process-product duality to process-object duality. In her framework, a mathematical conception, i.e., the set of internal representations and associations evoked by some mathematical concept, often consists of two complementary types, namely, (1) an operational conception in which the mathematical concept is seen as a process, algorithm, and/or (imaginable) action and (2) a structural conception in which a mathematical concept is treated as an object. For example, 2×3 has a clear process aspect of multiplication, but it can also be viewed as a prime factorization of the number 6, which on its turn can be seen as $3 + 3$. Three expressions for one and the same integer![3] Another example the expression $4/6$ has a clear process aspect of division, but at the same time, it can be one of the representations of the rational number $2/3$. This does not mean that once we have achieved the notion of $2/3$ as a fraction that we can forget $2/3$ as a division. Rather, the accomplished attitude toward $2/3$ is ambiguous in nature. Ambiguity is an essential characteristic of mathematical thinking and mathematical representation. This is something that students must gradually learn. For example, lower secondary students are surprised that a ratio such as $2 : 3$ is linked to the fraction $2/3$. During the learning process about ambiguity in representation, they may still be at work within an incomplete or even incorrect orientation base and rely on improper solution methods. For example, application of the thin lens equation $\frac{1}{f} = \frac{1}{d_o} + \frac{1}{d_i}$, relating the focal length f, the object distance d_o, and the image distance d_i, and the magnification equation $\frac{h_i}{h_o} = \frac{d_i}{d_o}$ for object height h_o and image height h_i, students often arrive in exercises at an equation like $\frac{4}{h_o} = -\frac{3}{5}$. Solving this equation is then often done by the trick of swapping the numerator and denominator to the equation $\frac{h_o}{4} = -\frac{5}{3}$, but many a student does not really understand why this can be done.

The process aspects of a function such as $f(t) = 10 - 4.9t^2$ are among others the rule for calculating function values and the solving of equations such as $f(t) = 0$. The function is a prescription for a calculation process. Meanwhile, also structural aspects can be identified, such as membership of the family of quadratic functions, being the solution of the initial value problem $f''(t) = -9.8$, $f(0) = 10$, $f'(0) = 0$, and being an object to which one can apply calculus operations like differentiation and integration. In general, the operational conception comes before the structural conception, and the structural conception is viewed as more abstract and advanced. The objectification of processes toward an integrated, object-like whole is called reification. Many difficulties in doing mathematics can be explained as a too strong focus of the students on the operational aspects of mathematical concepts. Again and again, researchers have found that much effort must be put into mathematical instruction for process-object development: for example, all years of secondary school mathematics are used (and needed) to let students develop an adequate understanding of the concept of mathematical function.

---

[3] Karam, Uhden and Höttecke (this volume) use the same example to illustrate that in physics and mathematics, different meanings are assigned to multiplication.

Arcavi et al. (2017) argued in their research-based pedagogical framework for the teaching and learning of algebra that the seemingly simple algebra of school mathematics requires a long path from arithmetic to algebra when adequate algebraic expertise is aimed at. The action-process-object-schema (APOS) framework developed by Dubinsky and others (Arnon et al. 2014) and the three worlds of mathematics framework developed by Tall (2013), in which the proceptual[4] symbolic world is one of the worlds, are two of the commonly used frameworks in mathematics education that take this process-object duality strongly into account in the development of mathematical understanding.

Students must learn to cope with the process-object duality and develop proceptual thinking, or otherwise they will have difficulties to see structure in algebraic questions. For example, 89% of the 73 first year psychobiology students who took in 2015 a resit of Basic Mathematics failed on the following multiple choice question:

To which of the following expressions can one simplify $\sqrt{(x^2-1)^2+4x^2}$?

(a) x^2+2x-1
(b) x^2+1
(c) $3x^2+1$
(d) $2x^2-2x$

The vast majority of students (61) selected the first option and apparently could not resist the visual salience of the erroneous statement $\sqrt{a^2+b^2}=\sqrt{a^2}+\sqrt{b^2}=a+b$, linked to over-linearization of expressions as a result of the feeling to be compelled to do something with the square root in combination with the squares. However, when explicitly asked about the correctness of this statement in a and b, they all knew that it is wrong. They only made the mistake when the symbols were replaced by algebraic expressions: maybe the cognitive load of covering up the subexpression x^2-1 and seeing $4x^2$ as $(2x)^2$ was so high that they could not deal well anymore with these eye-catching elements. But maybe we are here facing the difference between fast and slow decision-making, terms coined by Kahneman (2011), and students are triggered by the structure of the formula to think fast and follow their intuition, instead of thinking slowly and analytically, what would be wiser in this example.

Kirshner and Awtry (2004) concluded that many errors in algebra are not the result of conceptual misunderstanding, but of an overreliance on visual salience. At least we can conclude from the above example that the ability of the students to read through algebraic expressions and to foresee the effects of manipulations were limited. They could recognize certain features of algebraic expressions but failed in a flexible choice of an algebraic manipulation strategy. Bokhove (2011) added the notion of pattern salience to the gestalt view on a formula, that is, the notion of elements that seemingly scream for action. For example, many a student is triggered

---

[4]Procept, a contamination of process and concept, is a term introduced for the combination of symbol, process, and concept, to make clear that a mathematical object never completely loses its process nature.

by the brackets and powers in the equation $(x - 1)^2 + 2 = 6$ to expand brackets in the left-hand side as a first step to rewriting of the quadratic equation into standard form. But an algebraically skilled person would recognize that after subtracting 2 both sides are squares, namely, of $x - 1$ and 2. Again, students seemingly cannot resist fast thinking in case where slow thinking is more appropriate.

Van Stiphout (2011) used Sfard's framework to assess students' algebraic proficiency in pre-university education and found that, although students make progress during their school career, they struggle at the end of school still with tasks that require structure sense and that the difficulties increase according to the number of transitions that have to be made between an operation and a structural approach. Students are said to display structure sense if they can

- recognize a familiar structure in its simplest form;
- deal with a compound term as a single entity and through an appropriate substitution recognize a familiar structure in a more complex form;
- choose appropriate manipulations to make best use of a structure (cf. Hoch and Dreyfus 2006; Novotná and Hoch 2008).

Van Stiphout concluded that the process-object duality is inherently difficult for students and that Dutch mathematics education apparently does not pay enough attention to fostering the process of reification. Looking at the vast literature on learning and teaching algebra (see, e.g., Kieran 1992, 1997, 2007; Stewart and Reeder 2017), we have no reason to doubt that this is different in educational systems in other countries.

3.5 Formula and Equation

When we wrote that a variable gets meaning through the context in which it is used, we meant both the context of "doing school mathematics" and the context of "doing mathematics in school physics," which both have their own conventions. For example, the word formula has a special meaning in school mathematics, and the role of the letters in the *formula* $y = x^2$ is not the same as in the *equation* $y - x^2 = 0$. The words formula and equation are used to distinguish between the case of a functional relationship between the isolated variable that depends on the other variable and the case of a more general relationship between unknowns. For students, it is important to make a clear distinction between these different notions, even though the textbook may not be so explicit about it. A mathematician or scientist, however, is much used to applying the same algebraic symbolism for many purposes: $y = x^2$ may stand for an equation, a function definition, a process of computing the value of y from the value of x, and so on.

Note however that equivalent mathematical expression may elicit different mathematical thoughts: for example, $y = \frac{1}{2}x + \frac{1}{2}$ brings the idea of a straight line with slope $\frac{1}{2}$ more readily in mind than the equivalent $y = \frac{1}{2}(x + 1)$ and $y =$

$(x + 1)/2$, which connect more to the theory of rational functions. In physics, this meaning making of a formula also depends on the symbolism used. For example, a formula like $E = \frac{1}{2}mv^2$ elicits other thoughts in the head of a physicist than just its bare use of symbols: (s)he will probably think, in a glimpse of a second, of kinetic energy E, with m standing for mass of a moving object and v for its velocity. Redish (2005) noted that loading meanings onto symbols leads to differences in how physicists and mathematicians interpret formulas and equations and that labelling of constants and variables differs in these fields. Where variables in calculus are almost always an x, y, z, or t, and constants or parameters are commonly labelled by a, b, c, or d, physicists use many symbols in formulas and equation and load meanings to symbols like m, c, v, V, e, E, and so on, depending on the context (e.g., Q can stand for the amount of heat and the amount of electric charge). In physics, and science in general, it is mostly forbidden to change names in an expression because it will ruin its meaning: for example, the variables in Newton's 2nd law $F = ma$ stand in a unique way for physical concepts, and renaming it as $p = mv$ would turn it into the definition of momentum, a different physical concept. Replacing m by p in Newton's law would ruin it because m denotes mass by default in physics, and any other letter (besides M) does not have this extra meaning. It is as if Newton's law is in fact an equation combined with a set of rather fixed definitions, where one cannot do without the other. In contrast, there is in mathematics more freedom to select names of variables, for example, $\{(a, b) \in \mathbb{R}^2 \mid ab = 1\}$ is the same set as $\{(x, y) \in \mathbb{R}^2 \mid xy = 1\}$. The strong sense of notational conventions in physics and science in general, compared to the notational freedom or algebraic ambiguity in mathematics, occurs on many occasions.

3.6   The Equal Sign

Research (see, e.g., Godfrey and Thomas 2008; Knuth et al. 2006) has found that the meaning of the equal sign causes difficulties at all student levels. In arithmetic, the equal sign is mostly used as a sign for action ("do something; work it out now"): exercises like $2 + 3 = \square$ and $2 + \square = 5$ ask for the result of some calculation. When lower secondary students first practice with physics problems like "what is the distance covered by a cyclist when he bikes for two hours at a speed of 20 km/h?" and later on encounter the word formula *speed* × *duration* = *distance*, then they will interpret the equal sign as a sign for doing a computation. For them, it is not obvious that *distance* = *speed* × *duration* can mean something else. Replacing words by symbols does not change this perspective. The operational, process-oriented perspective of the sign suffices during early years at school, but students must develop an object-oriented understanding of equality as equation and equivalence relation so that they can think algebraically. They must also start recognizing when the equal sign is used in the sense of "is equal to," "is supposed to be equal to," "corresponds with," "is assigned the value," "is defined as," and "is equivalent to." In common language, the term "is equal to" or "corresponds with" is mostly used to express that for some useful purpose two things are swappable and that it makes no real difference which one to use. We may say that 100 centimeters equals 1 meter and may write down 100 cm $=$ 1 m. It is a matter of convenience which unit is used, but one will notice a difference in practice when doing measurements or calculations. The entities are swappable, but not the same. Written in the form 1 cm $= \frac{1}{100}$ m, it looks more like a definition or conversion rule. The equal sign has lost its notion of balance between the sides of the equality and is now used in a unidirectional way.

Thus, the dual nature of an algebraic expression to indicate both a process and an object is never lost: statements like

$$\frac{1}{x+1} + \frac{1}{x+2} = \frac{2x+3}{x^2+3x+2}, \quad \text{for every real number } x$$

and

$$\frac{2x+3}{x^2+3x+2} = \frac{1}{x+1} + \frac{1}{x+2}, \quad \text{for every real number } x$$

are in formal sense equivalent, but it cannot be denied that the first statement is about computing the sum of two rational expressions, whereas the second one represents a partial fraction decomposition. In the first statement, the equal sign is not read as "is formally equivalent to", but as "yields": it is about addition, triggered even more by the plus symbol between the two terms on the left-hand side. In the second statement, the equal sign indicates equivalence by referring to a process of simplification, which is not immediately triggered by the expression on the left-hand side but has more to do with the context in which it is used (say, integration or decomposition in similar terms). Students must become aware of the dual nature of an algebraic expression, or in the terminology of Tall (2013) become proceptual thinkers, in order to acquire algebraic expertise. It would help if teachers would explain their students why they sometimes read a formula from right to left instead of from left to right. Algebraic expertise of students would be increased if teachers would explain why they sometimes expand brackets in a mathematical expressions or why they do the opposite and factorize an expression.

It is interesting to notice that in programming language, different symbols are used for assignment (or definition), equality, and sameness: the mathematical software *Mathematica* uses =, ==, and === for these three purposes, respectively. In mathematical texts, the symbols := and $\overset{\text{def}}{=}$ are often used to make a definition; the symbols $\equiv$ and $\sim$ denote equivalence. But just using different symbols for different meanings does not help students develop algebraic insight or symbol sense,[5] but

---

[5] Arcavi (2005) described symbol sense as the ability to give meaning to symbols, expressions, and formulas and to have a feeling for their structure. Drijvers (2011, p. 22) confined the interpretation of symbol sense to the understanding of the meaning and structure of algebraic formulas and expressions, which involves (1) the strategic abilities to arrive at a problem approach and to

may only confuse them more. Instead, we are of opinion that teachers and textbook authors should be more explicit to their students and readers in explaining the ambiguous notations that they use so that students can familiarize themselves with purposeful ambiguity in mathematical notation and can start appreciating the mathematical symbolism. For example, teachers could talk with their students more about the distinction between an identical equation (or identity), which holds for all values of the variables involved, and a conditional equation, which is only true for certain values of the variables. Adding a few words to a mathematical expression may already help: taken by itself, $a + b = b + a$ is meaningless, but "for all real numbers a and b, $a + b = b + a$" describes the commutativity of addition more clearly. Instead of just writing down the equation $x^2 = 2$, one could add a few words to arrive at "for some number x such that $x^2 = 2$." When one writes "the function $y = x^2$," it is immediately clear that one means something else than "the equation $y = x^2$." Too easily in mathematical texts and in explanations on the blackboard, the words that support the meaning of mathematical expressions disappear.

Research has found that as a consequence, many students at all levels hold misconceptions about the equal sign and cannot distinguish formulas and equations. For example, some young students believe that the equal sign cannot be used in an equation that does not have an operator symbol (i.e., $2 = 2$) and that all operator symbols must be at the left-hand side of an equation. Older students, even freshmen at university (Godfrey and Thomas 2008), may believe that an equation must at least contain one variable or emphasize the solution aspect of an equation (and thus do not see $a + b = b + a$ as an equation).

3.7   Variables in Physics

In physics, a variable is mostly used as a name for a quantity that can vary (often with respect to time) and that in many cases can be measured. The name of a physical quantity is often an abbreviation of the full name: T for temperature, V for volume, m for mass, v for velocity, and so on. Measured values are always floating-point numbers, possibly with margins of error, or natural numbers (in counting processes). Physicists mostly compute with quantities as with real numbers except that they take into account the accuracy of floating-point arithmetic. A value of a physical quantity actually consists of three parts, namely, the numerical value (a number), the precision (the number of significant decimals or the margins of error), and the unit that is used to measure the quantity. This makes quantity arithmetic more difficult to learn and to use than reference-free number arithmetic. On the other hand, the units of physical quantities can also hint whether the formula into which numeric value numbers have been plugged is correct: when one keeps track of

---

maintain an overview of this process, (2) the capacity to view symbolic expressions globally, and (3) the capacity of algebraic reasoning.

the units during a computation, one should end with a numerical value and a valid unit for the computed quantity. This however requires algebraic skills that lower secondary students may not have sufficiently developed yet.

Quantity arithmetic is also often connected to a particular perspective on the context situation. The following example illustrates this. First, compare the given answers to the following problem:

A red and blue car move slowly in a straight line in the same direction from the same starting point, with the same speed of 2 m/s. The red car starts first, at time $t = 0$, and the blue car 3 s later. Give the formula for the distance s that the blue car has moved after t seconds, for $t > 3$ s.

$$\text{Answer 1}: \ s = 2(t - 3) \qquad \text{Answer 2}: \ s = 2t - 6$$

In generalized arithmetic, the two expressions are equivalent: a factored and expanded form. In quantity arithmetic, which takes dimensions into consideration, the first answer represents a time approach of the problem via a *distance* = *speed* × *time* formula, and the second answer represents a distance approach via a *distance* = *distance* − *distance* formula. In the terminology of Sherin's framework (2001, 2006) for understanding of physics equations, this means that two different symbolic forms are invoked, namely, intensive-extensive ($\Box \times \Box$) and opposition ($\Box - \Box$) and that the reasoning is different. But students are not always aware why to switch to an alternative equivalent form, for example, why to expand brackets or to factorize an expanded form. It takes much time, effort, and experience to develop symbol sense in both mathematics and physics context.

Whereas in mathematics, variables often occur in isolation (e.g., in a quadratic expression $x^2 + 2x + 3$), in physics, they are mostly used in relationships that combine two or more quantities. As example may serve Boyle's law $pV =$ constant, for pressure p and volume V. But the variables in this law are not only used as varying quantities; depending on the problem to solve, the symbols may also stand for a finite set of values or a single value. Thus, in Boyle's law, the variables may actually be functions of time, viz., $p \mapsto p(t)$ and $V \mapsto V(t)$, and the law says that the product of these functions is a constant function, i.e., $p(t) \cdot V(t) =$ constant, at every time t. In a problem like "suppose that $V = 10$ mL when $p = 3$ bar, how much is V when $p = 6$ bar?" p and V do not represent functions anymore but function values, namely, the pressure and volume at a certain fixed time. Actually, we solve such problem via a relationship between two states characterized by two physical quantities: $p_1 \cdot V_1 = p_2 \cdot V_2$, where $(p_1, V_1) = (3, 10)$.

Proportionality is another type of relationship between several states involving two or more physical quantities. For example, Charles's law (also known as Gay-Lussac's 1st law), $V/T =$ constant, for volume V and temperature T, can also be written as $V_1/T_1 = V_2/T_2$. In school physics problems, three of the four quantities are often given, and a student is asked to use the proportionality to compute the value of the fourth quantity. Again, the variables are not treated anymore as varying quantities but as momentary values or constants.

The focus of mathematical thinking may differ radically from the one in a study of a real-world problem. For example, when one encounters in mathematical work a rational expression of the form $\frac{ax}{x+b}$, the attention goes almost automatically to the singular behavior as x approaches $-b$. Compare this with the study of enzyme kinetics, where the Michaelis-Menten expression for the initial rate of transformation of a substrate, S, by an enzyme, is $v = \dfrac{V_{max}[S]}{K_M + [S]}$, where $[S]$ is the concentration of S, V_{max} is the maximum rate, and K_M is the so-called Michaelis-Menten constant. Because the parameters and concentrations are positive, the singularity is never encountered, and the common mathematical analysis is irrelevant. Sometimes a mathematical conflict corresponds with a problem in the context situation. An example is electric short circuit, in which the resistance becomes zero, and Ohm's law gives an infinitely high current. In reality, the resistance is never equal to zero, but it can be small enough to lead to a very high current. A physicist will often ignore the mathematical problems and use Ohm's law whenever it is convenient. In other case, a physicist may ignore the mathematical problems by circumventing them via practical reasoning.

In physics, or more generally in science, words like "big," "small," and "relatively small" can be used while talking about physical quantities, and the absolute values of the physical quantities are often used for this purpose. A force F_1 of -10 N is stronger than a force F_2 of 3 N. But if students have learned in mathematics lessons that $3 > -10$, they might be tempted to write $F_2 > F_1$ and call F_2 greater than F_1 in the sense of strength. The symbol $\ll$ signifies that the left-hand side is much smaller than the right-hand side. This notation is not used in mathematics. A small change of a quantity Q is also given a name in science, such as ΔQ, and one manipulates it as any other variable, except that one often ignores higher-order term like $(\Delta Q)^2$ to get a simpler model description. In this way, many physics laws are derived. Going from calculus of small changes to infinitesimal change and calculus of differentials is then a natural step. In a system dynamics approach to mathematical modelling, Van Buuren (2014) showed that difference equations are already within reach of lower secondary students. Actually, in his design experiments, when students were confronted with a change in notation for the formula for distance, velocity, and time from $s = v \cdot t$ to $\Delta x = v \cdot \Delta t$, they argued that it would be easier for them to use Δ-notation from the beginning (with ΔQ denoting a finite difference of the quantity Q).

3.8   Mathematical Modelling

Redish and Kuo (2015) remarked that not only the mathematics is different in physics but also its purpose is different: in physics, one tries to make sense of the real world, not of an abstract field built upon axioms. Equations often come from theories and laws of physics, and they are used in models to describe and explain the real world. As Redish (2016) argued, equations are used to organize and pack

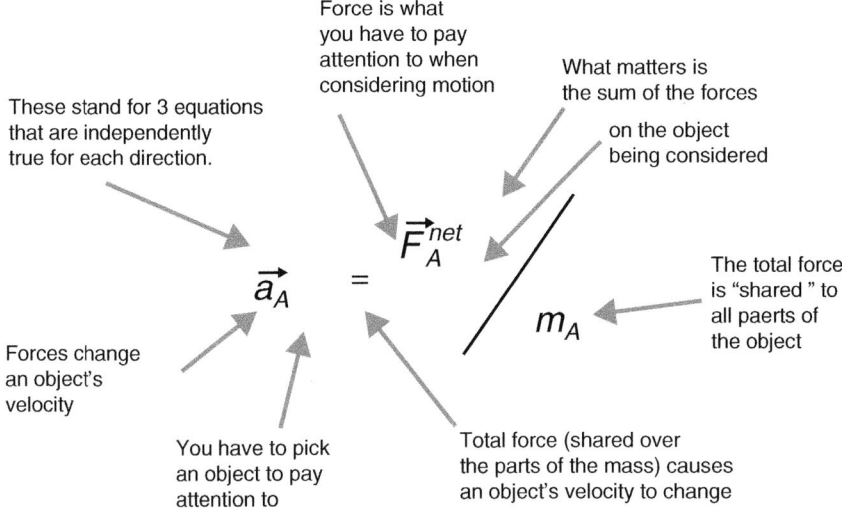

Fig. 3.1 Conceptual knowledge packed in a single equation, viz., Newton's second law for the 3-dimensional motion of object A. (Picture taken from Redish 2016)

conceptual knowledge. He illustrated this with Newton's second law of motion as shown in Fig. 3.1.

Torigoe (2015) argued that many students struggling with symbolic equations activate a web of resources appropriate for solving numeric problems and are not sensitive to the cues for resources appropriate for symbolic problem solving. Kanderakis (2016) argued that in physics, the continual interplay between physics concepts, models, mathematical symbolic expressions, and the real world causes students' difficulties in modelling activities, because the mathematics of secondary school physics is more complex semantically and conceptually than secondary school mathematics. He also used the history of mathematization of physics to show that mathematical concepts, structures, and operations are indispensable parts of modern physics and consequently of secondary school physics and to exhibit that the use and construction of models are essential for a quantitative mathematical study of the physical word, also when models are created to apply known physics theory to real physical systems. On the other hand, we are of opinion that the real world actually helps making physics models concrete, tangible, and understandable. Students just need a helping hand in the mathematization. A nice aspect of physics is that one can often act and think in a non-mathematical way as well, and this offers a broader perspective and extra means to check one's findings.

Redish (2005) noted that traditional physics education does not give enough emphasis to many of the critical steps in mathematical modelling: students are mostly provided with a ready-made model that they only have to manipulate mathematically in order to get an answer to a posed problem. This explains why many students find it difficult to create models themselves from verbal descriptions of problem situations. As Duggan et al. (1996) identified as a critical point in investigative work, selecting and defining relevant variables (such as independent, dependent, and control variables) and relating them to each other in a qualitative sense of cause-effect or in a quantitative way for a problem at hand are already difficult steps for lower secondary students to make at the beginning of inquiry-based investigative work. Cause-effect is not always as easy as it may look at first sight: for example, many a substance expands when the temperature increases. In this perspective, it is the temperature change that causes the expansion. But in a mercury thermometer, the temperature is read off from the height of the mercury in the glass tube. Here, the temperature depends on the height. So it is not straightforward on many occasions which variable to select as independent variable and which one as depending variable. In physics problems in which mass, density, and volume play a role or in physics problem involving distance, speed, and time, it depends on the concrete task which variable is best seen as the dependent one.

3.9 Proportionality, Ratio Table, and Cross-Multiplication Table

Two related notions stand out in mathematics and physics education, namely, proportionality and ratio table, and again mathematics and physics education treat them differently, even when it concerns only the notion of cross-multiplication table. Van der Valk and Broekman (2001) pointed out that in mathematics textbooks, a ratio table has in most cases the following characteristics:

- it consists of two rows in no particular order and a variable number of columns;
- the rows have a label to indicate the meaning of the numbers and to specify, if needed, the units used;
- the ratio between the numbers in the columns is the same for all columns;
- proportional arithmetic operations in the vertical and/or horizontal direction are used to calculate empty places in the ratio table.

There is still room for extension: the function table shown in Table 3.1 originates from a Dutch mathematics textbook and is used herein as an example to verify that the ratio between a certain power of x and y is the same for all columns. In this case, y is directly proportional to x^3, and in particular $y = 0.3x^3$. Note that in this example, the independent and dependent variable are in the first and second row, respectively. In a real context, swapping the rows is often useful: for example, in a

Table 3.1 Example of a function table in a mathematics textbook at upper secondary level, taken from Reichard et al., *Getal & Ruimte* [Numbers & Space] wi havo D deel 3, 2007, p. 74

x	1.8	2.3	3.7	4.1	5.3	6.1
y	1.75	3.65	15.2	20.7	44.7	68.1

Table 3.2 Example of a ratio table in a science textbook at lower secondary level, taken from Hogenbirk et al., *Natuur- en Scheikunde Overal* [Physics and Chemistry Everywhere], 2 mHV, 1997, p. 111

		× 3600
Duration	1 s	1 h
Distance	7.2 m	25,920 m ≈ 26 km
		× 3600

table of salary as a function of working hours, one can better put amount of salary in the first row and number of working hours in the second row because the ratio has then a meaning as hourly wages. This layout of a ratio table would better match with the distance-time table often used in physics textbooks.

In science textbooks, Van der Valk and his colleagues found several versions of ratio tables, including Table 3.2 about the conversion of the constant speed of a moped from m/s into km/h.

In this example, there are in fact two ways to compute the distance travelled in 1 h: using the multiplicative relationship within or between the contextual magnitudes (duration, distance). In Table 3.2, the symbol × 3600 suggests the first approach, i.e., multiply within the variable "distance." But one can also use the multiplicative relationship between the contextual magnitudes and multiply the duration of 3600 s by the unit rate of 7.2 m per 1 s. In mathematics education research literature, the first approach is called the scalar perspective on proportional reasoning, and the second one is called the functional perspective. In addition, there is the cross-multiplication algorithm. A high level of proportional reasoning involves the flexible use of either the scalar or functional relationship depending upon the ease of calculation with the numbers in the problem.

In the above example, the ratio table differs in four aspects from a mathematical ratio table:

- only four cells are present, which allow the cross-multiplication algorithm to be used instead of row and column operations;
- labels are missing near quantities and instead information like units is placed inside cells;
- the ratio of the numbers changes due to the use of units and rounding (1:7.2 becomes 1:26 from mathematical point of view);
- rows are changed compared to a mathematics-like table: The ratio expressed in the table changes from 1 s :7.2 m to 1 h :26 km, but the task linked to this ratio table is actually about speed, i.e., distance divided by duration; so why not an ordering of rows that reflects this?

Van der Valk and Broekman (2001) advocated that the subject of ratio tables is one of the areas in which mathematics and science teachers could collaborate and gear activities for one another to let their students develop good proportional reasoning skills. Otherwise, the ratio table may remain an algorithm instead of a tool for promoting understanding. This cooperation between mathematics and physics in teaching and learning about proportionality is not as straightforward as it may seem because proportional reasoning goes beyond the ratio table and includes verbalization of proportionalities; conversion of the verbalization to formulas, tables, and graphs; and recognition of behavior of these tables and graphs. About graphing in mathematics and physics, Ellermeijer and Heck (2002) pointed at technical, linguistic, and contextual differences. As an example of different perspective on proportionality formulas in mathematics and physics, we mention that one can often find in mathematics texts the characterization of proportionality and inverse proportionality of variables P and Q in the format of formulas $P/Q = c$ and $P \cdot Q = c$ for some constant c, respectively. In physics, the mainstream perspective of characterizing quantities P and Q as proportional and inversely proportional apparently is $P = c \cdot Q$ and $P = c \cdot Q^{-1}$, respectively. These two perspectives (constant ratio or constant multiplier) are reflected in the different ways the ratio tables are treated in textbooks in these disciplines and whether the scalar or functional perspective on proportional reasoning is promoted.

3.10   Conclusion

In this chapter, we have illustrated that math in physics is not the same as math in mathematics. Physicists and mathematicians do mathematics for different purposes, load meanings onto symbols differently (i.e., link equations with physical contexts and let symbols carry extra information not present in the mathematical formulas), and seemingly communicate in a different language. Redish and Kuo (2015) came to the same conclusion and also provided a lot of examples. These differences, when not explicitly discussed in classroom, confuse many students and often bring them into situations where they do not know how to proceed. What also happens in school practice is that students do not recognize similarities between math in mathematics and math in physics because they are obscured by the context. Here, it helps if a teacher points at similarities and discusses the students whenever possible. We are of the opinion that knowledge about the differences and possibly obscured similarities of use of math in physics and mathematics helps teachers understand students' obstacles in doing math. Once physics and mathematics teachers have more insight into students' difficulties in math caused by epistemological differences between the fields, they can jointly make steps toward coherent teaching. This requires an open mind, time and willingness for discussion with other teachers, goodwill from school or faculty directors, and instructional materials that are jointly written for coherence. It is not easy but doable if one sets achievable goals. One such goal is to be consistent between mathematics and physics wherever possible: it is unhelpful to have arbitrary differences in approaches and terminology between the subjects; students benefit if they are pointed at similarities in the use of mathematics in the various fields.

As a concrete example of what can be achieved by gearing representations in mathematics and physics lessons for one another, we take the proposal of Weijers

Table 3.3 A ratio table with extra rows for unit conversion; compare this table with Table 3.2

Duration (s)	1	3600
Duration (h)		1
Distance (m)	7.2	25,920
Distance (km)		25.92

and Van der Valk (2001) for extending the ratio table to include unit conversion. The main idea is to extend the number of rows of a ratio table for unit conversion, which in most cases also requires proportional reasoning. Table 3.2 about a moped going at a speed of 7.2 m/s is in the proposed representation equal to Table 3.3. At the end, the speed can be approximated to 26 km/h.

How small this change may look like at first sight; yet it enables the use of similar representations and operations in ratio tables in mathematics and science disciplines. When teachers in these disciplines look in each other's textbooks, discuss notations and methods with each other, and collaborate to bring more coherence in the use of mathematics in their subjects, they will find other opportunities to improve teaching and learning. In the recent reports of Bohan (2016) and Needham (2016), written to support UK teachers of 11–16 science in the use of mathematical ideas in the new science curriculum, one can find examples and recommendations. Concrete suggestions can be as small but as helpful as choosing a common notation for derivatives and denoting physical quantities, units, and symbols in the same manner with the same typography. Also it helps if mathematics teachers like physics teachers work with formulas that are independent of the units used: for example, they better avoid wordings like "the length of L centimeters" and "L is the length in centimeters" and then construct a formula in L, but instead do what physics teachers do, stating just "L is length." Coherence between mathematics and science can also be improved by sharing lesson planning formats where a common mathematical focus is identified or by having linked lessons on a common theme and comparing approaches from the various disciplines.

But as argued in this paper, essential differences between the use of math in mathematics and physics remain unavoidable. The best thing teachers can do is discuss with their students the differences and similarities of mathematical representations. One better points students at discipline-based conventions and how they relate to each other. For example, in mathematics, one prefers the lexicographic ordering in mathematical expressions and does not use labelled names if they can be avoided; physics often uses labelled names (if only to label the object under investigation or the source of a force) and has its own conventions of ordering. Teachers better clarify the meaning of algebraic terms and notations in their lessons and help students relate them to what they already know from earlier mathematics and physics lessons. For example, Van Buuren (2014) clarified in his modelling learning path for lower secondary physics student what he meant with terms like equation, formula, and a simple calculation. Even though these were operational definitions, which would be extended at later stages of education, they helped his students to make the first steps in algebraization of context situations and in doing computations that would prepare them for creating models later on in their school career.

In summary, the main recommendations for mathematics and physics teachers at all levels of education are that they discuss with their colleagues how mathematics is used in their fields, try to agree on consistent use of mathematics in their lessons wherever possible, and explicitly discuss in their classrooms with their students what the similarities and differences between mathematics use in their fields are. Concrete examples will help their students better grasp the mathematical methods and techniques and transfer them from one subject to another.

References

Arcavi, A. (2005). Developing and using symbol sense in mathematics. *For the Learning of Mathematics, 25*(2), 42–47.

Arcavi, A., Drijvers, P., & Stacey, K. (2017). *The learning and teaching of algebra: Ideas, insights, and activities.* London: Routledge.

Arnon, I., Cottrill, J., Dubinsky, E., Oktaç, Roa-Fuentes, S., Trigueros, M., & Weller, K. (2014). *APOS theory: A framework for research and curriculum development in mathematics education.* New York: Springer.

Bohan, R. (2016). *The language of mathematics in science. A guide for teachers of 11–16 science.* Hatfield: Association for Science Education. Retrieved April 13, 2017 from http://www.ase.org.uk/documents/language-of-mathematics-in-science-1/

Bokhove, C. (2011). *Use of ICT for acquiring, practicing and assessing algebraic expertise.* Doctoral theis, University of Utrecht. Utrecht: CD-β Press. Retrieved February 25, 2017 from http://dspace.library.uu.nl/handle/1874/214868

Drijvers, P. (Ed.). (2011). *Secondary algebra education: Revisiting topics and themes and exploring the unknown.* Rotterdam: Sense Publishers.

Duggan, S., Johnson, P., & Gott, R. (1996). A critical point in investigative work: Defining variables. *Journal of Research in Science Teaching, 33*(5), 461–474.

Ellermeijer, T., & Heck, A. (2002). Differences between the use of mathematical entities in mathematics and physics and consequences for an integrated learning environment. In M. Michelini & M. Cobal (Eds.), *Developing formal thinking in physics* (pp. 52–72). Selected contributions to the first international GIREP seminar, Udine, September 2001. Udine: Forum. Retrieved February 25, 2017 from www.fisica.uniud.it/URDF/articoli/ftp/2001/Imp%20Developing%20Formal.pdf

Freudenthal, H. (1983). *Didactical phenomenology of mathematical structures.* Dordrecht: Reidel Publishing Company.

Godfrey, A., & Thomas, M. O. J. (2008). Student perspectives on equation: The transition from school to university. *Mathematics Education Research Journal, 20*(2), 71–92.

Heck, A. (2001). Variables in computer algebra, mathematics, and science. *The International Journal of Computer Algebra in Mathematics Education, 8*(3), 195–221.

Heck, A., Kedzierska, E, & Ellermeijer, T. (2009). Design and implementation of an integrated computer working environment for doing mathematics and science. *Journal of Computers in Mathematics and Science Teaching, 28*(2), 147–161.

Hoch, M.,& Dreyfus, T. (2006). Structure sense versus manipulation skills: An unexpected result. In J. Novotná, H. Moraová, M. Krátká, & N. Stehlková (Eds.), *Proceedings of the 30th Conference of the International Group for the Psychology of Mathematics Education,* Prague (Vol. 3, pp. 305–312). Retrieved February 25, 2017 from http://files.eric.ed.gov/fulltext/ED496933.pdf

Kahneman, D. (2011). *Thinking, fast and slow*. New York: Farrar, Straus and Giroux.

Kanderakis, N. (2016). The mathematics of high school physics. *Science & Education, 25*(7–8), 837–868.

Kieran, C. (1992). The learning and teaching of school algebra. In D. A. Grouws (Ed.), *Handbook of research on mathematics teaching and learning* (pp. 390–419). Reston: National Council of Teachers of Mathematics.

Kieran, C. (1997). Mathematical concepts at the secondary school level: The learning and teaching of algebra and functions. In T. Nunes & P. Bryant (Eds.), *Learning and teaching mathematics: An international perspective* (pp. 133–158). Hove: Psychology Press.

Kieran, C. (2007). Learning and teaching of algebra in the middle school through college levels: Building meaning for symbols and their manipulation. In F. K. Lester (Ed.), *Second handbook of research on mathematics teaching and learning* (pp. 707–762). Charlotte, NC: Information Age Publishing.

Kirshner, D., & Awtry, T. (2004). Visual salience of algebraic transformations. *Journal for Research in Mathematics Education, 35*(4), 224–257.

Knuth, E. J., Stephens, A., McNeill, N., & Alibali, T. (2006). Does understanding the equal sign matter? Evidence from solving equations. *Journal for Research in Mathematics Education, 37*(4), 297–312.

Kücheman, D. E. (1981). Algebra. In K. M. Hart (Ed.), *Children's understanding of mathematics: 11–16* (pp. 102–119). London: John Murray.

Malisani, E. (2006). *The concept of vairable in the passage form the arithmetic language to the algebraic language in different semiotic contexts*. Doctoral thesis, Comenius University of Bratislava. *Quaderni di Recerca in Didattica Del G.R.I.M.*, 16(1). Retrieved February 25, 2017 from http://math.unipa.it/~grim/quad_16_suppl_1.htm

Malisani, E., & Spagnoloi, F. (2009). From arithmetical thought to algebraic thought: The role of the "variable". *Educational Studies in Mathematics, 71*(1), 19–41.

Needham, R. (2016). *The language of mathematics in science. Teaching approaches*. Hatfield: Association for Science Education. Retrieved April 13, 2017 from http://www.ase.org.uk/documents/ase-the-language-of-mathematics-in-science-teaching-approaches/

Novotná, J., & Hoch, M. (2008). How structure sense for algebraic expressions or equations is related to structure sense for abstract algebra. *Mathematics Education Research Journal, 20*(2), 93–104.

Radford, L. (1996). The roles of geometry and arithmetic in the development of elementary algebra: Historical remarks from a didactic perspective. In N. Bednarz, C. Kieran, & L. Lee (Eds.), *Approaches to algebra: Perspectives for research and teaching* (pp. 39–53). New York: Springer.

Redish, E .F. (2005). Problem solving and the use of math in physics courses. *Invited Talk Presented at the Conference, World View on Physics Education in 2005: Focusing on Change*, Delhi, August 21–26, 2005. Retrieved February 25, 2017 from arXiv:physics/0608268 [physics.ed-ph].

Redish, E. F. (2016). Analysing the competency of mathematical modelling in physics. In G. Tomasz & E. Dbowska (Eds.), *Key competences in physics teaching and learning* (pp. 25–40). Switzerland: Springer International Publishing. Preprint available at arXiv:physics/61604.02966 [physics.ed-ph].

Redish, E. F., & Kuo, E. (2015). Language of physics, language of math: Disciplinary culture and dynamic epistemology. *Science & Education, 24*(5), 561–590.

Schoenfeld, A. H., & Arcavi, A. (1988). On the meaning of variable. *Mathematics Teacher, 81*(6), 420–427.

Sfard, A. (1991). On the dual nature of mathematical conceptions: Reflections on processes and objects as different sides of the same coin. *Educational Studies in Mathematics, 22*(1), 1–36.

Sherin, S. (2001). How students understand physics equations. *Cognitive Science, 19*(4), 479–541.

Sherin, S. (2006). Common sense clarified: The role of intuitive knowledge in physics problem solving. *Journal of Research In Science Teaching, 43*(6), 535–555.

Skemp, R. R. (1986). *The psychology of learning mathematics* (2nd ed.). Harmondsworth: Penguin Books Ltd.

Stewart, S., & Reeder, S. (2017). Algebra underperformances at college level: What are the consequences? In S. Stewart (Ed.), *And the rest is just algebra* (pp. 3–18). Cham: Springer International Publishing.

Tall, D. (2013). *How humans learn to think mathematically: Exploring the three worlds of mathematics*. New York: Cambridge University Press.

Tarski, A. (1956). Symbolic logic. In J. R. Newman (Ed.), *The world of mathematics* (Vol. 3, pp. 1901–1931). New York: Simon& Schuster. Original work published 1941.

Torigoe, E. T. (2015). Unpacking symbolic equations in introductory physics (preprint). arXiv: 1508.00535 [physics.ed-ph].

Trigueros, M., & Ursini, S. (2003). First-year undergraduates difficulties in working with different uses of variable. In A. Selden, E. Dubinsky, G. Harel, & F. Hitt (Eds.), *CBMS issues in mathematics education* (Vol. 12, pp. 1–29). Washington, DC: American Mathematical Society.

Uhden, O. (2012). *Mathematisches Denken im Physikuntericht* [Mathematical thinking in physics education]. Doctoral Thesis, Technical University of Dresden. Logos Verlag: Berlin. Retrieved February 25, 2017 from http://d-nb.info/106773242X/34

Uhden, O., & Pospiech, G. (2010). Translating between mathematics and physics: Analysis of student's difficulties. In W. Kaminsky & M. Michelini (Eds.), *Teaching and learning physics today: Challenges? Benefits?* (pp. 102–106). *Proceedings of Selected Papers of the GIREP-ICPE-MPTL International Conference 2010, Reims*. Udine: University of Udine. Retrieved February 25, 2017 from http://iupap-icpe.org/publications/proceedings/GIREP-ICPE-MPTL2010_proceedings.pdf

Uhden, O., & Pospiech, G. (2012). Mathematics in physics: Analysis of students difficulties. In C. Bruguire, A. Tiberghien, & P. Clément (Eds.), *Science learning and citizenship, E-Book Proceedings of the ESERA Conference 2011*, Lyon (Part 3, co-edited by M. Michelini & R. Duit, pp. 218–222) Lyon: ESERA. Retrieved February 25, 2017 from http://www.esera.org/media/ebook/ebook-esera2011.pdf

Ursini, S., & Trigueros, M. (2001). A model for the uses of variable in elementary algebra. In M. van den Heuvel-Panhuizen (Ed.), *Proceedings of the 25th Conference of the International Group for the Psychology of Mathematics Education* (Vol. 4, pp. 327–334), Utrecht. Retrieved February 25, 2017 from http://files.eric.ed.gov/fulltext/ED466950.pdf

Ursini, S., & Trigueros, M. (2004). How do high school students interpret parameters in algebra? In M. J. Hoines & A. B. Fuglestadt (Eds.), *Proceedings of the 28th conference of the International Group for the Psychology of Mathematics Education*, Bergen (Vol. 4, pp. 361–368). Retrieved February 25, 2017 from http://files.eric.ed.gov/fulltext/ED489597.pdf

Usiskin, Z. (1988). Conceptions of school algebra and uses of variables. In A. F. Coxford & A. P. Shulte (Eds.), *The ideas of algebra, K-12* (pp. 7–13). Reston: National Council of Teachers of Mathematics.

Van Buuren, O. (2014). *Development of a modelling learning path*. Doctoral thesis, University of Amsterdam, Amsterdam: CMA. Retrieved February 25, 2017 from http://hdl.handle.net/11245/1.416568

Van der Valk, T., & Broekman, H. (2001). Science teachers' learning about ratio tables. In M. van den Heuvel-Panhuizen (Ed.), *Proceedings of the 25th Conference of the International Group for the Psychology of Mathematics Education*, Utrecht (Vol. 4, pp. 351–358). Retrieved February 25, 2017 from http://files.eric.ed.gov/fulltext/ED466950.pdf

van Stiphout, I. M. (2011). *The development of algebraic proficiency*. Doctoral thesis, Technical University Eindhoven. Eindhoven: EsOE. Retrieved February 25, 2017 from http://hdl.handle.net/11245/1.416568

Weijers, M., & Van der Valk. (2001). Aandachtspunten bij het gebruik van verhoudingstabellen in de natuurwetenschappen [Point of attention in using ratio tables in the natural sciences]. *Nieuwe Wiskrant, 20*(4), 23–27. Retrieved April 13, 2017 from http://www.fi.uu.nl/wiskrant/artikelen/artikelen11-20/204/204juni_wijers-vdvalk.pdf

Chapter 4
Mathematical Representations in Physics Lessons

Marie-Annette Geyer and Wiebke Kuske-Janßen

4.1   Introduction

Communication in science is characterized by the use of specific types of representations. Physicists use different ways to represent their knowledge and to explain and analyze phenomena. In the same way, these specific representations occur in physics lessons and play an important role in the teaching and learning of physics. Different media and representations are used as the following example of a typical situation in physics class illustrates: the design of an experiment is sketched, the experiment is conducted, the relation between the measured quantities is firstly represented in a table and then in a graph, a formula is derived, and conclusions are formulated. During all these steps, verbal language in an oral or written form occurs as well.

Particularly mathematical representations play an important role in physics and physics lessons. It is alarming that learners show problems especially in understanding and handling mathematics in physics (e.g., Uhden 2015; Pospiech and Oese 2014; Bagno et al. 2011; McDermott et al. 1987). Therefore, the interest of this article lies in the description and classification of different mathematical representations in physics lessons. This helps to understand specific characteristics of these different representations and is a first important step toward investigating students' handling with these representations and identifying their difficulties as well.

Beginning with a general description of representations from a cognitive sciences' and semiotics' view, the article presents the state of theory about representations in physics education with focus on mathematical ones. Based on this,

---

The authors contributed equally to this work.

M.-A. Geyer (✉) · W. Kuske-Janßen
Faculty of Physics, TU Dresden, Dresden, Germany
e-mail: marie-annette.geyer@tu-dresden.de; wiebke.kuske-janssen@tu-dresden.de

© Springer Nature Switzerland AG 2019
G. Pospiech et al. (eds.), *Mathematics in Physics Education*,
https://doi.org/10.1007/978-3-030-04627-9_4

an adapted model for the classification of representations is presented. Proceeding from theoretical considerations and empirical findings about the relevance of representations for learning and understanding physics and students' difficulties, two theoretical models are presented. These models are part of current research and offer an approach to analyze different changes of representations in physics classes. Ultimately several implications for teaching will be derived.

4.2 Representations

The Encyclopedia of Science Education defines representations as "notions or signs or symbols that stand for something in the absence of that thing, a thing which typically is a phenomenon or an object in the external world but can be just in our imagination" (Dolin 2016, p. 836 f.). This definition needs specification for our purpose of describing mathematical representations in physics lessons.

First of all the definition differentiates between the external world and our imagination. This links to a description of representations either as *external* or as *internal*. External representations are used to communicate meaning. These external representations are perceived by our senses and converted to an internal representation; some authors (e.g., Greca and Ataide in this book) also speak of mental models of the external representation. Gilbert defines: "The meanings attached in the brain to a given external representation may therefore be called a personal internal representation of it and is a visualization of it" (Gilbert 2016, p. 122).

Bruner (1974) specifies internal representations and how they evolve with increasing age and the development of cognitive abilities from birth till adulthood. He differentiates between enactive, iconic, and symbolic representations (cf. Fig. 4.1). As a child we are only able to represent the world internally by acting. We discover the world by movement and the sense of touch, and these actions are our mental representation of the world. An example for an enactive representation is the manual palpation of different substances. After a while learning progresses toward representing the world by iconic signs. We look at a cat and construct an internal picture of the cat. The picture has similarity with the original cat. Because of these similarities, we are able to connect our internal representations with the objects we

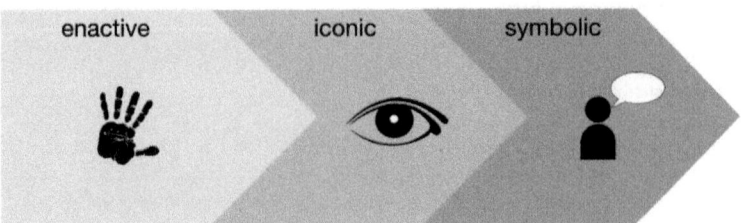

Fig. 4.1 Bruner's classification of representations

represent. A third and most complex way to represent objects is using symbolic representations. A symbol does not have any similarity with the object it represents. Frequently the mapping between representations and objects follows socially shared principles; the connection of a symbol and its content is an arbitrary convention. An example for symbolic representations is the human language. We see a cat and represent it internally by thinking the word "cat." This word does not have any similarity relations to the cat but is an abstract sign for a group of objects, in this case for animals walking on four paws, with a tail, that make "mew." In physics lessons, all three categories, enactive, iconic, and symbolic representations, play an important role. Conducting experiments enables students to make enactive representations, drawing a sketch encourages iconic representations, and all kinds of language and also mathematics are symbolic representations.

Because we think it is important to investigate the observable macro-level of physics lessons (e.g., how does the teacher act in class, what representations does he use, how do students handle different representations), we will concentrate on external representations initially. This detailed description is an important prerequisite for a better understanding of individual learning processes. Theoretical approaches for internal representations will be used in further research to derive implications for the use of different external representations in teaching.

Representations are always embedded in communicational processes. Because we are concentrating on representation as signs that contain information, we will focus on the informational side of communicational processes. With reference to the communicational model of Schulz von Thun, we ignore the self-revelation, relationship, and appeal sides of a message and concentrate on the factual information side (Schulz von Thun n.y.). Considering this focus communication intends a transportation of meaning from one person to another. Since the construction of meaning is always embedded in a person's prior knowledge, the outgoing and the incoming information of two communication partners will not be identical. This is important for the context of teaching and learning. Focusing on the information that a representation encodes, representations can be seen as signs. The definition above uses sign as synonym for representation. The use and meaning of signs is part of semiotics. As Peirce (1960) points out, each (external) sign is connected to an (external) object and to an (internal) interpretant: "A sign [. . .] addresses somebody, that is, creates in the mind of that person an equivalent sign, or perhaps a more developed sign. That sign which it creates I call the interpretant of the first sign. The sign stands for something, its object" (Peirce 1960, p. 135). The interpretant always depends on the person who is reading the sign and his or her individual knowledge. This triad of a sign is shown in Fig. 4.2. A good illustration of this from a physics lesson is the term "electric voltage." When a student uses the words "electric voltage," his/her speech is the (external) representation for the object, in this case the physical concept of electric voltage. His internal representation might differ from this object and how it is understood by another person, e.g., a physicist or a physics teacher. In physics education it is well known that many concepts that students have diverge from the established concepts of the community of the discipline. These different individual concepts of an object constitute the interpretant.

Fig. 4.2 Peirce's triad of a sign (cf. Peirce 1960): every sign, respectively, representation is connected to an object and an interpretant

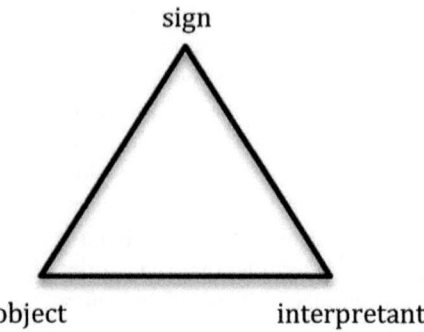

Another important characteristic of a representation is that it represents not every aspect of an object but focuses on a collection of characteristics that are relevant for the communicational purpose. In this way a representation is related to a model of an object. Lemke (1998) emphasizes this characteristic and therefore states that it is always important to use different representations to understand the meaning of an object or concept: "It is sometimes argued that the various representations of a 'concept' are entirely 'redundant' with one another, that they can be placed in one-to-one correspondence, so that meanings that can be made in one semiotic modality can be equally well made in the others. This is not the way scientific communication appears to work: meanings are made by the joint codeployment [co-deployment: combination and connection, authors' note] of two or more semiotic modalities, and such codeployment of resources is needed for canonical interpretation. In my opinion, semiotic modalities (e.g. language, depiction) are essentially incommensurable" (Lemke 1998, p. 110). This aspect will be deepened in the section about multimodal representation in this article.

4.2.1 Specific Nature of Mathematical Representations

Mathematical representations, for instance, formulas and graphs, are in some aspects different from representations in everyday life or in other disciplines. The reason is the nature of mathematical objects. Duval (2006) asks, "How can the represented object be distinguished from the semiotic representation used when there is no access to mathematical objects apart from semiotic representations?" (Duval 2006, p. 115). This citation describes the complexity of mathematical representations as consequence of the complexity of mathematical objects. A mathematical object is not something we can see or even touch, but it is an abstract cognitive construct. Therefore, enactive and iconic representations do not play an essential role for mathematical concepts in the first place. However, they may occur in the context of teaching.

Furthermore, it has to be considered that mathematical representations are ambiguous by nature. Hence, they are frequently interpreted in different ways. Heck and van Buuren (in this book) illustrate this particularity of mathematical representations with the help of different examples.

Based on Schnotz and Bannert (2003), Bauer (2015) describes the internal representation of a mathematical object. The internal representation includes possible external representations; the relationship with other mathematical terms and possible mathematical operations (Bauer 2015, p. 24). Redish and Kuo (2015) follow a comparable approach by using the concept of encyclopedic knowledge for describing the meaning of a physical formula. This approach leads to the relevance of multimodal representations for learning and teaching physics and will be described in this article in Sect. 4.4.

4.3 Classification of Representations in Physics Lessons

Examining a physics classroom, a huge variety of representations can be found. Students have to learn how to interpret, use, construct, or change between them. For analyzing the related teaching and learning processes, it is necessary to describe the different kinds of representations, their similarities and differences, and their cross-linking. Therefore, in this section a detailed classification of external representations in physics is presented. Because of the specific role of mathematical representations, these are highlighted in this classification. For this reason first a classification of representations from mathematics education is introduced which later is adapted to physics.

4.3.1 Representations in Mathematics

Prediger and Wessel (2011) developed a model to distinguish between different registers of representations ("Darstellungsregister") in mathematics lessons that is related, for instance, to Bruner (1974). It draws a distinction between the registers:

Objective
Pictorial
Verbal
Symbolic-numerical and
Symbolic-algebraic

Furthermore, the verbal register can be divided into everyday language, academic language, and special language.

The different representation registers in their model are basically ordered according to the degree of abstractness whereby the symbolic registers are the most abstract ones. However, this can be different for different situations depending on the context. For instance, abstract pictures can act as a mediator between academic and special language.

4.3.2 Representations in Physics

Physics involves its own representations which are characteristic for the discipline. The Encyclopedia of Science Education (Dolin 2016, p. 837) and other authors (e.g., Gilbert 2016, pp. 123–131) present different ways to classify them. As the particularity of mathematical representations in physics should be highlighted in this article, the classification by Prediger and Wessel (2011, p. 167) was adapted as it focuses on representations which include mathematical symbolism and concepts. To meet the conditions of the discipline of physics, some adjustments were made.

Figure 4.3 shows the adapted classification of representations in physics lessons that contains the following categories.

Objective Representations: This category relates to the enactive level by Bruner (1974) described in Sect. 4.2. It includes representations of objects or physical concepts that exist as objects. They can not only be seen but also touched or moved. In physics lessons, they occur, for example, during experimental situations and also in the form of objective models, e.g., a movable model of the solar system. This category has two subcategories: On the one hand, there are *iconic objective representations* in which similarities to the represented object exist, e.g., an objective model of an engine. On the other hand, there are *symbolic objective representations* in which a certain symbolism has to be known to understand and work with these kinds of representations, e.g., an objective model of the axes of a three-dimensional coordinate system (also in form of gestures, e.g., using three fingers for symbolizing the three axes).

Pictorial Representations: Photographs, pictures, drawings, sketches, etc. are classified as pictorial representations. Two subcategories are distinguished as well: First there are *iconic pictures* that still show similarities to the objects they represent. Examples are photographs and sketches of experiments. Second there are *symbolic pictures* that expose no or almost no similarity between the picture and the represented object or physical concept. These representations can only be interpreted and used if the symbolism is known and understood. For instance, a circuit diagram can only be interpreted if the circuit symbols are known. Also energy flow charts belong to this subcategory.

Verbal Representations: This category comprises written as well as spoken language. According to Prediger and Wessel (2011) and Prediger (2013), three levels with no clear boundaries are distinguished:

Everyday language includes predominantly expressions of spoken language. Frequently there is a lack of explicitness. (e.g., I use the switch. The light turns on.)
Academic language includes primarily expressions of written language. It is explicit, with more complex grammar and impersonal. To some extent words of special language occur. (e.g., the electric circuit is closed. The LED is on.)
Special language is characterized by high precision, conciseness, clearness, and high occurrence of specialized terminology. The communication is optimized

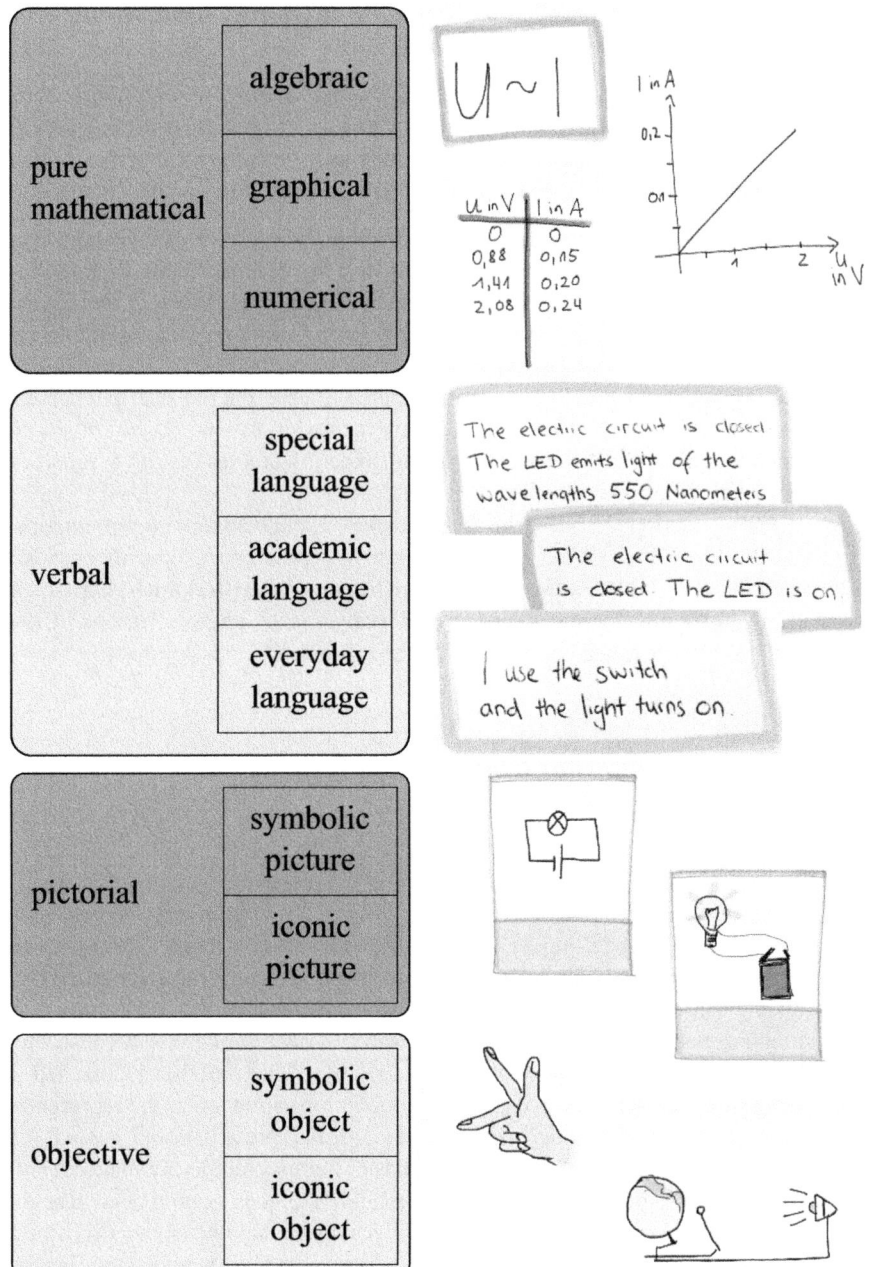

Fig. 4.3 Classification of representations in physics lessons (Bear in mind that the term "object" in the subcategories does not mean the object that is represented but its representation in form of an object)

toward efficiency and clarity. (e.g., the electric circuit is closed. The LED emits light of the wavelength 550 nanometers.)

It can happen that other representations than verbal ones occur within these three types of languages, e.g., formulas within a written text or gestures within a talk. In contrast to Prediger (2013, p. 175), we see this as a blending of or cross-linking between different kinds of representations and not as part of the verbal categories.

Pure Mathematical Representations: As mathematical objects are abstract constructs, the mathematical representation of physical objects inherits the high level of abstractness. For instance, data that was measured in an experiment is represented in form of a table. This kind of representation is called *numerical*. Line diagrams or graphs that are drawn or interpreted are categorized as *graphical*. And mathematical terms, equations, and formulas which are derived to describe physical phenomena or that are used to solve physical problems are located in the category of *algebraic* representations. These are often used to sum up experimental results and to represent a general relation between physical quantities.

As already mentioned above, this classification of mathematical representations in physics is based on a classification used in mathematics education. Some adjustments had to be made to meet the conditions of the discipline physics and to make it applicable for the planning and analyzing of physics lessons. These modifications are deepened and justified in the following.

4.3.3 Different Kinds of Objective and Pictorial Representations

Both the objective and the pictorial categories in the presented classification consist of two subcategories, whereas Prediger and Wessel (2011) have only one category for each of them in their model. The distinction between iconic and symbolic objects, respectively, and pictures refers to the idea of Schnotz and Bannert (2003). They distinguish between depictive and descriptive representations. It is important to distinguish between these two different sorts of representations because they relate to different levels of abstractness. A depictive representation is attached to the content in a way that makes it easy to extract information about the represented object. Structural characteristics of the object and its representation are equal (cf. Schnotz and Bannert 2003, p. 143). These kinds of representations are named *iconic* here. On the other hand, a descriptive representation includes symbols that describe the represented object in an abstract way. Only with the help of conventions these symbolic representations can be interpreted or constructed (cf. Schnotz and Bannert 2003, p. 143). In the presented classification, the term *symbolic* is used to describe these kinds of representations.

The meaning and the relevance of this distinction will be illustrated by the following examples of objective and pictorial representations.

Objective representations in physics can be of different levels of abstractness. For instance, a physical model of a combustion engine represents the structure and functionality of it. It can be moved to illustrate different phases of the process. Such a model shows similarities to the real object that it represents and hence is classified as an *iconic objective representation.*

But many objective representations in physics lessons include symbolism. Similarities to the objects that they represent are missing. This is due to the fact that most physical theories describe objects or constructs that cannot be observed directly, e.g., the electric current. Because this implies extra challenges for students who have to handle them, they should be seen in a different way. For instance, when the physics teacher demonstrates three of his fingers explaining the relations of electric current, magnetic field, and force, a lot of information is embedded. Only if the rules of the symbolism and the physical concepts are known this representation can be useful to students. The described gesture is classified as a *symbolic objective representation.*

The example of an experiment in which the electrical conductivity of different materials is investigated helps to illustrate the difference between iconic and symbolic pictures. A picture of the experimental set-up is less abstract than a circuit diagram. A student has to be able to relate the circuit symbols to concrete objects to understand the latter. This means students have to face different requirements when they interpret these different sorts of pictorial representations. The circuit symbols show almost no similarity to the objects they are representing. They are symbols and therefore at a higher degree of abstractness compared to photographs or sketches.

The same applies to flow charts in physics that are used, for example, to represent energy transformations. Leisen (2004, 2005) sees them coequal to line graphs. However, if they are not focusing on quantifiable aspects, it seems more appropriate to classify them at a lower level of abstractness compared to mathematical representations. The different degree of symbolism and abstractness justifies why these *symbolic pictures* have to be distinguished from *iconic pictures* (e.g., photographs).

Especially for physics teachers, it is important to consider this difference between iconic and symbolic objects or pictures. Using different kinds of representations creates different challenges for students. Students first have to get used to the way in which physics is communicated. They have to learn how to interpret, use, and interrelate different symbolism and physical concepts that are represented in pictures and objects. Using objective and pictorial representations in physics lessons is not necessarily easy for students although they are meant to be clear, descriptive, and demonstrative (especially in comparison to mathematical representations).

4.3.4 Graphical Representations

In the classification by Prediger and Wessel (2011), (line) graphs are located in the pictorial register, whereas in our classification, they belong to one subcategory of pure mathematical representations, *graphical representations* (cf. Fig. 4.3).

Fig. 4.4 Variety of
classifying a vector in physics

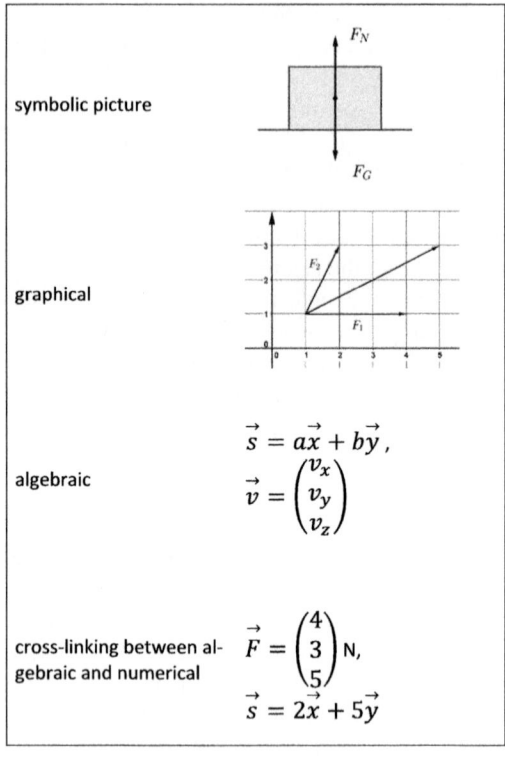

Thereby, they are considered as different from pictures and are assigned to a higher
level of abstractness, in a greater distance to the level of objective representations.
There are several reasons for it: Within graphs symbolic features occur (e.g., labels
of the axes, trend lines, error bars) which increase abstractness and semantic density
of the representation. Furthermore, graphs contain quantifiable aspects; they are
representing a functional dependency. For these reasons they are mathematical
representations and have to be seen differently to pictures.

It should be noted that the mapping of a representation to the presented
classification in Fig. 4.3 could depend on the situation. The example in Fig. 4.4
that examines a vector in physics is illustrating this.

The presented classification was predominantly developed to analyze the particularity of mathematical representations in physics lessons. Therefore, it focuses on
representations that occur in textbooks, at the black- or whiteboard, in the students'
notes, and in the teachers' and students' speech. Nowadays technology allows a
dynamic variation and connection of mathematical representations as well. This
can be seen as a blending of the presented categories. For instance, in simulations
pictorial and different kinds of pure mathematical representations are connected.
The article by Euler and Gregorcic in this book illustrates how this blending of
different representations can be used in a fruitful way to provide learning of physics.

4.4 Relevance of Different Representations for the Learning and Understanding of Physics

Physicists use many different representations for their work and communication. Thus, representations are an important part of physical methods and techniques that students should be introduced to. Teachers can use a multimodal approach in their physics lessons to emphasize this and to enable their students to understand and work on physical topics. In this section it is highlighted as well that students should not only learn how to use and interpret representations in physics but in addition gain some meta-knowledge and meta-competences about representations in physics.

4.4.1 Representations as Part of Physical Methods

Understanding of physics includes that students should know about physical concepts, theories, structures, terms, and methods and apply them appropriately, for instance, to investigate, describe, and explain a physical phenomenon. This means, using the words of Airey and Linder (2009), that the students should get access to a *way of knowing*, "the coherent system of concepts, ideas, theories, etc. that have been created to account for observed phenomena in a discipline" (Airey 2009, p. 11). For instance, talking about different states of matter and phase transitions in physics, the related *way of knowing* includes among others the concepts of energy and heat, latent heat, physical quantities like temperature and pressure, and physical terms like melting, evaporating, condensing, and freezing. Following Airey and Linder (2009), these different aspects of the *way of knowing* are represented by different edges of a polygon in Fig. 4.5. Access to these aspects can be gained through different representations,[1] e.g., images and formulas. It may occur that some of the aspects of a *way of knowing* are not addressed or even not known (represented as a question mark in Fig. 4.5).

The representations which are important to understand a *way of knowing* are in general of different types. For the presented example of states of matter, some of them are shown in Table 4.1.

Each of these representations has specific potentials for uncovering a particular aspect of the regarded *way of knowing*. This has been described by Fredlund et al. (2012) as the *disciplinary affordance* of the representation.[2] Many of these representations "overlap" and can link to other representations, while others have quite specific *disciplinary affordances* (i.e., meanings that cannot be made by other means). Only when there is a combination of different representations a

---

[1] Airey and Linder (2009) use the term *semiotic resources* and distinguish between tools, activities, and representations. In our argumentation this distinction is not important; tools and activities can be seen predominantly as handling with objective representations.

[2] Fredlund et al. (2012) use the terminology of *semiotic resources* as well; therefore they call it the *disciplinary affordance* of a *semiotic resource*.

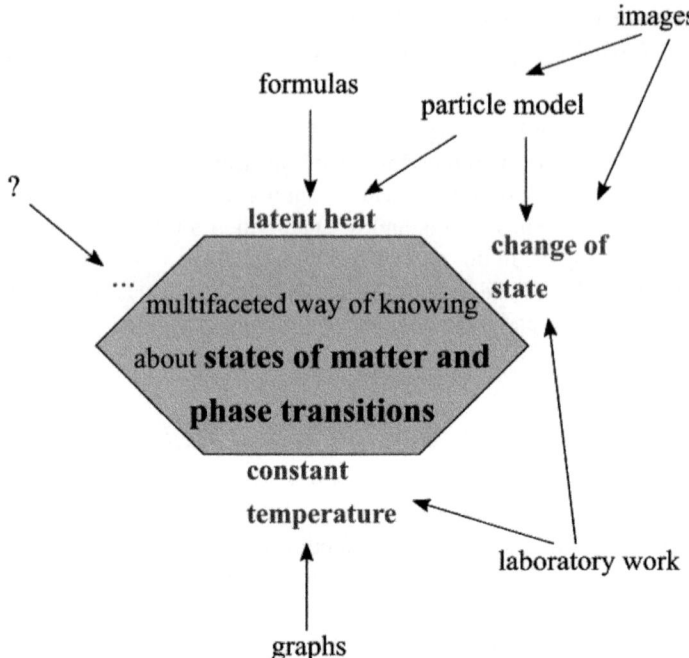

Fig. 4.5 Different representations give access to different aspects of the physical concept that should be understood (cf. Airey and Linder 2009). For instance, exploring the phenomenon of constant temperature during a phase transition could take place with the help of laboratory work and graphs

holistic experience of the *way of knowing* (i.e., understanding) can be reached. Therefore, students should be taught to become fluent in changing between different representations. This fluency is necessary but not sufficient for an understanding. Students might only imitate the use of different representations and thus still not experience the corresponding aspects of the *way of knowing* (cf. Airey and Linder 2009). In that way understanding different representations and using them correctly is part of the methods in physics. These different representations emphasize the importance of multimodal representations from a perspective that focuses on an understanding of the discipline itself more than on the specific content: "a set of carefully selected multimodal representations, each with their own (unique, supplementary or complementary) affordances, is needed in order to generate a collective disciplinary affordance" (Linder 2013, p. 47).

Airey and Linder (2009, p. 40) claim that a critical constellation of semiotic resources is necessary to reach a holistic understanding of a physical concept. Students should become fluent in this certain constellation of resources first. Teachers can follow a multimodal approach to support this.

Table 4.1 Examples of different representations giving access to the way of knowing about states of matter and phase transitions

Pure mathematical representations	$Q = m \cdot c \cdot \Delta T + Q_{latent}$
	$Q\_latent = q \cdot m$

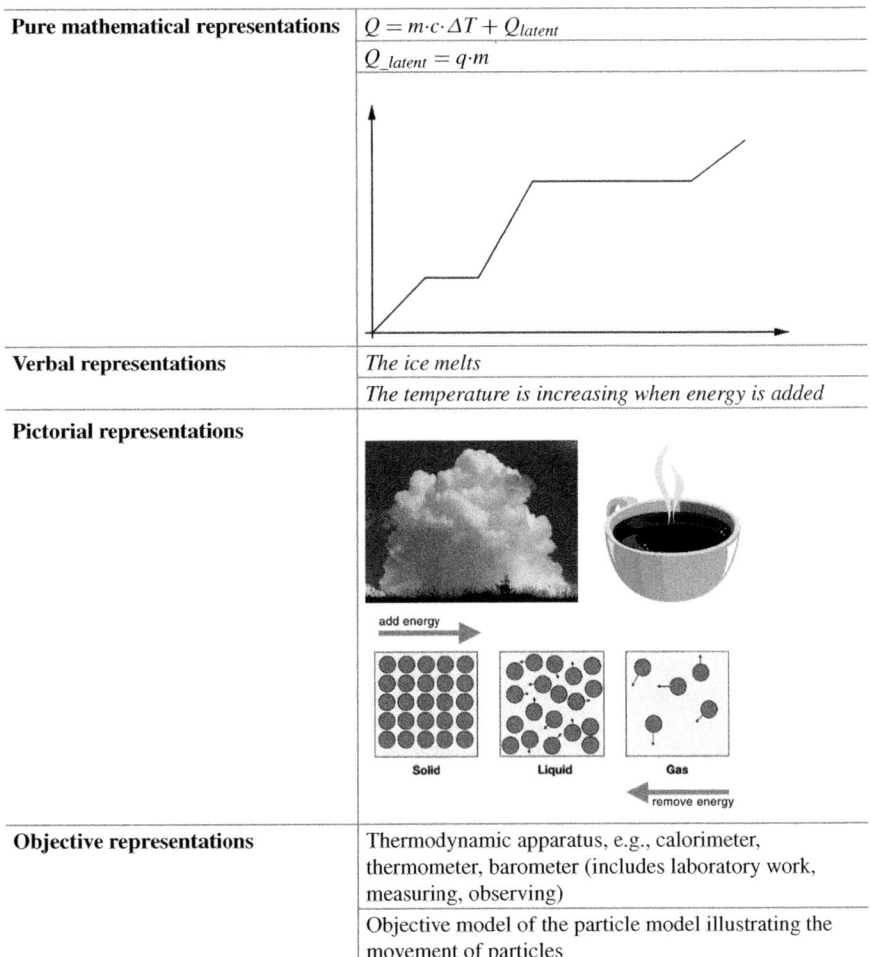

Verbal representations	*The ice melts*
	The temperature is increasing when energy is added
Pictorial representations	
Objective representations	Thermodynamic apparatus, e.g., calorimeter, thermometer, barometer (includes laboratory work, measuring, observing)
	Objective model of the particle model illustrating the movement of particles

4.4.2 Multimodal Representations

In Sect. 4.2, one of the characteristics of representations was specified: they focus on some aspects of the object they represent and leave out others. This means that to achieve a holistic experience of an object, e.g., a physical object, concept, or law, different representations have to be included. Considering different representations, we get a holistic impression of the physical context. On the other hand, using different representations enables different approaches to one concept and is therefore important for understanding and learning the physical concept itself. Lemke (2004) argues that science itself is multimodal and that students need to learn all "languages of science." Students should be able to interpret, connect, and integrate different representations. He states that "Scientific literacy is not just the knowledge of scientific concepts and facts; it is the ability to make meaning conjointly with verbal concepts, mathematical relationships, visual representations, and manual-technical operations" (Lemke 2004, p. 38).

Multimodal representations can fulfill different functions for learning and thus help to support learning and understanding of physics.

Different Information and Processes: As different representations display different (complementary) information, multimodal representations support learning different aspects of a phenomenon. In addition different representations promote different processes, e.g., a table will mainly be used for reading single values, a graph will help to recognize trends, while an equation invites to calculate.

Understand One Representation with Another One: Moreover, multimodal representations help students to understand a certain representation better by using a second representation to constrain the interpretation of the first one. Either students can relate an unknown representation to a well-known form of representation (constrain by familiarity) or by using a more specific representation to give more detailed information about the first ambiguous representation (constrain by inherent properties).

Construct Deeper Understanding: And finally multimodal representations can help students to construct deeper understanding, either relational understanding (relating different representations) or extending knowledge from one representation to another (e.g., transfer knowledge about a velocity-time graph to understand an acceleration-time graph) or allowing abstraction by referencing different representations, and thereby gain insight in the underlying structure of a topic (Ainsworth 2008; Ainsworth 2006). Furthermore, Waldrip et al. (2010) argue that "unless learners can represent their understandings in diverse modes, then their knowledge is unlikely to be sufficiently robust or durable" (Waldrip et al. 2010, p. 69).

As mathematical representations are highlighted in this article, a multimodal approach has to be stressed especially for them. Redish and Kuo (2015) reason from a cognitive semantics approach. Whereas Airey and Linder (2009), as described above, derive the necessity of using multimodal representations from within physics itself, Redish and Kuo (2015) focus more on the question what helps a student to understand a physical concept. They discuss what it means to make meaning of a mathematical expression in physics. For this, they apply three concepts from cognitive semantics: embodied cognition, encyclopedic knowledge, and contextualization. All three concepts imply that it is necessary to connect the mathematical expression with other cognitive resources to understand its meaning. On the one hand, the mathematical expression is connected with a physical experience (embodied cognition) or on the other hand with a network of different concepts that are related to it (encyclopedic knowledge). Furthermore, the interpretation of a mathematical expression depends on the context in which it is used (contextualization). For instance, if the sign Q is used in physics lesson, it can mean either heat or electric charge. Redish and Kuo (2015) do not mention multimodal representations explicitly, but all presented concepts focus on the connection of different resources. Considering that physical knowledge is represented in different representations and thereby multimodal, it is always necessary to connect, e.g., an equation with other representations related to it. This means connecting different representations can help to develop a better conceptual understanding.

4.4.3 Successful Learning with and about Representations

The theoretical considerations we described above let us assume that multiple representations effect the learning of science positively. Bauer (2015) quotes several studies that seem to report these positive effects for learning: van der Meij and de Jong 2006; Kozma et al. 1996; Ainsworth et al. 1998; Mayer and Sims 1994. However, he underlines that there is the possibility of cognitive overload. If students are not able to handle the multimodal representations that are used, they can interfere successful learning. He summarizes studies that could not show any effect of multiple representations on learning or even showed a negative effect: Chandler and Sweller 1991; Ainsworth et al. 1998; van Someren 1998.

This leads to the question what competencies students need to use (multimodal) representations successfully. Lemke (2004) emphasizes the importance of not only using different representations for teaching and learning, but in addition students should learn *about* different representations (their characteristics, their meaning, etc.) explicitly. Ainsworth (2008, pp. 199–204) specifies what students need to know for a successful learning by using representations. They should understand:

- The form of a representation (e.g., how to read a graph)
- The relation between the representation and the domain (e.g., what the symbols in a formula in a given physical context mean)
- How to select an appropriate representation (e.g., what representation can help to solve a specific problem)
- How to construct an appropriate representation (e.g., how to draw a graph or design a table)
- How to relate representations (e.g., how to connect a graph with a formula)

A similar approach can be found in the work of diSessa (2004). He developed a concept of metarepresentational competence (MRC) to describe higher levels of representational competence and enhances the importance of teaching MRC explicitly. The concept of MRC helps to understand how students design new representations, discuss the adequacy and suitability of representations, understand the purposes of representations, explain representations, and learn new representations quickly.

4.5   Students' Difficulties with Mathematical Representations in Physics

In the previous sections, it has been discussed why different representations are important in physics and which particularities mathematical representations have. A multimodal approach and learning about representations might improve learning and understanding of physics. Nevertheless it has to be considered that it challenges students to handle mathematical representations as several studies in mathematics and physics education research demonstrate.

For instance, Leinhardt et al. (1990) and Hattikudur et al. (2012) describe that students of lower secondary school struggle a lot with graphs. They have different misconceptions and difficulties with both interpretation and construction of graphs (e.g., graph-as-a-picture confusion, tendency toward linearity). A similar situation is reported regarding algebraic representations: students have difficulties to construct, adapt, and interpret even simple equations that describe everyday situations (Malle et al. 1993).

Studies that investigated how students change between different representations in mathematics demonstrate that it matters between which representations they translate. The difficulty depends on which kinds of representations are involved (Nitsch 2015). For instance, most difficulties occur during a change from a table, graph, or formula to a verbal representation (Bossé et al. 2012). Furthermore, novices have greater problems to connect different representations compared to experts (Kozma 2003).

With respect to mathematical representations in different disciplines (e.g., in physics), research results show that even if students have high mathematical skills, they are sometimes not able to apply them in these contexts (e.g., Rebello and Cui 2008; Planinic et al. 2013). Furthermore, they tend to neglect the context and focus on surface features of the representations. A conceptual understanding is mostly missing (e.g., Kim and Pak 2002; Tuminaro and Redish 2007; Uhden 2015; Bagno et al. 2011; Strahl et al. 2010; Pospiech and Oese 2014; Eriksson et al. 2014).

These multiple difficulties of students demonstrate the importance of further research in this field. There is still much to be learned about how students handle mathematical representations in physics, for example:

- How do students construct, adapt, and interpret mathematical representations in physics?
- How do they connect different mathematical representations and change between them in physics?
- What can be expected from students at which age with respect to the handling of mathematical representations in physics?
- To which extent do the students imitate their teachers in handling mathematical representations in physics?
- How can teaching be changed to promote a conceptual understanding of mathematical representations in physics?

The importance of multimodality and the described students' difficulties empha-size the importance of learning explicitly about representations. Two approaches that could help students to link different representations and to get a better insight in students' difficulties translating from one representation to another are presented in the next section.

4.6 Change of Representations in Physics Lessons

As described in Sect. 4.4, it is necessary to work with different kinds of represen-tations to achieve a holistic understanding of a physical concept or phenomenon. Therefore, the translation between different representations in physics becomes a part of the skills that students should learn.

A change of representations can be seen as a translation from a source represen-tation to a target representation (cf. Janvier 1987, p. 29). The elements and structure of the source have to be transferred into the elements and the structure of the target. As described above, different kinds of representations focus on different aspects of a given physical content, some of the represented information is lost, and other is added during this process.

The direction of the translation should be taken into consideration as well. For instance, it is a different task for students to construct a graph referring to a given formula compared to derive a formula from a given graph (cf. Duval 2006, p. 122).

In the following, two theoretical models are presented that focus on the change of representations in physics. They illustrate the particular role of mathematical representations in this process and its connection to verbal representations.

4.6.1 Changing Between Representations of Functional Dependencies in Physics

To describe the process of changing between different representations of functional dependencies, a model was developed (cf. Fig. 4.6). It was derived from a model in mathematics education (cf. Adu-Gyamfi et al. 2012) which characterizes the change between representations of functional dependencies, i.e., it covers only numerical, graphical, algebraic, and verbal representations as source or target representations.

During a translation from the source representation to a target representation, different categories of activities (A, B, C) can occur. For reaching the target repre-sentation, not every category has to occur. Furthermore, the order and frequency of applying them is not fixed.

- Activities A: stepwise realization
    The origin of this category is the construct *implementation verification* by Adu-Gyamfi et al. (2012). It brings out the use of algorithms and step-by-step

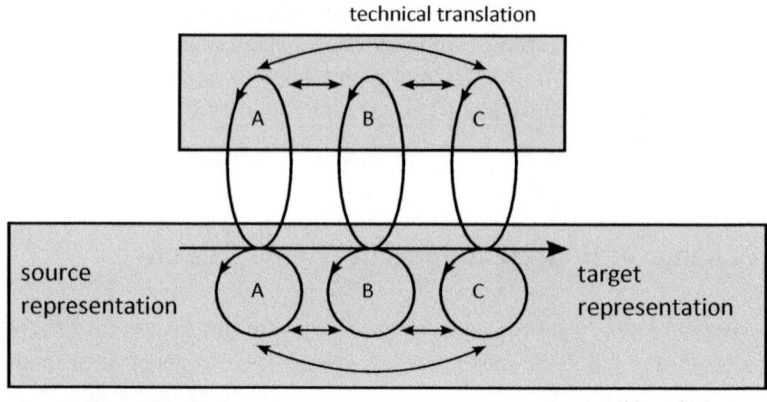

A stepwise realization
B use of characteristics
C verification of consistency

Fig. 4.6 Model of changing between representations of functional dependencies in physics (cf. Geyer and Pospiech 2015)

activities during a translation process. The target representation is built, e.g., by calculating values or plotting points.

- Activities B: use of characteristics

    Like in the *attribute verification* by Adu-Gyamfi et al. (2012), these activities focus on using key characteristics for the translation. Explicitly or implicitly given properties of the source representation are used to build the target representation, e.g., the kind of dependency between the related quantities.

- Activities C: verification of consistency

    This category was derived from the construct *equivalence verification* by Adu-Gyamfi et al. (2012) and concerns the verification of the consistency of the source and target representation. Both should convey consistent information. For example, single steps of the translation process are examined.

Each one of A, B, or C can be part of a technical or structural translation. These terms refer to the technical and structural role of mathematics in physics (cf. Pietrocola 2008; Karam and Pietrocola 2010). The possible approaches to a solution can focus on pure mathematics including remembered algorithms and routines (technical translation) or on an argumentation in which mathematics and physics are strongly intertwined (structural translation).

The task in Fig. 4.7 illustrates a characteristic example for a change of representations in a mechanics lesson.

Using this example, Table 4.2 describes how students could solve the task and maps possible steps to the categories of the model in Fig. 4.6.

In an explorative laboratory study with students aged about 14 years, the model in Fig. 4.6 will be used to describe and structure the process of the students working on physical-mathematical tasks which include changes of representations. This will be a step toward a first validation of the model.

> *Draw an appropriate graph which shows the relation between time and distance of the given data for the first 7 minutes.*
>
> Source
> representation:
>
> | t in min | 0 | 1 | 2 | 3 | 4 | 5 |
> |---|---|---|---|---|---|---|
> | s in km | 0 | 1.7 | 3.3 | 5.0 | 6.7 | 8.4 |
>
> Target representation: graph between 0 min and 7 min

Fig. 4.7 Example for a change between a table and a graph in physics

In the following, we will focus on the change between algebraic and verbal representations. This case shows that each of the kinds of representations can also occur in different modifications, e.g., different types of verbalizations.

4.6.2 Verbalization of Physical Formulas

Students frequently fail at describing conceptual meaning of formulas (e.g., Bagno et al. 2011). Formulas in physics are used to sum up physical and mathematical meaning, and physicists are able to connect the symbols used in the formulas to different knowledge resources (e.g., physical terms, theories, experiments, and mathematical relations and applications). Thus formulas construct a complex meaning; Redish describes that formulas are used "to organize and pack our conceptual knowledge" (Redish 2016, see also the example in the article of Heck and van Buuren in this book). Meanwhile students use formulas frequently for calculations. Therefore, it is interesting to have a closer look on how formulas can be translated into natural language and connected with verbal descriptions.

To be able to describe different forms of speaking about a formula, a model was developed that is based predominantly on linguistic theory and reflections. Referring to Hoffmann (1987, pp. 64–71) linguistics describes specialized language on the one hand in horizontal separations and on the other hand with vertical layers. Horizontal layers of specialized language describe the differences of language in different scientific fields, for example, language used by mathematicians compared to the language used in the field of physics or on a more detailed scale language in experimental nuclear physics compared to language in experimental solid-state physics. The interest for vertical layers of specialized language on the other hand focuses on differences within one scientific field. Hoffmann describes five vertical

Table 4.2 During a translation from a table to a graph, different activities can occur during a technical or a structural translation. Some examples are presented

There are only positive numbers in the table. The x-axis and y-axis are drawn for the first quadrant	A technical
The table shows distance in kilometer and time in minutes. Every minute the distance of a moving object was measured. The distance is a function of time. The x-axis is labeled with *t in min*, and the y-axis is labeled with *s in km*	A structural
The axes are scaled (arbitrarily)	A technical
All six points that are given in the table are plotted	A technical

All points seem to be in a line. Hence they could be connected by a straight line which shows the relation between s and t	A technical
The graph shows a uniform motion because the distances between every minute are the same. There is a proportional relationship between distance and time. This is represented by a straight line through the origin. If there is no change in velocity, the line can be extended up to 7 min	B structural

If the direct proportional relationship is recognized without plotting all the points, the graph could be drawn by extracting the slope with the help of the given point (1|1.7) or calculating the quotient s/t that is approximately constant and around 1.7	B technical
If the uniform motion is recognized without plotting all the points, the graph could be drawn by extracting the velocity of the moving object out of the pair (1 min|1.7 km). It moves (approximately) 1.7 km/min which is represented by the slope of the graph	B structural
The slope of the graph represents the velocity of the constant motion. 1.7 km/min is 102 km/h. The represented relation can be an idealized description of a car moving with 102 km/h at a straight part of a highway	C structural
The s-value for $t = 7$ can be calculated: $s = 1.7 \cdot 7 = 11.9$. The point (7|11.9) is (approximately) part of the curve	C technical

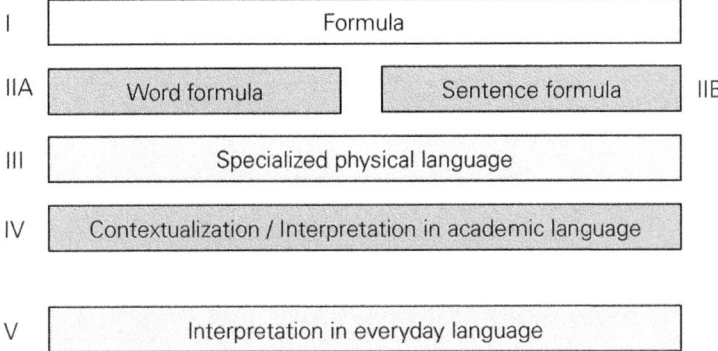

Fig. 4.8 Levels of verbalization of a physical formula (cf. Janßen and Pospiech 2015)

levels differentiated according to the level of abstraction manifested in linguistic characteristics of the language. A high level of abstraction is characterized by the use of artificial symbols for elements and relations; a low level of abstraction is characterized by a high proportion of natural, everyday language (cf. Hoffmann 1987 p. 64–71).

This in German Linguistics well-established theory of specialized language was applied to physics education and led to the model represented in Fig. 4.8. The model describes different levels of verbalization of a formula. The levels do not describe a preferred order in which the levels should be used in physics classes, but they describe in reference to Hoffmann from I to V decreasing semantic density. The bigger the number of the level in the model, the less dense the meaning is described, and the closer it approaches to everyday life and speech.

The model was developed to describe how teachers speak in class about formulas. However, it can also be used to reflect how it is spoken about formulas in general and thus to investigate the connection between verbal and mathematical representations. Fig. 4.9 presents an example for each level.

Level I On top of the model, the formula stands as a representation with a very high density of semiotic. Following Bruner (1974) "the possibility of consolidation – the quality that allows to condense meaning for example $F = MA$" (Bruner 1974, p. 18) is a characteristic of symbolic representations. A formula condenses a high amount of meaning on very little space.

Level II A first step to reduce the semantic density is to translate either the physical symbols (IIA) or the mathematical symbols in the formula (IIB). Thus, either a word formula or a sentence formula is built.

Level III At this level of verbalization, all symbols are translated to their technical terms of special language. It leads to a sentence with physical and mathematical terms. The sentence contains exactly the same information as the formula, but all symbols are verbalized.

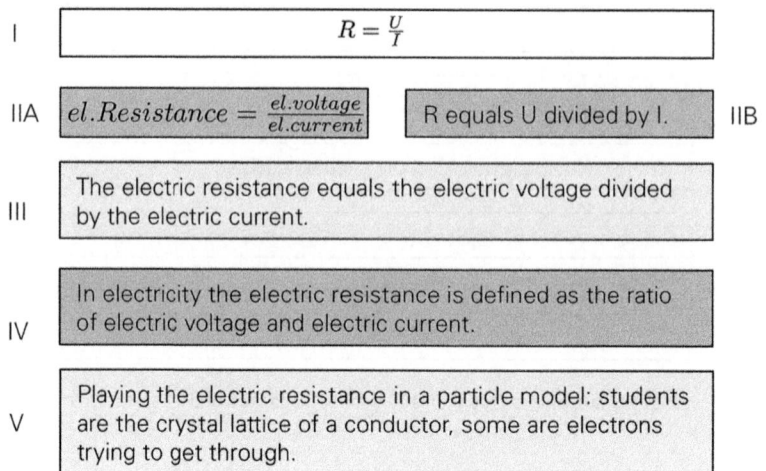

Fig. 4.9 Example for different levels of verbalization of a physical formula

Level IV Here the formula is connected to physical theories and mathematical implications. Only at this level any external information is added to the formula itself. The information added can be the following: contextualization of the formula in a physical theory (at least by naming the theory), referring the formula to an experiment (e.g., comparing experimental and theoretical values), mathematical implications (e.g., proportionalities: if the resistance increases at constant voltage, the current decreases), and a "categorization" of the formula (e.g., as a definition, a natural law, an empirical principle). All information that is added contributes to the conceptual understanding of a formula. There are many different possibilities to connect the formula to other knowledge. The kind of information that is connected with a formula depends on the individual person and its individual knowledge. Accordingly to this, the verbalization is a kind of an interpretation. The academic language used at this level is characteristic for school. Special terms still occur, but speaking with the word of Hoffmann (1987), the language is at a lower vertical level of specialized language.

Level V Level V connects the formula with didactical analogies, models, or everyday life. The formula could be associated with analogies from everyday life (e.g., "Imagine you walk on a very crowded street and you are not able to move forward very fast"), with models (e.g., "playing" particle model with students) or with applications of the formula to a problem (e.g., "The current in this circuit is too high. Because of that we have to integrate a bigger series resistor"). The language at this level is very close to everyday language. Furthermore, analogies and models used at this level always focus on some aspects of the formula; others are left out. This corresponds to the definition of representations in general presented before: different representations give different information about the represented thing.

On the one hand, this model helps to describe the way teachers speak about formulas. On the other hand, it helps to demonstrate how much information is contained in a formula. In this way, it can help to foster the sensibility for the large amount of decoding that students have to cope with when they want to understand a physical formula.

4.7 Implications for Teaching

Representations, either objective, pictorial, verbal, or pure mathematical ones, are an important part of physics and physics lessons as they are embodied in physical methods, techniques, and communication processes. The previous chapters show that if students are able to interpret, adapt, construct, and connect different representations, a better understanding of physics is possible. Thus, we come up with three main conclusions for teaching physics including a special approach for representations.

4.7.1 Use of Multimodal Representations

Generally students should learn how to gain information from different representations and how to use multimodal representations as a part of learning physics. This includes understanding the characteristic of representations to only focus on some aspects of an object or topic and that different representations are necessary to see a complete picture of it. Therefore, a variety of different modes of representations should be introduced and worked with in physics class.

4.7.2 Change of Representations

There are two general principles to embed change of representations in physics lessons. On the one hand, the teacher could present different representations and connect them with each other. On the other hand, students themselves should learn actively how to use different representations and how to translate between them. For this Prediger and Wessel (2012, p. 30) recommend different activities for mathematics lessons that can similarly be performed in physics lessons:

- Change from one representation to another. (e.g., Describe an experimental situation which the given graph could be a result from.)
- Organize different representations content wise. (e.g., Group all pictures, descriptions, graphs, and formulas according their type of motion.)

- Check/correct two representations for the same content. (e.g., Tina drew the following graph related to the given formula. Change the graph that it fits to the formula.)
- Explain why two representations do not refer to the same content. (e.g., Paul found the following formula that describes the given situation. Explain why it does not fit. How must it be changed?)
- Investigate relations or characteristics by changing to another representation. (e.g., Which of the cars whose movement is described with the help of the following table has a higher acceleration? Find out with the help of a graph.)
- Explain how to recognize structures or characteristics in different representations. (e.g., How do you recognize which material is a better electrical conductor with the help of these graphs?)
- Collect and reflect different representations of a content. (e.g., Collect all kinds of representations which illustrate Ohm's law.)
- Change in one representation and describe consequences for other representations. (e.g., Investigate with the help of the formula how the refraction angle changes when the arrival angle of the light is doubled. Which consequences follow in the picture that shows the path of the light?)

In this context of changing between representations, especially the connection to verbal language should be stressed. Waldrip et al. (2010) sum up and emphasize the consequences of Lemke (2002, 2004): "students need repeated opportunities to translate disciplinary understandings into natural language, even if such translations can only ever be partial rather than complete, because of the abstractness of the scientific forms of representation" (Waldrip et al. 2010, p. 68). This claim is enhanced by the cited studies about students' difficulties to understand formulas conceptually (cf. Sect. 4.5).

4.7.3 Explicitly Learning About Representations

As described above some authors demand explicit teaching about representations. This includes not only knowing how to construct representations and gain information from representations but also to reflect about it: "Students need to be more actively engaged in constructing and interpreting representations by actively discussing the properties of representations, including their strengths and limitations" (Waldrip et al. 2010, p. 70). This claim leads to a metarepresentational competence students should achieve. Based on how diSessa (2004, p. 293) describes the elements of a metarepresentational competence, activities during class can be the following:

- Invent or design own representations.
- Judge the adequacy and suitability of representations, also concerning different tasks.
- Discuss the purposes of representations generally and in specific contexts.
- Explain what specific representations mean and how they transport this meaning.

4.8 Conclusion

To sum up, researchers and teachers should pay particular attention to representations in physics lessons. On the one hand, they play an important role in specific communicational processes within the discipline. On the other hand, they have the potential to improve learning and to gain a differentiated view of a topic. Therefore it is important for educators firstly to decide consciously which representations they use for teaching a specific topic, which means which kind of representation fulfills the purpose of their teaching the best. Secondly they should think about how to connect representations to each other and thirdly have a lot of situations in lessons in which they discuss the characteristics, the advantages, and the disadvantages of representations with their students.

References

Adu-Gyamfi, K., Stiff, L. V., & Bossé, M. J. (2012). Lost in translation: Examining translation errors associated with mathematical representations. *School Science and Mathematics, 112*(3), 159–170.

Ainsworth, S. (2006). DeFT: A conceptual framework for considering learning with multiple representations. *Learning and Instruction, 16,* 183–198.

Ainsworth, S. (2008). The educational value of multiple-representations when learning complex scientific concepts. In J. K. Gilbert et al. (Eds.), *Visualization: theory and practice in science education* (pp. 191–208). Dordrecht: Springer.

Ainsworth, S., Bibby, P. A., & Wood, D. J. (1998). Analysing the costs and benefits of multi-representational learning environments. In M. Someren (Ed.), *Learning with multiple representations* (pp. 120–134). Pergamon, as cited in Bauer (2015).

Airey, J. (2009). *Science, language and literacy. Case studies of learning in Swedish University physics.* Uppsala: Uppsala University.

Airey, J., & Linder, C. (2009). A Disciplinary Discourse Perspective on University Science Learning: Achieving Fluency in a Critical Constellation of Modes. *Journal of Research in Science Teaching, 46*(1), 27–49.

Bagno, E., Eylon, B.-S. & Berger, H. (2011). How to promote the learning of physics from formulae? In: *GIREP-EPEC & PHEC 2009 International Conference* (=Physics Community and Cooperation Vol. 2), pp. 77–83.

Bauer, A. (2015). *Argumentieren mit multiplen und dynamischen Repräsentationen.* Würzburg: Würzburg University Press.

Bruner, J. S. (1974). *Entwurf einer Unterrichtstheorie.* Berlin: Berlin Verlag.

Bossé, M. J., Adu-Gyamfi, K., & Cheetham, M. R. (2011). Assessing the difficulty of mathematical translations: Synthesizing the literature and novel findings. *International Electronic Journal of Mathematics Education, 6*(3), 113–133.

Chandler, P. & Sweller, J. (1991). Cognitive load theory and the format of instruction. *Cognition and Instruction, 8*(4), 293–332, as cited in Bauer (2015).

diSessa, A. A. (2004). Metarepresentation: Native competence and targets for instruction. *Cognition and Instruction, 22*(3), 293–331.

Dolin, J. (2016). Representations in science. In R. Gunstone (Ed.), *Encyclopedia of science education* (pp. 836–838). Springer Science+Business Media Dordrecht 2015, Corrected Printing 2016.

Duval, R. (2006). A cognitive analysis of problems of comprehension in a learning of mathematics. *Educational Studies in Mathematics, 61*(1). Springer, 103–131.

Eriksson, U., Linder, C., Airey, J., & Redfors, A. (2014). Introducing the anatomy of disciplinary discernment: an example from astronomy. *European Journal of Science and Mathematics Education, 2*(3), 167–182.

Fredlund, T., Airey, J., & Linder, C. (2012). Exploring the role of physics representations: An illustrative example from students sharing knowledge about refraction. *European Journal of Physics, 33*, 657–666.

Geyer, M. –A., & Pospiech, G. (2015). Darstellungen funktionaler Zusammenhänge im Physikunterricht. Darstellungswechsel in der Sekundarstufe 1. In *PhyDidB –Didaktik der Physik – Beiträge zur DPG-Frühjahrstagung Wuppertal 2015*.

Gilbert, J. K. (2016). Chapter 7. The contribution of visualisation to modelling-based teaching. In J. K. Gilbert & R. Justi (Eds.), *Modelling-based teaching in science education* (Models and modeling in science education) (Vol. 9, pp. 121–148). Cham: Springer.

Hattikudur, S., Prather, R. W., Asquith, P., Alibali, M. W., Knuth, E. J., & Nathan, M. (2012). Constructing graphical representations: Middle schoolers' intuitions and developing knowledge about slope and Y-intercept. *School Science and Mathematics, 112*(4), 230–240.

Hoffmann, L. (1987). *Kommunikationsmittel Fachsprache. Eine Einführung*. 3., durchgesehene Auflage. Berlin: Akademie-Verlag (=Sammlung Akademie-Verlag Bd.44. Sprache).

Janßen, W., & Pospiech, G. (2015). Versprachlichung von Formeln und physikalisches Formelverständnis. In S. Bernholt (Hrsg.), *Heterogenität und Diversität – Vielfalt der Voraussetzungen im naturwissenschaftlichen Unterricht. Gesellschaft für Didaktik der Chemie und Physik, Jahrestagung in Bremen 2014* (S. 636–638). Kiel: IPN.

Janvier, C. (1987). Translation process in mathematics education. In *Problems of representation in the teaching and learning of mathematics* (pp. 27–32). Hillsdale: Lawrence Erlbaum.

Karam, R., & Pietrocola, M. (2010). Recognizing the structural role of mathematics in physical thought. In M. F. Tasar & G. Çakmakci (Eds.), *Contemporary science education research: International perspectives* (pp. 65–76). Ankara: Pegem Akademi.

Kim, E., & Pak, S.-J. (2002). Students do not overcome conceptual difficulties after solving 1000 traditional problems. *American Journal of Physics, 70*(7), 759–765.

Kozma, R. (2003). The material features of multiple representations and their cognitive and social affordances for science understanding. *Learning and Instruction, 13*, 205–226.

Kozma, R. B., Russell, J., Jones, T., Marx, N., & Davis, J. (1996). The use of multiple, linked representations to facilitate science understanding. In S. Vosniadou (Ed.),. (Hrsg.) *International perspectives on the design of technology-supported learning environments* (pp. 41–60). Mahwah: L. Erlbaum Associates, as cited in Bauer (2015).

Leinhardt, G., Zaslavsky, O., & Stein, M. K. (1990). Functions, graphs, and graphing: Tasks, learning, and teaching. *Review of Educational Research, 60*(1), 1–64.

Leisen, J. (2004). Der Wechsel der Darstellungsformen als wichtige Strategie beim Lehren und Lernen im DFU. *Fremdsprache Deutsch, 30*, 15–21.

Leisen, J. (2005). Wechsel der Darstellungsformen. Ein Unterrichtsprinzip für alle Fächer. *Der Fremdsprachliche Unterricht Englisch, 78*, 9–11.

Lemke, J. L. (1998). Multiplying meaning: Visual and verbal semiotics in scientific text. In J. R. Martin & R. Veel (Eds.), *Reading science: Critical and functional perspectives on discourse of science* (pp. 87–113). London: Routledge.

Lemke, J. L. (2002). Mathematics in the middle: Measure, picture, gesture, sign, and word. In M. Anderson, A. Saenz-Ludlow, S. Zellweger, & V. Cifarelli (Eds.), *Educational perspectives on mathematics as semiosis: From thinking to interpreting to knowing* (pp. 215–234). Ottawa: Legas Publishing.

Lemke, J. L. (2004). The literacies of science. In E. Wendy Saul (Ed.), *Crossing borders in literacy and science instruction* (pp. 33–47). Newark, DE/Arlington, VA: International Reading Association/NSTA Press.

Linder, C. (2013). Disciplinary discourse, representation, and appresentation in the teaching and learning of science. *European Journal of Science and Mathematics Education, 1*(2), 43–49.

Malle, G., Wittmann, E. C., & Bürger, H. (1993). *Didaktische Probleme der elementaren Algebra.* Wiesbaden: Springer.

Mayer, R. E. & Sims, V. (1994). For whom is a picture worth 1000 words - Extensions of a dual-coding theory of multimedia learning. *Journal of Educational Psychology, 86*(3), 389–401, as cited in Bauer (2015).

McDermott, L. C., Rosenquist, M. L., & van Zee, E. H. (1987). Student difficulties in connecting graphs and physics: Examples from kinematics. *American Journal of Physics, 55*(6), 503–513.

Nitsch, R. (2015). *Diagnose von Lernschwierigkeiten im Bereich funktionaler Zusammenhänge. Eine Studie zu typischen Fehlermustern bei Darstellungswechseln.* Wiesbaden: Springer Spektrum.

Peirce, C. S. (1960). *Collected papers of Charles Sanders Peirce* (2. Aufl.). Cambridge, MA: Belknap Press of Harvard University Press.

Pietrocola, M. (2008). Mathematics as structural language of physical thought. In M. Vicentini & E. Sassi (Eds.), *Connecting research in physics education with teacher edu-cation,* Bd. 2. ICPE.

Planinic, M., Ivanjek, L., Susac, A., & Milin-Sipus, Z. (2013). Comparison of university students' understanding of graphs in different contexts. *Physical Review Special Topics – Physics Education Research, 9*(2), 020103, 1–9.

Pospiech, G., & Oese, E. (2014). The use of mathematical elements in physics – View of grade 8 pupils. In L. Dvořák & V. Koudelková (Eds.), *ICPE-EPEC 2013 Conference Proceedings. Active learning – In a changing world of new technologies* (pp. 199–205). Charles University in Prague, MATFYZPRESS publisher.

Prediger, S. (2013). Darstellungen, Register und mentale Konstruktion von Bedeutungen und Beziehungen – mathematikspezifische sprachliche Herausforderungen identifizieren und bearbeiten. In M. Becker-Mrotzek, K. Schramm, E. Thürmann, & H. Vollmer (Eds.), *Sprache im Fach – Sprachlichkeit und fachliches Lernen* (pp. 167–183). Waxmann, Münster u.a.

Prediger, S., & Wessel, L. (2011). Darstellen – Deuten – Darstellungen vernetzen. Ein fach- und sprachintegrierter Förderansatz für mehrsprachige Lernende im Mathematikunterricht. In S. Prediger & E. Özdil (Eds.), *Mathematiklernen unter Bedingungen der Mehrsprachigkeit – Stand und Perspektiven der Forschung und Entwicklung in Deutschland* (pp. 163–184). Waxmann, Münster u. a.

Prediger, S., & Wessel, L. (2012). Darstellungen vernetzen. Ansatz zur integrierten Entwicklung von Konzepten und Sprachmitteln. *Praxis der Mathematik in der Schule. Sekundarstufen I und II, 45*(Jg. 54), 28–33.

Rebello, S., & Cui, L. (2008). *Retention and transfer of learning from mathematics to physics to engineering.* American Society for Engineering Education.

Redish, E. F. (2016). Analysing the competency of mathematical modelling in physics. In G. Tomasz & E. Dbowska (Eds.), *Key competences in physics teaching and learning* (pp. 25–40). Cham: Springer International Publishing.

Redish, E. F., & Kuo, E. (2015). Language of physics, language of math: Disciplinary culture and dynamic epistemology. *Science & Education, 24*(5–6), 561–590.

Schnotz, W., & Bannert, M. (2003). Construction and interference in learning from multiple representation. *Learning and Instruction, 13,* 141–156.

Schulz von Thun, F. (n.y.). *Das Kommunikationsquadrat.* https://www.schulz-von-thun.de/die-modelle/das-kommunikationsquadrat (downloaded on 2018-06-11).

van Someren, M. (Hrsg.). (1998). Learning with multiple representations. Pergamon, as cited in Bauer (2015).

Strahl, A., Schleusner, U., Mohr, M., & Müller, R. (2010). Wie Schüler Formeln gliedern – eine explorative Studie. *PhyDid, 1*(9), 18–24.

Tuminaro, J., & Redish, E. F. (2007). Elements of a cognitive model of physics problem solving: Epistemic games. *Physical Review Special Topics-Physics Education Research, 3*(2), 020101.

Uhden, O. (2015). Verständnisprobleme von Schülerinnen und Schülern beim Verbinden von Physik und Mathematik. *Zeitschrift für Didaktik der Naturwissenschaften, 22,* 1–12.

van der Meij, J., & Jong, T. d. (2006). Supporting students' learning with multiple representations in a dynamic simulation-based learning environment. *Learning and Instruction, 16*(3), 199–212. https://doi.org/10.1016/j.learninstruc.2006.03.007, as cited in Bauer (2015).
Waldrip, B., Prain, V., & Carolan, J. (2010). Using multi-modal representations to improve learning in junior secondary science. *Research in Science Education, 40*, 65–80.

Chapter 5
What Is Learned About the Roles of Mathematics in Physics While Learning Physics Concepts? A Mathematics Sensitive Look at Physics Teaching and Learning

Olaf Krey

5.1 Introduction

Physics is part of the human attempt to find the laws that govern nature. Doing physics implies a lot of activities, e.g.:

- Observing and experimenting
- Describing systems (identifying relevant entities and the relations between them)
- Establishing cause and effect relationships
- Defining observables and taking measurements
- Analysing relationships between observables in a lab
- Quantifying and mathematizing these relationships (thereby expressing them in a more precise way)
- Analysing the (mathematical) expressions and producing testable statements about the real or at least lab world or even predicting "new" entities in the real world
- Trying to systematize a body of knowledge, presented mathematically, etc.

This list is not exhaustive, but one would easily agree that all the items on this list are part of the "physics game". There are very different conceptions about the epistemology of physics (e.g. Losee 2001; Wenning 2009). Depending on a specific author's background, the field of physics, the physicist or the historical time under consideration, there are, of course, different emphases, but all of these approaches consider mathematics and experiments as core elements of today's physics endeavour.

O. Krey (✉)
Universität Augsburg, Augsburg, Germany
e-mail: olaf.krey@physik.uni-augsburg.de

© Springer Nature Switzerland AG 2019   103
G. Pospiech et al. (eds.), *Mathematics in Physics Education*,
https://doi.org/10.1007/978-3-030-04627-9_5

In science education the role of experiment in physics has been considered extensively (e.g. Hegarty-Hazel 1990; Leach and Paulson 1999; Lunetta et al. 2007), and aspects from the philosophy and history of physics concerning the use and value of experiments have found its way into the mainstream discussions of the science education community (e.g. the explorative and affirmative style of experimentation (Steinle 1998)). Unfortunately, this is not the case for the role of mathematics, which has simply not been a popular field of research until recently. The interplay between experiments and mathematization in physics and the implications for teaching and learning physics – despite a few exceptions (see contribution by Mäntyla and Proonen) – are even less investigated.

For decades the nature of science has been a highly valued field of science education research and is today considered to be an important aspect of scientific literacy. However, in general the discussion culminates in lists of items that summarize important characteristics learners should know about the nature of science (Lederman et al. 2002; McComas et al. 1998; Osborne et al. 2003; Rutherford and Ahlgren 1990). None of them includes the role of mathematics in (any acceptable) depth (cf. Krey 2012). Accordingly, well-developed physics units that help teaching about the role of mathematics are rare to find, although there are thoughtfully developed teaching units offering great potential for teaching physics content and aspects of the nature of physics at the same time (e.g. the HIPST case studies (Höttecke 2012)).

This is in line with many students' (and teachers') views of using mathematics in physics as a rather artificial, distant, algorithmic activity not leaving much space for creativity, controversial discourse or individual taste. It is not unlikely that the way we use mathematics in our physics lessons contributes strongly to the naïve realistic view of physical theories as simplified descriptions of reality that is well documented for students and teachers in the literature (e.g. Abd-El-Khalick 2012). From this perspective, the lack of research about the role of mathematics in learning and teaching physics may have covered an influential factor for an adequate understanding of the nature of physics. Over the years, research in the field of teaching and learning about the nature of science has validated the assumption that aspects of the nature of science had to be addressed *explicitly* and *reflectively* in order to be learnt efficiently (e.g. Abd-El-Khalick and Lederman 2000). It can be assumed that the same holds to be true for teaching and learning about the role of mathematics in physics.

The twofold aim of this paper is (a) to help clarify the roles of mathematics in (learning) physics, not by means of an abstract philosophical analysis but by two case studies about (i) uniform motion and (ii) image formation by thin lenses and (b) to exemplify how intentionally or unintentionally teaching about the roles of mathematics in physics occurs in the average physics classroom. I have chosen the two cases to illustrate an inductive and a deductive way of teaching. By doing so, the author wishes to draw attention to the fact that learning about the roles of mathematics in physics occurs (usually implicitly) at most of our physics lessons and most likely shapes our students' views on the nature of (learning) physics. To make this point, the case studies are used to illustrate certain characteristics of the use of mathematics in physics. Although they have been chosen to cover equations from different epistemological categories, the case studies are not everything that could or should be said about the use of mathematics in physics. It is not the aim of this chapter to come up with another list of possible characteristics of the use of mathematics in physics. Therefore, I will not even make an attempt to generate a complete list of those features. My modest aim is to illustrate a few of the messages (potentially and/or implicitly) inherent in average physics lessons. I thereby hope to help develop sensitivity towards potential learning opportunities for our students. Before considering the two cases, a more elaborated framework allowing to frame the chosen cases and to guide their analysis is necessary.

5.1.1   Thinking About the Roles of Mathematics in (Learning) Physics: A Basic Framework

As shown on the left side of Fig. 5.1, the author of this paper makes use of a more or less consensual distinction between reality (R) and mathematical theory (MT) (as suggested by, e.g. Ludwig 1985). This distinction is made for analytical purposes, and it can be discussed whether the separation of the two is adequate and empirically valid. Accepting this distinction allows to define a physical theory (PT) as a mapping ($\sim$) between reality and mathematical theory: $PT = R(\sim)MT$. (For details, see Ludwig 1985, 1990.) Obviously, mathematics can stand for itself, as an artificial creation by the human mind without a direct relation to the real world (pure mathematics). The world can also be considered independently of any mathematical theory, and humans can engage with this real world in many ways. Without going into further details, it becomes obvious that from this point of view, a large part of physics can be considered as the venture of building bridges between mathematics (MT) and the real word (R). This broad perspective allows to recognize otherwise isolated findings as aspects of a greater common – the attempt to build bridges between R and MT.

Ludwig's simple approach ($PT = R(\sim)MT$) serves well as an integrating perspective for many more elaborated descriptions of the roles of mathematics in physics. One of the limitations of this approach has been pointed out by philosophers (Cartwright 1983) and physicists, e.g. Einstein, who states: "As far as the laws of mathematics refer to reality, they are not certain; and as far as they are certain, they do not refer to reality" (Einstein 2002). That's why on the right side of Fig. 5.1, a pre-mathematical model (M) is included. Elements of this simplified (e.g. by selection, isolation, idealization) model are the entities to which mathematical structures are mapped. Refining Ludwig's approach I therefore suggest to consider two mappings ($-$) and ($*$), one between the real world and the

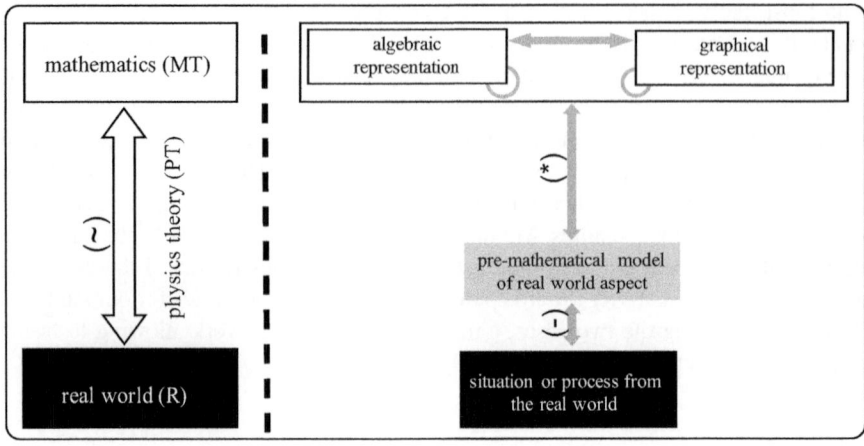

Fig. 5.1 Physics as building and walking on bridges between mathematics and (models of) the real world by the use of algebraic and graphical mathematical representation

pre-mathematical model ($-$) and a second one between the pre-mathematical model and mathematics (*). In total we have PT = R($-$)M(*)MT = R($\sim$)MT.[1] From the world of mathematics, the (arguably) two most important forms of representation (algebraic and graphical) have been selected to exemplify translations between them (straight arrows) as well as constructive and interpretive activities (straight and circular arrows) that go along with their usage.

This model is not meant to describe physics as a human activity as a whole. (I honestly doubt this is possible.) Its purpose here is to provide a framework for this chapter. Another weakness of this model is its lack to specifically address further important aspects, e.g. concerning the role of experiments or more precisely the great epistemological gap between experimental data and their algebraic representation (Falk 1990). This is something to keep in mind.

As mentioned before, this primary framework is a rather broad one, allowing to frame otherwise quite different attempts to describe how mappings between mathematics and the real word are established in science education. The mathematics-physics interplay can be described in many helpful ways, focusing on different aspects. (For a reconstruction of interplay patterns based on middle school physics classroom evidence, see Lehavi et al. 2017; for an overview on modelling concepts, see Pospiech in this book; for a related framework assuming a "physical-mathematical model" to be essential for an in-depth analysis of using mathematics in learning physics, see Uhden et al. 2012.)

---

[1]Ludwig is aware of this complexity and his approach is much more detailed than can be explicated here. (For details see Ludwig 1985, 1990.)

The way bridges between mathematics and the real world are built and used in physics instruction has been identified as a matter of concern quite early. Wagenschein likens the learners' empty recall of physics equations to "paper flowers" (Wagenschein 1995); others articulate their impression of physics being destroyed/decomposed/ripped by calculation ("zerrechnete Physik", Dittmann et al. 1989). More recently a distinction between a *technical* dimension and a *structural* dimension of mathematics in physics (Karam 2014) roughly corresponding with *low-level* and more *elaborated* mathematical activities on the student side (Krey 2012) has been suggested. The technical dimension refers to an instrumental role of mathematics, to mathematics as a calculation tool which implies a procedural focus on practices such as to blindly use equations for solving quantitative problems, to focus on algorithmic manipulation, etc. The structural dimension in contrary refers to an organizational role of mathematics, to mathematics as a reasoning instrument which suggests the focus on physical interpretation of mathematical representations and the derivation of equations from physical principles, etc. (cf. Karam 2014; Krey 2012).

A very simple and widely spread distinction (with far-reaching consequences) can be made between an inductive and a deductive approach to teaching physics content. An *inductive* approach starts from observations (real world), while a *deductive* approach starts from already established (e.g. algebraically represented) laws and principles. Although these are general teaching patterns not necessarily referring to a quantitative relationship to be taught, for this article the implications of their application are of interest as far as these patterns are used to frame ways to the laws of physics, usually represented mathematically (e.g. as an equation).

Working out more elaborated patterns of the mathematics-physics interplay in physics education is necessary to better understand how a misrepresentation of the roles of mathematics in physics can be avoided. On the one hand, this knowledge is likely to be more useful for (future) physics teachers than for actual physics learners, and at the same time, in-depth philosophical considerations about the role of mathematics in physics tend to become difficult and challenging, perhaps too difficult for the average high school learner. On the other hand, it is rather simple to define certain features that can also serve as fields of reflection for physics learners (Krey 2012). Here are a few questions that more or less directly lead to these features and provide first opportunities for learners to learn to appreciate the use of mathematics in physics: How do mathematical representations influence *cognitive load*? How does mathematics contribute to the *exactness* of physics? How can mathematics influence information exchange (*communication*) in physics? How is the *certainty* and *objectivity/intersubjectivity* of physics knowledge influenced? At what *cost* do we apply mathematics in physics? It has been shown that students have an inappropriate understanding of all the above questions (Krey 2012).

Karam and Krey (2015) suggest that physics equations can be categorized according to their (subjective) epistemological status. This leads to four categories:

principles (e.g. $\sum \vec{p} = const.$), definitions (e.g. $\vec{p} = m\vec{v}$), empirical regularities (e.g. Balmer's formula) and derivations (e.g. $a = v^2/r$). It is reasonable to assume that teaching and learning (about) physics equations implicitly or explicitly always go hand in hand with an underlying message about the roles of mathematics in physics. However, the message can be very different depending on what category a certain equation belongs to, e.g. whether it was introduced via an inductive or deductive way.

Since it must be assumed that most of the knowledge about the nature of science is taught and learned implicitly, the message sent by an inductive approach needs to be considered more carefully, and perhaps some explicit measures have to be taken to avoid a misrepresentation (see case study 1 on uniform motion). This becomes even more clear by (a) recognizing the fact that only a few equations can be obtained by a deductive approach (derivation) in secondary school physics and (b) knowing about a certain tendency by teachers to choose an inductive approach over a deductive one, even if a deductive one is easily accessible.[2] When physics equations are involved in the teaching of physics, deductions are often used in simple ways, e.g. to predict the final velocity of a falling object based on the equation for free fall. In contrast, the equation for free fall (or accelerated motion in general) itself is obtained by an inductive approach. From this, the introduced equation likely appears as the summarized and generalized result of a longer learning-teaching sequence, the only important "fact" of the lesson.

Although there is no empirical evidence (since a study focusing on the actual use of mathematics in the secondary classroom has not been conducted yet), it can be considered as common ground that the above-mentioned technical dimension of using mathematics in physics is by far over-represented. Calculation in the sense of given-sought problems calling for a plug and chug strategy and the corresponding low-level activities are assumed to be a widespread practice. For example, more recently learners' physics problem-solving has been observed, and their activities could often be described in a calculation frame (Bing and Redish 2009) or as following a recursive plug and chug strategy (Tuminaro and Redish 2007). Furthermore, analysing pupils' beliefs about the role of mathematics in physics supports the assumption of having experienced a lot of low-level activities in school (Krey 2012). Although this problem is known for decades, it has not been addressed efficiently so far. However, a few attempts to develop future teachers' awareness about the problem as well as their equation-related pedagogical content knowledge (Shulman 1986) have been suggested, implemented and evaluated (e.g. Karam and Krey 2015).

"We do students a disservice by treating conceptual understanding as separate from the use of mathematical notations" (Sherin 2001). This chapter is written in

---

[2]See Redfors et al. in this book for examples of deductive elements in physics textbooks that are not used by the teachers. Although there are many examples where equations could be deduced from theoretical or mathematical considerations in high school (Snell's law, centripetal acceleration, etc.), it is rather hard to find many examples for equations that could be dealt with in middle school (e.g. total resistance in series and parallel circuit).

the spirit of this quotation. The modest objective for this article is not to offer a solution for the mentioned problems, but rather provide a rather small contribution to a desired future solution, a small step on a long way to go. The author suggests that teaching physics, whether we are aware of it or not, at the same time means (at least implicitly) teaching about the role of mathematics in physics. In what follows, two cases covering content of most, if not all, secondary physics curricula are considered: uniform motion and image formation by thin (converging) lenses. These cases are used to identify a few of the messages usually sent implicitly to our students and to point out aspects of the mathematics-physics interplay that could be discussed explicitly in the physics classroom in some cases, but at least inform the teacher about a few implications and potentials of her or his everyday physics teaching. For this purpose, an inductive approach of introducing an equation (uniform motion) is complemented by a derivation of the thin lens equation (deductive approach). The author's hope is that the roles of mathematics in learning and teaching physics become clearer and a step towards a better balance between the technical and the structural role of mathematics can be envisioned. The way we use mathematics may play a decisive role in how our students perceive the nature of physics, and this is why I argue that any view about the nature of physics that does not consider the roles of mathematics is incomplete and missing one of the essential aspects.

By going through the two cases, typical, not necessarily observed, teaching scenarios are described from a meta-perspective. After considering a certain aspect of the teaching and learning about uniform motion or image formation, a potentially important or interesting insight about the roles of mathematics in physics is highlighted in a shaded box. The collection of these possible (and sometimes obvious) insights relates to the above-mentioned fields of reflection; by the help of which, one may find a way to help students learn about the roles of mathematics in physics explicitly and reflectively, as suggested for other aspects of the nature of science.

5.2 Case Study 1: Uniform Motion

Usually during their first year of science or physics in secondary school, pupils deal with motions. They learn to distinguish between straight line and curvilinear motion (rotary, circular, oscillatory motion) as well as between uniform and accelerated motion. The most simple (straight line uniform) motion is then analysed in more detail. While semi-quantitative approaches are possible and a helpful intermediate step, sooner or later, pupils start to measure distances and times and display and evaluate a functional dependency (see Box 5.1).

In Box 5.1 the widely used inductive approach to teaching physics at school (see Fig. 5.2) is exemplified. This inductive way of teaching usually starts with a real-world problem, phenomenon, a specific situation or process that is potentially

Fig. 5.2 Inductive way of
teaching physics

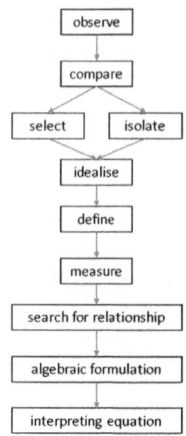

interesting to be investigated. The process under consideration (e.g. the motion of an air bubble) is observed in more detail; relevant aspects are identified by comparison and isolated. Usually an observable is identified or defined to allow measurement. In measurements numerical data are produced (quantification), which (often but not necessarily via a graphical representation) lead to an algebraic statement, usually an equation (mathematization). The specific situation is encapsulated in a diagram or equation. However, the equation stands for much more than just this specific situation from which it was obtained, e.g. represents all uniform motions (with velocity v). In total, a generalization is made and a mathematical statement usually is the end result.[3] Thereby, the gap between the real world and the world of mathematics has been bridged (cf., Fig. 5.1).

> Mathematisation comes with generalisation and downgrades specific situations to examples of a more general insight.[4]

---

[3]Further examples of equations that are likely to be introduced inductively include $x(t) = vt$ (uniform motion), $x(t) = \frac{1}{2}at^2$ (accelerated linear motion), $R = \frac{U}{I} = const.$ (Ohm's law), $R = \varrho \frac{l}{A}$ (Pouillet's law), $Q = mc\Delta T$ (heat equation), etc.

[4]Depending on your perspective, you may wish to emphasize the epistemological upgrade in terms of generality, width of applicability, elegance and simplicity that comes with the mathematical formulation of physical knowledge about the world. However, focusing on the concrete problem (e.g. the motion of an air bubble) from which this exploration started, this specific problem is downplayed and loses its relative importance as it becomes an example of a class of processes with many other processes in this class that are basically the same in that they are just one of many examples for a uniform motion. While none of these perspectives is more correct than the other, I doubt that many young learners can truly appreciate the (surely valid) upgrade at a first encounter.

Box 5.1: Uniform Motion

Task Using the given glass tube filled with coloured water, investigate the motion of the air bubble in the tube, when one end is lifted a bit.

Expectations and Objectives Write down what kind of motion you expect. Give reasons!

Materials/Equipment glass tube filled with coloured water, marker pen, stop watch, ruler

Procedure Lift one end of the tube a little, so that the air bubble takes at least 30s from one end to another. Mark the position of the air bubble every 5s (marker pen). Take measures of the total distances travelled (total displacement) after 5s, 10s, 15s, ...

Observations and Measurements

t in s	0	5	10	15	20	25	30	35	40
x in cm	0	5,4	11,2	16,9	22,7	28,6	34,3	40,2	46,1

observations: _____

Data Analysis and Conclusion
1. Plot the measured distances over time.

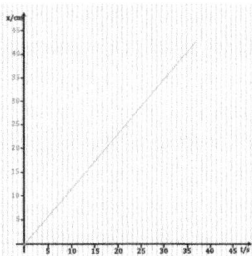

2. Summarize your findings mathematically.

$$x \sim t$$

For every time t and the related distance x we have

$$v = \frac{x}{t} = constant.$$

This can be described as a uniform motion, since the air bubble travels approximately the same distance every 5s. The velocity therefore is constant.

To experts this feature of generality that comes with equations is what makes them "icons of knowledge" (Bais 2005), important nodes in an expert's widely branching web of knowledge. For learners, however, equations can (and often do) become isolated facts that are considered to be the only really important message, which leads to an "obsession" with formulas (Schecker 1985). For example, $v = \frac{x}{t}$ (v – velocity, x – displacement, t – time) is what is usually remembered, after the teaching/learning path described above. This can already be seen as a seed for the common practice of the subsequent instrumental use of equations described in the literature as calculation framing (Bing and Redish 2009) and recursive plug and chug (Tuminaro and Redish 2007), which are part of the technical dimension (Karam 2014) of the use of mathematics in physics, characterized by low-level activities (Krey 2012).

The above-illustrated inductive approach certainly is a valid one that also has its parallels in the history of physics as well as today's research. However, in school physics this approach is probably over-represented. Reasons (among others) can be seen in the assumed learners' lack of mathematical prerequisites that would allow for alternatives (See Karam, Uhden, & Höttecke in this volume to clarify the validity of this assumption.) or the relatively easy planning by the teacher for given-sought scheme problems.

By teaching along this pattern, an implicit message about the role of mathematics in physics, and therefore the nature of physics, is delivered. This message may include the false (or at least one-sided) impressions that, e.g.:

- Experiments are the beginning of any knowledge generation process in physics.
- The main function of mathematics is to summarize experimental results and allow for calculations.
- Quantification and mathematization are more or less the same.
- Mathematics describes reality, using artificial signs and symbols.
- Making use of mathematics means to make statements in a largely arbitrary language to describe insights more easily to be expressed in our natural language.

That all the above-mentioned statements are problematic is well documented in the literature (Cartwright 1983; Koponen and Mäntylä 2006; Krey 2012).

However, there are valid lessons to be learned here as well. As mentioned before, in the example above, it is the formula $v = \frac{x}{t}$ (v – velocity, x – displacement, t – time) that is usually remembered by pupils. Quite often the equation stands for both the definition of velocity and the characterization of a uniform motion. From a teacher's perspective, this can be an intermediate step, but not a satisfying end result. The equation above is the definition of an average velocity for the time interval $[0; t]$, and for learners it is an additional step to understand that $v = \frac{x}{t} = const.$ is what makes this equation an appropriate description for a uniform motion. Even more precise would be a quantification like "for all t". This implies that the initially learned concept of velocity over the years becomes more refined, and formerly ignored or unseen details become accentuated and perhaps more meaningful. The language of mathematics in which statements about velocity are made allows (and forces) to express these refinements carefully. The distinction between momentary (limit concept) and average velocity (ratio concept) shows how this differentiation process occurs conceptually and mathematically.

> Mathematization requires and induces conceptual precision.

In the above example as well as in many other cases collecting data in table, representing them graphically and expressing regularities algebraically (and perhaps even retranslating into common language) are a quite common strategy. It develops pupils' abilities to deal with different types of representations. Learners can experience (often implicitly) that different representations have certain advantages and disadvantages. For example, a certain kind of relationship between observables may be recognized easier in a graph than in a table; an equation allows to chunk different dependencies between observables. The lesson to be learned:

> Different mathematical representations (of the same content) allow access to information in different ways and therefore make one aspect easier accessible than others.

To embrace this message fully, an understanding of (mathematical) representations, their different strength and weaknesses and their coordination in case multiple representations are used is required (Ainsworth 2006). However, at least in German physics lessons and especially in lab work activities, the three-step approach table-graph-equation to collect, visualize and summarize measurement readings often becomes a ritual instead of a rational choice. (See Geyer and Kuske-Janßen in this volume for a more detailed description of translation activities.) This again gives very good reason to students to think about the use of mathematics as something static, ritualized and algorithmic rather than a creative, question-driven and goal-oriented process.

The example of the introduction of the concept of constant velocity and uniform motion hopefully made visible a few of the usually missed learning opportunities about the role of mathematics in physics as a vital part of the nature of physics. Three non-trivial insights were emphasized. If made explicit to learners, those insights might be able to initiate thought processes that perhaps over several years in school may help to develop a more appropriate view on the role(s) of mathematics in physics.

In what follows, another basic example that is part of every secondary physics curriculum is considered. This example helps to illustrate other aspects of an attempt to establish a connection between the real world and the world of mathematics. As it is one of the rare cases in secondary school physics making use of a deductive approach, the example is meant to complement the view of mathematics as a mere language used to express experimental results more precisely and more efficiently as suggested implicitly by the inductive approach. For this purpose we consider a second field of physics, namely, ray optics and more specifically the magnification equation and the thin lens equation.

5.3 Case Study 2: Image Formation by Thin (Converging) Lenses

There are two equations that usually are presented to the students when teaching about ray optics and image formation by thin lenses. The magnification equation $\frac{h_i}{h_o} = \frac{d_i}{d_o}$ and the thin lens equation $\frac{1}{f} = \frac{1}{d_o} + \frac{1}{d_i}$ are usually considered for convex lenses (h – height, d – distance to lens, o – object, i – image, f – focal length). Although they are closely related to one another, for many students they are two isolated facts that have to be learned. In the case of the lens equation, an immediate interpretation is difficult, as the form of the equation is pretty unfamiliar to a learner of school physics. This may explain part of the difficulties to understand the full meaning of these equations and the relation between them.

5.3.1 Towards the Magnification Equation

A teaching sequence could start with an explorative experimental setting from which students learn that for a given object distance (d_o), a sharp image formation on the screen is only possible for a certain image distance (d_i). They may also come up with the hypothesis that the size of a sharp image (image height, h_i) is somehow related to the object distance (d_o). A more explanative experimental approach may follow in which the relation between object distance (d_o), object height (h_o), image distance (d_i) and image height (h_i) is investigated. Sooner or later pupils usually can be convinced that $\frac{h_i}{h_o} = \frac{d_i}{d_o}$ is a valid statement. So far everything that has been said about the role of mathematics in the case of an inductive approach before can be said here as well. A pattern is found within the measurements and students express that pattern more precisely by means of the magnification equation. However, something is different here already. The equation is valid only for a specific constellation, the case of a sharp image formation. That is why we usually write it as a proportion, knowing that the kind of continuity that can be found in the equation describing a uniform motion ($x = vt$) is not a part of the package so to speak. At a first glance however, it would be totally acceptable to write $h_i = \frac{h_o}{d_o} d_i$, and wishful thinking may suggest that for a given object height and object distance, any image height may be realized by adjusting the image distance (screen position). From a purely mathematical point of view, it is no problem to consider a function $h_i(d_i)$, and there is no reason why one would assume that the domain of this function would not be R^+, the positive real numbers. And given the formal analogy between the two equations $x(t) = v \cdot t$ and $h_i(d_i) = \left(\frac{h_o}{d_o} \right) \cdot d_i$, it is (from a mathematics point of view) surprising that the application for describing natural phenomena requires different constraints to more or less the same mathematical concept. For the physicist this may not be of much surprise, but for pupils this seems worth to be made explicit, as it points to an important feature of doing physics and the interplay between mathematics and physics.

> The often seemingly perfect mapping between real world and mathematics is not a guaranteed given and it is by far not self-evident.

In the example discussed, the mathematical structure is more comprehensive than the real-world situation does require. However, at least there is a mathematical structure that does fit our needs.

5.3.2   Towards the Thin Lens Equation

The second equation in the field of ray optics and image formation by thin lenses (lens equation) is usually motivated by the question about how image distance and object distance relate to one another for a given lens. A possible activity for students is presented in Box 5.2.

Box 5.2: Thin Lens Equation – Part 1

The photo shows a candle and a sharp inverted image of the candle appearing at a screen behind the converging lens.

Investigate for this lens (and two other converging lenses) how the image distance depends on the object distance.

Generalize the relationship you discovered in terms of a functional dependency.

By doing a few explorative experimental moves, which is an important part of finding a problem solution, pupils usually find a few regularities. For example, they will find that by placing the candle too close to the lens ($d_o \leq f$), a sharp image formation cannot be observed on the screen. They may remember from constructing ray diagrams that a virtual image is formed. For candle positions outside the focal length ($d_o > f$), one finds that when moving the candle (object) closer to the lens, the image location will move away from the lens. So both the location of the object and the image move in the same direction.

For further investigation there are at least two different strategies available – an inductive experimental strategy and a deductive mathematical strategy. From my experience most pupils and (also students at university) will follow the experimental approach, which is not much of a surprise, since as mentioned before, this is what happens in physics lessons quite often, when a physical law is introduced. (One could argue that it is simply more fun to do an experiment. Well, I think this could be true, but what we consider to be fun is learned, at least to a large extent.) They vary the object distance, perhaps in 1 cm steps, and make measurements of the related image distance. For a converging lens ($f = 5cm$), this may lead to the table of measurements and their graphical representation presented in Box 5.3.

This way, no doubt, a regularity is found, and by plotting the graphs for different lenses, one may obtain a more general statement about the relationship. A few students may perhaps be convinced quickly that this is a linear relation by only considering object distances not too close to the focal length, which is not surprising, since most relations we consider in school physics are linear ones.

Box 5.3: Measurement Readings and Plot Showing the Relation Between d_o and d_i

d_o in cm	d_i in cm	
6,0	30,0	
7,0	17,5	
8,0	13,4	
9,0	11,2	
10,0	10,0	
11,0	9,3	
12,0	8,5	
13,0	8,2	
14,0	7,8	
15,0	7,5	

relation between object distance and image distance (converging lens, f=5cm)

Those who make it to a more appropriate graph would be happy that they have found and documented a relation in an empirical way. How one could express the regularity found in an algebraic representation is a question that also bears an opportunity to learn *about* physics. Physicists strive for understanding – often by investigating specific phenomena in an experimental setting – and describe relations in mathematical terms, but in the end they want to have general laws from which they can form explanations and predictions. As Feynman puts it:

> First we have an observation, then we have numbers that we measure, then we have a law which summarizes all the numbers. But the real *glory* of science is that *we can find a way of thinking* such that the law is *evident*. (Feynman et al. 1964; 26-3)

This is a rather deep reason for why a structural role of mathematics is essential for physicists, implying that mathematical representations help to make logical deductions which provide "a way of thinking such that the law is evident" and therefore are essential tools for the analysis of natural phenomena. This rather philosophical approach can be reduced to a more pragmatic argument that perhaps is more suitable to convince learners of the helpfulness of a deductive mathematical strategy, at least in the context of our example, as new measurements have to be taken for every lens and for every new object distance that has not been investigated before. This also points to an interesting feature of the physics enterprise. Apparently, on one hand asking *"Why?"*, the search for reason and understanding is what keeps physicists going. On the other hand, however, their knowledge is a huge collection of *how* observables are related. And of course, this web of knowledge is structured, and there are statements of different epistemological and hierarchical status that allow to form explanations and predictions (cf. the different epistemological categories for physics equations described in the framework section), but whether this is the same as an answer to the question "Why?" can be argued.

Box 5.4 shows the main steps in deriving the lens equation from prior knowledge including mainly (a) the magnification equation,[5] (b) the ability to read and mentally manipulate ray diagrams, and (c) the fact that light travels in a straight line. The – in this case geometrical – representation of the situation allows an analysis and extended mental manipulation. Without visualizing the situation in a geometrically "correct" way, it is much more difficult to identify meaningful relations between the relevant observables.

> Mathematical representations (e.g. geometrical diagrams, tables, graphs, equations, symbols) materialise abstract ideas (entities and the relations between them). These ideas can be manipulated (literally) (Fischer 2006) and therefore serve as external cognitive tools.

As mentioned before, the derivation of the thin lens equation is one of the rather rare opportunities to demonstrate a deductive way to generate knowledge in the secondary physics classroom. Based on already established knowledge, a geometrical representation is generated, analysed and combined with already established pieces of knowledge represented algebraically (magnification equation). Through purely mathematical reasoning, a "new" equation is produced and interpreted. Perhaps predictions based on the derived equation are made and tested by experiment (Fig. 5.3).

Fig. 5.3 Deductive way of teaching physics

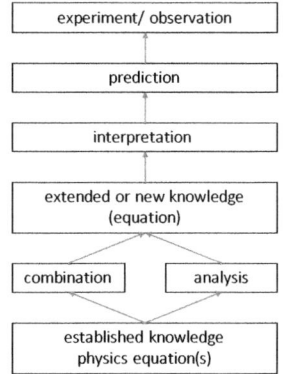

---

[5]Of course, the magnification equation can also be derived by using the intercept theorem, which shows the deep structural equivalence between geometry and light propagation based on the ray model of light, which allows the mapping between relevant aspects of the real world and geometric entities.

Box 5.4: Thin Lens Equation

Ray diagram

The necessary translation from the real world situation to this diagram is not an easy task. Physics knowledge and mathematical representational ability have to be coordinated in order to come up with a helpful representation.

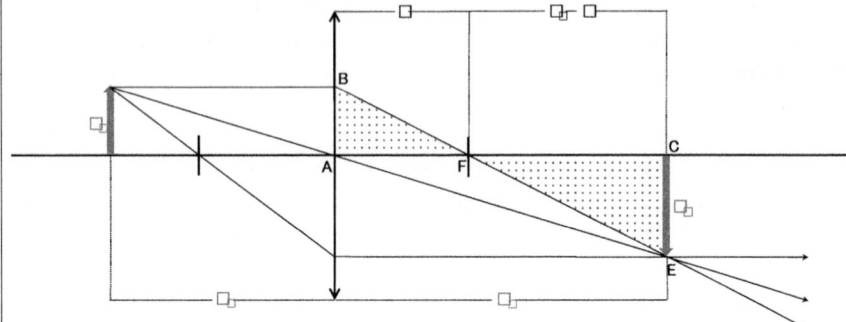

An analysis by geometrical means

1. Since the measure of each interior angle in ΔABF is equal to the measure of the corresponding angle in ΔFCE, the two triangles are similar.

2. The intercept theorem applies, since $\acute{A}B$ und $\acute{C}E$ are parallels. We can therefore establish a relation between $f, d_i - f, h_o, h_i$.

Finding an algebraic form of representation and combine it with prior knowledge

The intercept theorem gives $\frac{h_i}{h_o} = \frac{d_i - f}{f}$. With the magnification equation $\frac{h_i}{h_o} = \frac{d_i}{d_o}$ we get

$$\frac{d_i}{d_o} = \frac{d_i - f}{f} \therefore d_i f = d_o(d_i - f) \therefore d_i f - d_o d_i = -d_o f \therefore d_i = \frac{d_o f}{d_o - f}.$$

The result of the suggested derivation is $d_i = \frac{d_o f}{d_o - f}$. Most textbooks, however, will present $\frac{1}{f} = \frac{1}{d_o} + \frac{1}{d_i}$ as the lens equation. While the first is a direct answer to the problem/question that motivated the investigation, the latter seems to satisfy a purely aesthetic concern. It's a matter of taste to prefer a certain kind of simplicity here – and that actually is a bit of a surprise. Aesthetics and taste sometimes function as guiding principles or at least are influential to the physicist's work. This indeed is astonishing to most learners but also clearly indicates that physics is a human enterprise. The more so, as this aesthetic concern is not related to experimental equipment or phenomena that can be beautiful as well but to mathematical representations used in physics.

Aesthetical aspects are influential to the physicist's work.

Of course, going through this derivation can be a challenge for pupils – for some of them, it involves barriers too high to overcome by themselves. For most students the connection between the symbols and their meaning can get lost, and they blindly follow the manipulation of mathematical symbols. In a way that is not much of a surprise, since this is another great feature of using mathematics in physics. The signs themselves are meaningless; they only carry the meaning that has been explicitly ascribed to them. To be meaningful these signs need to be embedded in a natural language, and they and the results achieved by their manipulation cannot be more precise than the precision with which one can say what these signs actually mean. By the use of the mathematical mechanism, one only has to be diligent in executing the manipulations to keep the same level of precision without actually thinking about it. Interpreting the manipulated equation, however, does require deeper thought (cf. von Weizsäcker 2004, p. 90). It is this conservation of the real-world relations within the mathematical calculus that allows the tool-like use of mathematics in physics. Whether this can be understood at all or whether one simply has to get used to it is not my concern here. However, I argue that to know about the roles and functions that mathematics plays in physics gives a more complete understanding of the nature of physics.

The mathematical calculus that comes with a physics theory allows to find logical implications without content-related deeper thought (reduction of cognitive load). The derived statements are as precise as the definitions of all observables and operations allows them to be.

For the learner, lost in translation, the interpretation of the derived equation is of particular importance. Sometimes, and especially for younger pupils, it might be necessary or extremely helpful to demonstrate and check the validity of the new (intermediate) equation by prediction and experimental test.

In school we usually end here and one can wonder if all this is worth it. However, as far as I can see, by going through the whole process and pointing out the more or less obvious features of the use of mathematics in physics, by relating the world of mathematics to the experimental world and by bringing together experiences in manipulating the experimental setting and mathematical analysis, we create a learning opportunity worthwhile. This may help to make the thin lens equation not just another "paper flower" (Wagenschein 1995, S. 288) but the meaningful peak of a challenging learning path.

5.3.3   The Thin Lens Equation: Beyond the Secondary School Physics Curriculum

What follows goes beyond the average physics school curriculum, but illustrates another important feature of the use of mathematics in physics. This feature becomes visible by considering the implications of an already constructed mathematical representation, which sometimes helps to suggest new insights about the real world.

Above we only considered the case of $f < d_o$ which lead to a first quadrant branch of a hyperbola. By connecting the plotted measurement readings to a continuous graph, the idealization of "continuity" was already applied (and the usefulness of irrational values for the object and image distances was assumed, although those measurement readings obviously do not occur). But otherwise, the branch of the hyperbola conserves the observations made: $d_o \longrightarrow f \Longrightarrow d_i \longrightarrow \infty$, $d_o = 2f \Longrightarrow d_i = 2f$, $d_o \longrightarrow \infty \Longrightarrow d_i \longrightarrow f$. Exploring the derived thin lens equation $d_i = \frac{d_o f}{d_o - f}$ from a mathematical point of view reveals that there must be a second branch of the hyperbola for $f > d_o$. A first part of this branch ($f > d_o > 0$) turns out to describe the formation of virtual images, for which is $d_i < 0$ by convention. So the full consideration of the mathematical solution suggests the existence of a real world entity – virtual images in this case. And also the second part of this hyperbola branch ($d_o < 0$, dotted line in Fig. 5.4) turns out to have a meaningful interpretation for the image formation for virtual objects.

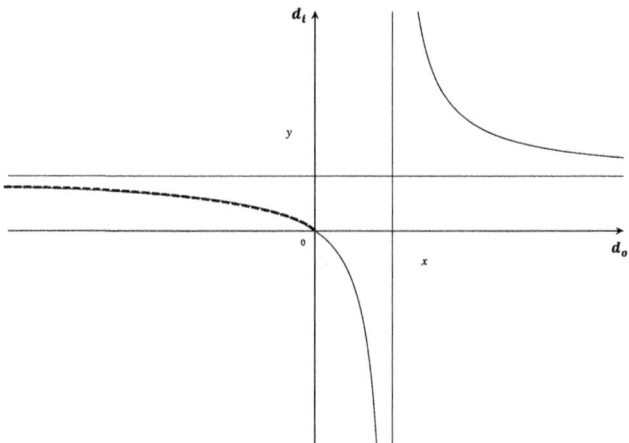

Fig. 5.4 Different parts of the hyperbola describing image formation for real objects and real images (first quadrant), but also for real objects and virtual images (fourth quadrant) and virtual objects (second quadrant)

> Mathematical descriptions can suggest the existence of real world entities or relations.

I am not suggesting that this has to become a necessary curricular content in high school physics or that this would correspond to a historical development; it does not. However, I consider this case to be a helpful example to illustrate an important feature of the mathematics physics interplay. While there are other correspondences that may help to point towards that feature (e.g. Snell's law of refraction and the critical angle), it is hard to find a historically valid and authentic example (such as the prediction of the positron) that can be discussed easily in a high school physics classroom.

5.4   Summary

Within an analytic framework based on Ludwig's conception of a physical theory that allows to view physics as an attempt to build bridges between the real world and the world of mathematics, two cases have been looked at in more detail, the uniform motion and the image formation by thin lenses. This helped to see more clearly which problems and potentials building and using these bridges may provide. The inductive way of introducing a physics equation and (even more so) the less usual deductive way (see Redfors et al. in this volume) are full of more or less implicit messages sent to our students (un)intentionally by the teacher.[6] The explicit consideration or even discussion of these messages bears great potential for learning about the role of mathematics in physics and therefore about the nature of science. It may be considered wishful thinking by some and a plausible hypothesis by others that using mathematics in physics is not necessarily a barrier for learning or limited to routine activities used in given-sought problems but actually a great cognitive tool that allows for analysis and understanding of physics. Perhaps we could even assume that the way we use mathematics in the physics classroom could have a direct influence on the content learning and conceptual understanding of physics. Physics is a difficult subject, and the use of mathematics does not make it easier, but

---

[6]It should be clarified that by no means the author intends to argue for a more deductive and against inductive teaching sequences. Both of them have their advantages and disadvantages. They are adequate to situations, outcomes and learners or they are not. For example, an arranged deductive way of introducing a law of physics (even the lens equation the way it was presented) can lead to lots of low-level activities on the learners' side and simply illustrate bad teaching. Furthermore, both approaches can help teach valuable lessons about the nature of science in general as well as the role of mathematics in (learning) physics more specifically. What the author wishes to say though, is that both approaches to teaching differ in the learning opportunities they can potentially provide with regards to the mathematics-physics-interplay.

it might well be possible that by learning about the nature of physics and about the role mathematics plays within this human endeavour, physics is perceived as more meaningful. If this article helped to generate or elaborate an idea of how the general roles of mathematics in physics can be found or identified in an everyday physics classroom – perhaps sometimes by verbalizing the obvious, the author would be delighted.

References

Abd-El-Khalick, F. (2012). Nature of science in science education: Toward a coherent framework for synergistic research and development. In B. J. Fraser, K. G. Tobin, & C. J. McRobbie (Eds.), *Second international handbook of science education* (pp. 1041–1060). Dordrecht/Heidelberg/London/New York: Springer.

Abd-El-Khalick, F., & Lederman, N. G. (2000). Improving science teachers' conceptions of nature of science: A critical review of the literature. *International Journal of Science Education, 22*(7), 665–701.

Ainsworth, S. (2006). DeFT: A conceptual framework for considering learning with multiple representations. *Learning and Instruction, 16*(3), 183–198.

Bais, S. (2005). *The equations. Icons of knowledge.* Cambridge, MA: Cambridge University Press.

Bing, T. J., & Redish, E. F. (2009). Analyzing problem solving using math in physics: Epistemological framing via warrants. *Physical Review Special Topics – Physics Education Research, 5*, 1–15.

Cartwright, N. (1983). *How the Laws of physics lie. English.* Oxford: Oxford University Press.

Dittmann, H., Näpfel, H., & Schneider, W. B. (1989). Die zerrechnete Physik. In W. B. Schneider (Ed.), *Wege in der Physikdidaktik. Band 1. Sammlung aktueller Beiträge aus der physikdidaktischen Forschung* (Vol. 1, pp. 41–46). Erlangen: Palm & Enke.

Einstein, A. (2002). Geometrie und Erfahrung. Zweite Fassung des Festvortrages gehalten an der Preussischen Akademie der Wissenschaften zu Berlin am 27. Januar 1921. In M. Janssen, R. Schulmann, J. Illy, C. Lehner, & D. K. Buchwald (Eds.), *The collected papers of Albert Einstein. Volume 7. The Berlin years: Writings, 1918–1921* (Vol. 7, pp. 382–388). Princeton: Princeton University Press (Reprint).

Falk, G. (1990). *Physik: Zahl und Realität. Die begrifflichen und mathematischen Grundlagen einer universellen quantitativen Naturbeschreibung.* Basel: Birkhäuser.

Feynman, R., Leighton, R., & Sands, M. (1964). *The Feynman lectures on physics, volume I. mainly mechanics, radiation, and heat.* Reading: Addison-Wesley. Retrieved from http://www.feynmanlectures.caltech.edu/.

Fischer, R. (2006). *Materialisierung und Organisation.* München/Wien: Profil Verlag.

Hegarty-Hazel, E. (Ed.). (1990). *The student laboratory and the science curriculum.* London/New York: Routledge.

Höttecke, D. (2012). HIPST—History and philosophy in science teaching: A European project. *Science & Education, 21*(9), 1229–1232.

Karam, R. (2014). Framing the structural role of mathematics in physics lectures: A case study on electromagnetism. *Physical Review Special Topics – Physics Education Research, 10*(1), 10119-1–10119-23.

Karam, R., & Krey, O. (2015). Quod erat demonstrandum: Understanding and explaining equations in physics teacher education. *Science & Education, 24*(5), 661–698.

Koponen, I. T., & Mäntylä, T. (2006). Generative role of experiments in physics and in teaching physics: A suggestion for epistemological reconstruction. *Science and Education, 15*(1), 31–54.

Krey, O. (2012). *Zur Rolle der Mathematik in der Physik. Wissenschaftstheoretische Aspekte und Vorstellungen Physiklernender.* Berlin: Logos Verlag.

Leach, J., & Paulson, A. (Eds.). (1999). *Practical work in science education. Recent research studies.* Roskilde: Roskilde University Press.

Lederman, N. G., Abd-El-Khalick, F., Bell, R. L., & Schwartz, R. S. (2002). Views of nature of science questionnaire: Toward valid and meaningful assessment of learners' conceptions of nature of science. *Journal of Research in Science Teaching, 39*(6), 497–521.

Lehavi, Y., Bagno, E., Eylon, B.-S., Mualem, R., Pospiech, G., Böhm, U., Krey, O., & Karam, R. (2017). Classroom evidence of teachers' PCK of the interplay of physics and mathematics. In *Key competences in physics teaching and learning* (pp. 95–104). Cham: Springer. https://doi.org/10.1007/978-3-319-44887-9.

Losee, J. (2001). *A historical introduction to the philosophy of science* (4th ed.). Oxford: Oxford University Press.

Ludwig, G. (1985). In Z. Raum (Ed.), *Einführung in die Grundlagen der Theoretischen Physik* (Vol. 1). Braunschweig/Wiesbaden: Vieweg.

Ludwig, G. (1990). *Die Grundstrukturen einer physikalischen Theorie.* Berlin: Springer.

Lunetta, V. N., Hofstein, A., & Clough, M. P. (2007). Learning and teaching in the school science laboratory: An analysis of research, theory, and practice. In S. K. Abell & N. G. Lederman (Eds.), *Handbook of research on science education* (pp. 393–441). New York/London: Routledge.

McComas, W. F., Clough, M. P., & Almazroa, H. (1998). The role and character of the nature of science in science education. In W. F. McComas (Ed.), *The nature of science in science education. Rationales and strategies* (pp. 3–40). Dordrecht: Kluwer Academic Publishers.

Osborne, J., Collins, S., Ratcliffe, M., Millar, R., & Duschl, R. (2003). What "ideas-about-science" should be taught in school science? A Delphi study of the expert community. *Journal of Research in Science Teaching, 40*(7), 692–720.

Rutherford, F. J., & Ahlgren, A. (1990). Science for all Americans. Retrieved from http://www.project2061.org/publications/sfaa/online/sfaatoc.htm

Schecker, H. (1985). *Das Schülervorverständnis zur Mechanik. Eine Untersuchung in der Sekundarstufe II unter Einbeziehung historischer und wissenschaftshistorischer Aspekte.* Bremen: Universität Bremen.

Sherin, B. L. (2001). How students understand physics equations. *Cognition and Instruction, 19*(4), 479–541.

Shulman, L. S. (1986). Those who understand. Knowledge growth in teaching. *Educational Researcher, 15*(2), 4–14.

Steinle, F. (1998). Exploratives vs. theoriebestimmtes Experimentieren: Ampères frühe Arbeiten zum Elektromagnetismus. In M. Heidelberger & F. Steinle (Eds.), *Experimental Essays – Versuche zum Experiment* (pp. 272–297). Basden-Baden: Nomos Verlag.

Tuminaro, J., & Redish, E. F. (2007). Elements of a cognitive model of physics problem solving: Epistemic games. *Physical Review Special Topics – Physics Education Research, 3*, 1–22.

Uhden, O., Karam, R., Pietrocola, M., & Pospiech, G. (2012). Modelling mathematical reasoning in physics education. *Science & Education, 21*, 485–506. https://doi.org/10.1007/s11191-011-9396-6.

von Weizsäcker, C. F. (2004). *Der begriffliche Aufbau der theoretischen Physik.* Stuttgart/Leipzig: S. Hirzel.

Wagenschein, M. (1995). *Die pädagogische Dimension der Physik.* Aachen: Hahner Verlagsgesellschaft.

Wenning, C. J. (2009). Scientific epistemology: How scientists know what they know. *Journal of Physics Teacher Education Online, 5*(2), 3–15.

Part II
Learning Mathematization

Chapter 6
Blending Physical Knowledge with Mathematical Form in Physics Problem Solving

Mark Eichenlaub and Edward F. Redish

6.1 Introduction

Physicists and educators have long held problem-solving to be one of the key tools to help students understand physics (Meltzer and Otero 2015). If problem-solving is a bridge to expert-like understanding, we should find ways to let students experience expert-like thinking in as many dimensions as possible while working problems. This includes learning new physical concepts and mathematical techniques, because experts and novices differ greatly in the amount of physics and math they know. But experts also diverge from novices in their problem-solving strategies, their patterns of metacognition (Schoenfeld and Sloane 2016), their epistemological stances towards their work (and abilities to negotiate between various stances), their conception of what mathematical entities are, and their expectations for how to derive meaning from their work. These differences between experts and novices are part of a "hidden curriculum" that students need to learn as they progress in physics, but which we rarely teach explicitly (Redish et al. 2010).

In particular, researchers have singled out math as a particular sticking point in problem solving in introductory physics. Much of the existing research seeks to document student understanding, or misunderstanding, of particular mathematical tools, such as differentiation or coordinate systems. Our teaching experience shows that even when students appear to have mastered the appropriate tools in previous classes, they may still struggle to use those tools effectively in physics problems. In previous work, one of us (Redish and Kuo 2015) laid out an argument that this is largely because the ways that physicists make meaning with mathematics are unfamiliar to students. Even if they are skilled with the manipulations of algebra

M. Eichenlaub (✉) · E. F. Redish
Department of Physics, University of Maryland, College Park, MD, USA
e-mail: meichenl@umd.edu; redish@umd.edu

© Springer Nature Switzerland AG 2019   127
G. Pospiech et al. (eds.), *Mathematics in Physics Education*,
https://doi.org/10.1007/978-3-030-04627-9_6

and calculus, students' expectations about how to interpret variables may lead them astray. For example, many students, given a problem about test charges and electric fields, will say that changing the magnitude of a test charge changes the magnitude of the electric field it measures. They reason from the equation $E = F/q$ that if q increases, E decreases. The students understand the math involved well, but don't account for the way the force on a charge changes with the charge – there was a hidden functional dependence they did not see because physics culture assumes the reader will associate every symbol (in this case, F) to its physical meaning. That would make the functional dependence of F on q clear, but students don't yet expect to have to find this physical meaning when solving problems. The challenge for educators is to create problems and problem-solving environments that encourage students to search for physical meaning in mathematics.

In creating problems, educators often separate "qualitative" problems that test and build intuition from "quantitative" problems to develop mathematical skills (Hsu et al. 2004), indicating an implicit assumption that these are separate faculties that are used and developed individually. We believe that for experts, intuition and mathematics are not insulated from each other, or even cleanly separable. Instead, they reinforce each other; intuition is often connected to mathematics and mathematics is understood partially via intuition. While solving a problem, an expert will blend mathematical forms such as equations (or abstracted properties of equations), with intuitive conceptual schema to create richer mental spaces than those derived from formal mathematics alone.

For an example of this blending of intuition and mathematical form, we look at Sherin (2001)'s description of "symbolic forms", a class of blended intuitive-formal conceptual structures that experts (and in Sherin's case, second-year physics students) use to understand equations. To introduce symbolic forms, we'll take an example from Sherin, who describes two students thinking about a ball falling through the atmosphere at terminal velocity. The students intuitively understand that air drag and gravity are both acting on the ball, but balance each other out, leaving no net acceleration. In Sherin's account, the students activate a conceptual schema for "balancing" of competing influences. This balancing schema could potentially match many different physical scenarios, or even everyday scenarios, such as expenses balancing out income when breaking even financially, but here it is called on to understand air drag and gravity. The students then associate the balancing schema with the abstracted symbol template for equations $\square = \square$, where each square represents one of the two balancing influences. The students know that they are looking for an equation with an expression related to gravity on one side and an expression related to air drag on the other. The students' work on a specific equation is then informed by this pairing of the intuition behind balancing with the symbolic template. The combined intuition and formal structure are collectively a symbolic form. Sherin identified 21 symbolic forms in his data corpus; our purpose here is to use them as one example of blended intuitive and formal thinking that is found in experts and potentially in students as well.

Symbolic forms are not a complete account of how physicists make meaning with equations. The example of failed meaning-making in the eq. $E = F/q$, cited earlier, involves the correct use of the symbolic form Sherin identified as "prop-", where a schema related to "if one goes up, the other goes down" is blended with the symbol template $\left[\frac{\cdots}{\ldots x \ldots} \right]$, but this symbolic form alone wasn't enough to lead students to the right answer.

We cannot give a full account of all the ways experts bring meaning into equations, but as a second example, we consider experts' ontology of equations, i.e., the types of objects equations are in experts' conceptual schemas. For example, here are a few examples of physicists writing about the relation between the Yukawa potential, $V(r) = \frac{qe^{-mr}}{r}$ and the Coulomb potential, $V(r) = \frac{q}{r}$.

In the limit of $m \to 0$ the Yukawa potential becomes the Coulomb or gravitational potential ... (Heile 2015)
    ... if we choose ... $m_0 = 0$, the potential reduces to the Coulomb potential energy ... (Townsend 2000) [source uses m_0 in place of m]
    We can take the limit $\alpha \to 0$ and recover the Coulomb potential. (Hassani 2013) (source uses α in place of m)
    The Coulomb potential of electromagnetism is an example of a Yukawa potential ... (Wikipedia 2016)
    We see ...that if the mass m of the mediating particle vanishes, the force produced will obey the $1/r^2$ law. If you trace back over our derivation, you will see that this comes from the fact that the Lagrangian density for the simplest field theory involves two powers of the spacetime derivative ... (Zee 2010)

In some cases, physicists see themselves as enacting a change in the Yukawa potential. They or their reader actively "take the limit" or "choose $m = 0$". Other times, the Yukawa potential changes, but there's no clear agent involved. It may "become" or "reduce to" the Coulomb potential and the mass may "vanish", but no entity is identified as enacting the change. In contrast to these dynamic descriptions, the relationship can also be described statically. Nothing in particular is happening when the Coulomb potential "is an example of" the Yukawa potential.

This is just a sampling of physicists' language on the topic. The details of how they describe the Yukawa potential-Coulomb potential relationship may depend on both the physicist and the context of what they're communicating in complicated ways. Our goal here is simply to illustrate that there is a significant diversity of ways to conceptualize of an equation.

These examples come from professional, graduate, and upper-division under-graduate material, where such a diversity of conceptualizations of equations is commonplace. By contrast, in introductory physics textbooks, equations are usually treated as static entities to be scrutinized.

Outside the nucleus the nuclear force is negligible, and the potential is given by Coulomb's law, $U(r) = +k(2e)(Ze)/r, \ldots$ (Tipler and Mosca 2007).
Coulomb's law can be written in vector form ... $\tilde{\mathbf{F}}_{12} = k \frac{Q_1 Q_2}{r_{21}^2} \hat{\mathbf{r}}_{21} \ldots$ (Giancoli 2000).
The electric force acting on a point charge q_1 as a result of the presence of a second point charge q_2 is given by Coulomb's Law: $F = \frac{kq_1q_2}{r^2} = \frac{q_1q_2}{4\pi\epsilon_0 r^2}$ (Nave 2017).

The main exception we have observed to this "equations as static entities" ontology is in descriptions of formal operations on equations. These descriptions come up during derivations (e.g. "differentiate with respect to t", "set them equal to each other", etc.). Also, equations are sometimes described as active entities, for example "Coulomb's law describes a force of infinite range which obeys the inverse square law" (Nave 2017) in that they "describe" things, but this does not represent the same diversity of conceptions we saw with regard to the Coulomb and Yukawa potentials.

This mostly-static view of equations stands in contrast to introductory physics sources' descriptions of the physical quantities the equations represent

We can divide up a charge distribution into infinitesimal charges ... (Giancoli 2000)

The force exerted by one point charge on another acts along the line joining the charges. It varies inversely as the square of the distance separating the charges and is proportional to the product of the charges. (Tipler and Mosca 2007)

In describing the force, field, or charges associated with Coulomb's law, introductory textbooks use both agentive language ("We can divide") and non-agentive ("The force ....acts...."). The second quotation here also mixes dynamic ("varies inversely...") with static ("is proportional to...") language in the same sentence. So while a diversity of ontological viewpoints are generally considered acceptable for thinking about physics in introductory settings, this seems to apply much more to physical quantities than to equations. As we move to more expert settings, the equations themselves take on the same diversity of ontologies.

Sfard (1991)'s notion of conceiving of functions as either objects or processes is similar to ours, but here we consider "process" views where the equation itself is changing, as opposed to Sfard's notion of a static function which describes change when inputs transform into outputs. Our point here is simply to illustrate one more small piece of the diversity in expert conceptual systems used to make mathematics physically meaningful. This piece, like symbolic forms, is never explicitly taught. It is a part of the hidden curriculum, and something we can try to find evolving in students as they progress towards expertise.

Based on an exploratory analysis of problem-solving interviews, we suggest that students, in the right circumstances, use a large and diverse arsenal of productive, sophisticated, and creative ways to conceptualize physics problem-solving. They do not always access these resources when they would be productive, and many of the difficulties students experience with using math in physics are not so much difficulties of having the appropriate tools, but of applying them appropriately. While much of the hidden curriculum will need to be learned via years of enculturation in the physics community, there are entire swaths of it that don't need to be explicitly taught so much as activated. Small interventions that encourage students to use specific problem-solving strategies, can, in some cases, greatly enhance students' access to productive ways of thinking about mathematical tools that are rarely explicitly taught.

The strategies we're investigating are commonplace, well-known to physicists, and generally well-regarded components of effective problem-solving. They include examining special and extreme cases, dimensional analysis, and estimation. Our contribution to understanding these strategies is to suggest that their scope can be very broad. They can be used at different stages of problem-solving and in different ways. We also give examples of how students use these strategies to construct meaning from mathematical expressions in ways similar to how experts do it.

6.2 Theoretical Framework: Resources, Framing, and Epistemic Games

Our analysis is situated in the resource model (Hammer 2000; Redish 2004). In this framework, students don't have monolithic conceptual understandings; they have many small pieces of knowledge, or resources, that they can call on while solving a problem. When solving a problem, students will activate various resources and construct a solution based on them. If students don't solve a problem correctly, it may be that they don't have the appropriate resources, or that they do, but aren't activating them in that context. In the previous example of a test charge and the measured electric field, students did activate resources relating to understanding inverse mathematical relationships (including the prop- symbolic form), but did not activate resources related to the functional dependence of force. Whether or not students activate a resource can depend on how they associate it with other resources they are using, so in a future problem, students might improve their performance if they've learned to activate resources related to functional dependence when they see questions about forces in electromagnetism.

The issue is not so simple, though. The students in question were all able to recite the mantra "the electric field is independent of the test charge". In this sense, they knew the answer to the problem, but they didn't call on this knowledge, or if they did, didn't apply it. In addition to resources related to manipulating mathematical equations and resources related to intuitive understanding of physics, students also have "epistemological resources", resources related to how they seek to obtain and justify knowledge (Hammer and Elby 2003).

A student who uses an equation because it makes intuitive sense may come to the same answer as a student who uses an equation they found in a textbook they consider authoritative, but the way they are thinking about knowledge is very different; they are using different epistemological resources. The students who answer the test charge problem incorrectly are probably not activating epistemological resources related to interpreting each variable physically, or resources related to finding concordance between memorized facts (such as the electric field being independent of the charge) and the results of reasoning based on equations.

To understand why students sometimes use one set of epistemological resources and sometimes another, we use the lens of epistemological framing (Bing and Redish 2009). Because we could potentially use any resource at our disposal (i.e. every fact, technique, or type of reasoning we can conceive of) on a given problem, the space of problem-solving strategies we have to search through to find one effective approach is extremely large. We begin by narrowing the problem down to a certain type of problem, and then search through the resources we associate with that type. Calling on a physical principle to solve a problem requires activating different epistemological resources than using an equation does, and those resources often are associated with different epistemological framing (Gupta and Elby 2011; Kuo et al. 2013). Students who answered that changing the magnitude of a test charge changes the magnitude of the measured electric field may have entered a "calculation" frame, and didn't remember or pay attention to their knowledge that the electric field is independent of the test charge because they didn't frame the task as one in which physical principles are relevant.

Moving towards expertise in problem solving is as much about using what resources you have effectively as it is about picking up new resources. As students work physics problems, they need to learn not only new content, but new ways of relating to the content. They need to be able to choose productive epistemological frames and activate appropriate resources. All of these are difficult tasks that live mostly in the hidden curriculum.

Analyses of problem solving often break the task down into a series of steps. Sometimes this is prescriptive, as when textbooks list a series of steps to make in solving a problem. For example, Redish et al. (2010) describes a textbook with the following scaffold for problem solving

Model! – Make simplifying assumptions.

Visualize! – Draw a pictorial representation.

Solve! – Do the math.

Assess! – Check your result has the correct units, is reasonable, and answers the question

and gives an example where the method failed. The textbook posed a question asking us to find the volume occupied by the water evaporated after sweating during exercise. The solution manual followed each step, finding that the volume was simply the volume of an ideal gas with the appropriate number of molecules, ignoring that the evaporated water will, by convection and diffusion, spread out over a very large volume. The textbook's solution manual follows each individual step, but nonetheless comes to a nonsensical answer to a problem by failing to "tell the story of the problem". From this example, Redish finds

Tying the analysis to a rubric – a formal set of mapped rules ... does not help if it does not also activate an intuitive sense of meaning by tying the problem to all we know and recognize about a system

We also view problem-solving as a series of steps, but not as steps for students to follow. Instead, the steps are a framework for researchers to understand how students solve problems. This approach is common in physics education research. For example, in analyzing student difficulties using math in physics, Wilcox et al. (2013) proposed the ACER framework, which consists of Activation of the tool, Construction of the model, Execution of the mathematics, and Reflection on the result.

Whereas a prescriptive problem-solving script tells students to follow precise steps in a given order, Wilcox et al. write, "…we are not suggesting that all physics problems are solved in some clearly organized fashion, but a well articulated, complete solution involves all components of the ACER framework." That is, having the framework allows the researchers to narrow their focus and identify specific tasks students are struggling with, rather than simply bemoaning that they can't apply math appropriately. In that paper, Wilcox et al. found that students' resources for the technique of taking a Taylor expansion weren't activated by the appropriate signal, which was one variable of interest being very much smaller than another, and suggested that problems be written to focus on building this particular association for students between signal and mathematical technique.

Frameworks like ACER are effective at picking out specific technical steps that students don't take in problem-solving. Our interest here is broader, including student epistemologies, attitudes towards mathematics, conceptualization of the entities involved, and other aspects of the hidden curriculum. The framework of epistemic games is a flexible one that allows analysis of both problem-solving moves and the motivations behind them.

We have previously discussed epistemological frames in problem-solving. Framing is a general feature in psychology, and when we work in a particular frame it often cues a script for how that type of activity typically goes, which sets expectations for what will happen next and what sorts of actions are appropriate (Goffman 1974).

An epistemic game is a script (with additional structure to be described below) that allows us to understand the moves students make in problem solving (Tuminaro and Redish 2007). As we watch students solving problems, we assign their problem-solving to some particular epistemic game, which we take to structure the types of resources they call on and the order in which they use them. An epistemic game will generally have a particular epistemological frame associated with it, but adds additional structure. The viability of epistemic games as an analysis framework stems from its psychological plausibility via the connection to psychological scripts and that, when Tuminaro and Redish (2007) analyzed student problem solving, they found that certain epistemic games were repeated many times on different problems and in different circumstances. The term "epistemic game" comes from Collins and Ferguson (1993), although the version we use here is that of Tuminaro and Redish (2007).

In an epistemic game, as in games like solitaire or chess, one or several players make moves. These moves might be mathematical moves, such as *add the same quantity to both sides of the equation*, conversational moves, such as *offer a reason supporting your position*, or physical moves, such as *draw a picture of the situation*. Because players can make various types of moves, analyzing the moves lets us focus on different aspects of the hidden curriculum in problem-solving.

As the players of an epistemic game make moves, they gradually fill out an epistemological form, a template for what the solution to the problem should look like, which may be physical or verbal. Finally, players either reach the e-game's stopping condition and decide they are done, or else switch to a different game or give up on the solution attempt.

Tuminaro and Redish identified six common games that students play during problem solving, such as *recursive plug-and-chug*, in which students identify a formula and put values into it without interpreting the results, and *mapping meaning to mathematics*, which describes the problem-solving process in which students analyze the physics of a situation, turn their analysis into equations, manipulate the equations, and then turn the result into a new physical understanding.

Students use e-games to guide their inquiry, and their (generally unconscious) choices for what e-game to play have large effects on their problem-solving process. Different games have different rules about what sort of evidence is salient, what sort of moves are allowed, what type of arguments to give, and what it means to be done with a problem. When students get stuck on a problem or come to answers that don't make sense from the viewpoint of experts, they often have resources that would allow them to solve the problem, but never access them because they are not included in the current frame (Tuminaro and Redish 2007; Bing and Redish 2012).

We do not consider playing an epistemic game favorable or unfavorable; that depends on which epistemic game and how appropriate it is to the situation. Epistemic games also aren't confined to students; experts play them as well, and do it very effectively. For example, in his short paper "A Model of Leptons" Weinberg [1967] searches for an equation to describe leptons and their interactions. The method is to list various properties the equation should have—what symmetries it has, what types of solutions to avoid, etc. Each such consideration can be translated into a particular feature that the final equation should have, and by combining a sufficient number of features, only one equation is left that satisfies them all— the final equation derived for leptons and their interactions. Weinberg is playing an epistemic game we call "significant features". This is a game used to generate solutions to a given problem (as opposed to evaluating a proposed solution). To play, one lists relevant significant features a solution ought to have, such as a maximum at a certain place, or matching a certain symbolic form. Each feature is translated into a formal constraint or piece of the sought solution, such as the derivative being zero at the maximum or a symbolic template which matches the symbolic form appearing in the equation. As the player discovers more features and their associated forms, they gradually fill out the equation (or plot or other form) they are seeking. The game ends when they either decide they have completely specified the answer to the problem or decide that they don't know enough features to do so.

In Sherin (2001)'s work, two students decide that under constant acceleration the equation for velocity as a function of time is either $v(t) = v_0 + at$ or $v(t) = v_0 + \frac{1}{2}at^2$, but cannot decide between the two. Sherin analyzes this as using the "base plus change" symbolic form. Students conceptualize the situation as velocity starting at some given value, then changing to a new value, and realize that this maps onto the symbolic template $\square + \Delta$. The symbolic form doesn't distinguish between the terms at and $\frac{1}{2}at^2$ as "changes" to map onto Δ in the symbolic template. Both are positive (for positive acceleration and time) and indicate an object speeding up. Sherin's analysis is that using only a symbolic form isn't enough for students to determine the correct equation. We agree, and add that the students are playing the same "significant features" game that Weinberg did in building a model of leptons. They begin with a feature they want to the solution to have – matching the conceptual schema of base + change, and translate that into a mathematical form – the $\square + \Delta$ symbol template. Although they ran out of features to finish constraining their answer to the one correct answer, they were nonetheless playing the same epistemic game, just with very different material and at different levels of expertise.

6.3   Data and Analysis

The students we interviewed were enrolled in an introductory physics for life science course at the University of Maryland. Most are juniors, with some sophomores and seniors. The course prerequisites include one semester each of calculus, probability, chemistry, and two semesters of biology. Students are mixed between having taken physics in high school and not.

This is a population of relative novices in physics, but who have taken from 5 to 12 college science courses before taking this one; they generally have strong expectations about how science courses and problem-solving in them work, which the instructor (Redish) routinely challenges. (See (Redish et al. 2014) for more details on the creation and principles behind the course.) All interviews used a think-aloud protocol, encouraging students to write and articulate their thoughts at all times as they solved problems. Some interviews were one-on-one with the interviewer (Eichenlaub) and others were group interviews in which the interviewer was present but participated minimally, with occasional small interventions designed to prompt use of specific problem-solving strategies. We conducted a total of 24 hour-long interviews with 23 different students enrolled in the first of two semesters of this course.

With these interviews, we were interested in the breadth of approaches and conceptualizations students take in problem solving, including whether and how they blend physical intuition with mathematical formalism and how they conceive of variables, parameters, and entire equations. We chose problems and problem-solving strategies that we hoped would elicit epistemic games with a strong interplay of intuition and formalism. Our goal was to bring out a diversity of interesting conceptual systems in students' solution attempts. The strategies we investigated were *examine extreme or special cases*, *dimensional analysis*, and *estimation*, chosen especially because they are all familiar parts of an expert physicist's toolkit, but are not always taught explicitly at the introductory level.

We wanted to make fine-grained analysis of small, interesting incidents in our interviews, so we took video of the interviews ensuring that the field of view captured all students (for group interviews) or student and interviewer (for one-on-one interviews) so that we could reference speech, gesture, and other expressions. Students wrote on a whiteboard, which we photographed at the interview's conclusion.

Our goal in analyzing these interviews was to generate hypotheses about cognitively-rich ways that students can interact with math and physics. This was exploratory analysis, not confirmatory, so the results we present here are case studies to be examined in more detail in the future. Our focus was on finding particularly interesting moments throughout the problem-solving sessions, including moments of blended mathematical/intuitive sensemaking and moments that show how students conceive of the mathematical entities they're working with. To that end, we reviewed the videos highlighting incidents that stood out to us, then discussed them together to generate hypotheses regarding student conceptualizations that interested us. Here we present those hypotheses along with descriptions of the incidents that we watched while generating them.

Below, we describe each strategy and report briefly on how students in our interviews took up the strategy before discussing, through the lens of epistemic games, specific cognitive aspects of problem-solving that these strategies elicited.

6.3.1 Extreme and Special Cases

Most physical systems we examine in problem solving have one or more free parameters that enter the problem. For example, in trying to find the effective spring constant of two springs connected in series to form a single combined spring, the individual spring constants are such parameters. If we set one of these parameters to its largest or smallest possible value, we're looking at an extreme case. So for springs in series, we could set the second spring constant to be infinite, in which case it is completely rigid, does not contribute at all to the stretching of the combined spring, and the effective spring constant would simply be that of the other spring. Using this fact to try to understand something about the general situation is a strategy we call "extreme case" reasoning. We might also consider the case where the two spring constants are equal. Then each spring stretches the same amount, the total stretch is twice as much as the stretch of an individual spring, and the effective spring constant is half that of an individual spring. We call this "special case" reasoning. The two are almost the same, but extreme cases have been discussed independently in the literature, so we identify them as separate but closely-related reasoning strategies.

Clement and Stephens (2009) studied extreme cases in a grade school setting, finding that looking at the extreme case helps students build vivid, dynamic mental imagery, consistently leading to better intuitive understanding of physics scenarios. Used in quantitative problem solving, extreme cases not only boost our intuition, but also allow us to connect that intuition to equations we've generated or are considering. Our accuracy and intuition for thinking about extreme cases has led physicists to make their study a standard problem-solving tool (Morin 2008). Nearing (2003) elaborated on why extreme cases lead to better intuition in his undergraduate textbook on mathematical physics

> How do you learn intuition?
>
> When you've finished a problem and your answer agrees with the back of the book or with your friends or even a teacher, you're not done. The way to get an intuitive understanding of the mathematics and of the physics is to analyze your solution thoroughly. Does it make sense? There are almost always several parameters that enter the problem, so what happens to your solution when you push these parameters to their limits? In a mechanics problem, what if one mass is much larger than another? Does your solution do the right thing? In electromagnetism, if you make a couple of parameters equal to each other does it reduce everything to a simple, special case? When you're doing a surface integral should the answer be positive or negative and does your answer agree?
>
> When you address these questions to every problem you ever solve, you do several things. First, you'll find your own mistakes before someone else does. Second, you acquire an intuition about how the equations ought to behave and how the world that they describe ought to behave. Third, It makes all your later efforts easier because you will then have some clue about why the equations work the way they do. It reifies the algebra.

Extreme cases, to Nearing, are not about the physics situation alone or the mathematical expression alone, but a way of bridging the two into a unified qualitative and quantitative understanding of physics.

In a prototypical use of the extreme or special case reasoning, students first derive an expression, in terms of parameters of the problem, that is a potential solution to the problem. For example, they might find the acceleration of a block in terms of various masses, angles, and coefficients of friction involved. They then use their physical intuition for extreme cases to evaluate this potential solution.

This evaluative use can be analyzed as a "sanity check" epistemic game. This game begins after students generate a candidate solution to a problem, and is used to test whether the solution makes sense. The prototypical moves of the game are

1. Identify a feature which the candidate solution intuitively ought to have.
2. Check whether the candidate solution has this feature.
3. If it does, identify a new feature the solution ought to have. If it does not, either reject the solution and start over, or enter a new epistemic game to determine why the solution and feature do not match.
4. Continue playing the game until you can't think of any more features or are satisfied with your confidence in the candidate solution.

When playing the sanity check game with the extreme case strategy, these moves could look like this:

1. Identify a physical variable in the problem.
2. Imagine it becoming extremely large, extremely small, or some special value that stands out.
3. Intuitively identify the behavior of the system in this case.

4. Analyze the same limit of expression in the potential solution.
5. Compare the results of (3) and (4) for consistency. If they are consistent, confidence in the solution increases. If they are inconsistent, choose a new e-game to figure out whether it is your intuition or the mathematical expression that is incorrect.
6. Repeat for other variables in the problem.

This game encourages students to repeatedly compare a mathematical expression with a physical intuition, and so promises to be a good place to learn about how students use math to inform physical understanding and vice versa.

Although we've outlined a canonical version of the game above, physicists use extreme cases in many other ways. The snippets from physicists discussing the relation between the Yukawa and Coulomb potentials in Sect. 6.1 discuss sending a parameter (α) to an extreme (zero), but instead of examining the physical behavior of a system in this limit, they discuss an equation itself simplifying to a different equation.

Further, in many cases beyond the introductory classroom, we can only find analytic solutions for the limiting cases of an equation, so studying the asymptotic behavior of otherwise intractable physical systems has become the most common analytical approach in modern mathematical physics (Bender and Orszag 1999). As a result, extreme cases and special cases lead to a host of useful tools, resources, and intuitions for physicists, including for example perturbation theory and the WKB method. The power of this game is one of the reasons that the predilection of introductory students to "put numbers in right away" (thereby reducing the problem to one that looks more like "just math") is often counter-productive.

In interviews, we gave students several problems where we expected the extreme cases game to be useful: the half-Atwood machine (Fig. 6.1), the electric field on the axis of a ring of charge, springs in series and parallel, and the area of an ellipse.

In every case, we found that students have strong and accurate physical intuitions for the extreme or special cases. In some circumstances, students consistently spontaneously play the sanity check game using special case reasoning. For example, every student interviewed on the ellipse problem (Fig. 6.2) considered the

Fig. 6.1 The half-Atwood problem: A block of mass M is attached to a block of mass m via a massless string strung over a pulley as shown. The setup is frictionless. What is the acceleration of the block m?

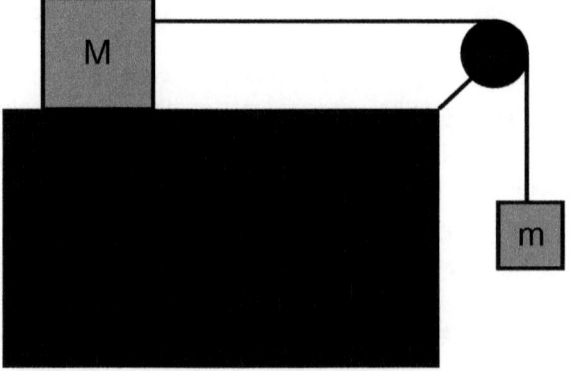

Which of these could be a formula for the area of the ellipse shown?

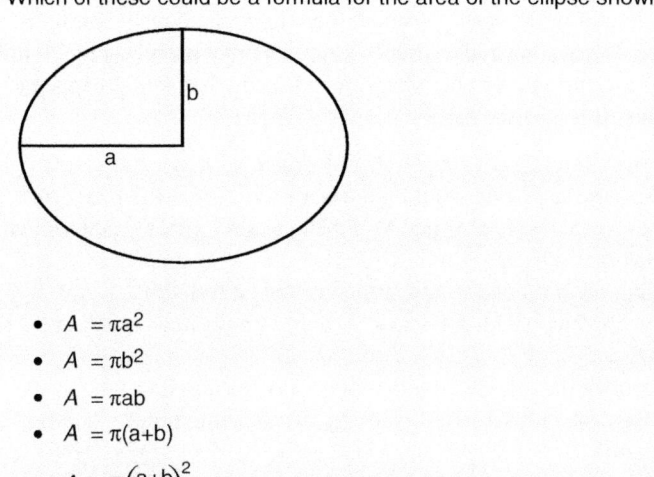

- $A = \pi a^2$
- $A = \pi b^2$
- $A = \pi ab$
- $A = \pi(a+b)$
- $A = \pi\left(\dfrac{a+b}{2}\right)^2$

Fig. 6.2 The ellipse problem

special case $a = b$, a circle, and used it to evaluate the given answers. No students, on the other hand, spontaneously checked the extreme case $b \to 0$, however, when prompted by the interviewer to consider "a long, skinny ellipse", most did use this extreme case to answer the question correctly.

Extreme/special case reasoning also proved consistently valuable to students answering the half-Atwood problem (Fig. 6.1) and to students finding the electric field on the axis of a ring of charge.

The students in our interviews found this strategy less effective when asked to determine the effective spring constant of two springs connected in series. Asked to consider this problem without being prompted to think of extreme cases, Lizzie, Myra, and Lelia (pseudonyms) had the following discussion

1. Lelia: What's Hooke's law again? Oh yeah, T is this. [writes an equation for Hooke's law] So in this. The length would technically be twice as long.
2. Lizzie: oh for the two.
3. Lelia: technically this k coefficient would be twice as long as one of them.
4. Lizzie: yeah [erases board and writes $T = k\Delta L$].
5. Lelia: so I think k-series would be them added together. Cause I remember I remember from.
6. Lizzie: the homework.
7. Lelia: yeah there's two connected the new k coefficient is twice as much, I think.
8. Lizzie: we have two k's. [all writing equations involving k, T, and ΔL].
9. Lizzie: k-series would be k-one plus k-two.
10. Lelia: yeah, that's what I'm thinking.

Lizzie, Lelia, and Myra (did not speak above) associate higher spring constants with more length of the spring, leading them to conclude that springs in series have an a spring constant that adds. After working on other problems for 20 minutes, they returned to the springs, and the interviewer asked what would happen if one spring were much stiffer than the other

1. Lizzie: the stretch, the easy one would stretch a lot.
2. Lelia: and the hard one would stretch a little bit, so the total stretch would be mostly due to the softer spring. So i mean again I guess k-constant would be the softer one.
3. Lizzie: but the hard one would still contribute a little bit.
4. Lelia: yeah, but we don't know. I don't know how much, you know what percentage.
5. Myra: can we like divide it by the number of springs?
6. Lelia: like k-one plus k-two divided by two or something?
7. Lizzie: or n?
8. Myra: cause I'm thinking because if one is way easier to stretch and the other one is not stretching at all, but each spring is still contributing some stretching, so then you divide it by the number of springs.

Their physical intuition is correct, but in the remaining time, they are unable to match their intuition to an equation, and ultimately revert to their original answer of $k_{eff} = k_1 + k_2$. Although their effort to play extreme cases didn't result in a correct equation, they did make correct conclusions about the mathematical form of the answer, specifically that the effective constant should be (very nearly) the same as that of the softer spring, and they consistently attempted to match physical intuition to equations. However, without a clear mapping from spring constants onto physical stiffness, it was difficult for them to find a correct equation.

6.3.2 The Dimensional Analysis Game

There are several strategies based on the idea that if two physical quantities are equal, they must have the same dimensions. We refer to these strategies collectively as "dimensional analysis", and they are taught extensively at the introductory level (Robinett 2015), while also remaining of professional interest to physicists for more than a century (Bridgman 1922). A prototypical example of playing the sanity check epistemic game for evaluating a formula using dimensional analysis would be

1. Find an equation that may be a solution to a given problem.
2. Evaluate the physical dimensions of each term on the left side of the equation.
3. Multiply the dimensions of all terms on the left hand side together to get the dimensions of the entire left hand side.
4. Repeat (2) and (3) for the right hand side.
5. Compare the dimensions of each side of the equation. If they are the same, the equation may be correct. If they are not, the equation is incorrect.

This game allows students to catch some mistakes in their answers. Students in our sample played dimensional analysis readily on questions that specifically asked about dimensions, for example asking which of a set of four formulas could be the surface area of an object, but also occasionally used it productively in questions aimed at understanding functional relationships. For example, when asked,

> Sixteen students are sharing N large cheese pizzas. Assuming that the students share the pizza evenly, which expression gives the number of students each pizza must feed?

many students had difficulty choosing between the expressions $N/16$ and $16/N$, among other distractors. Two interviewees noted that the number 16 had units of students, and because the answer they were looking for had units of students, the choice must be $16/N$.

Our data set was not set up to investigate the more elaborate dimensional analysis game in which students are asked to use the dimensions of relevant variables to explicitly construct formulas, or pieces of formulas, in cases where the full analytical derivation is too long, complicated, or intractable to be useful (Robinett 2015), although we believe this game would be interesting to research in the future. Constructing a formula from elemental pieces, as well as understanding an incomplete formula which contains scaling information but cannot be numerically evaluated, may lead to rich student cognition.

6.3.3 Estimation

By estimation, we mean integrating personal knowledge, a corpus of memorized numbers, and approximation heuristics to obtain order-of-magnitude estimates of interesting quantities, either in physics or in everyday scenarios. Like dimensional analysis and examining extreme cases, estimation is a highly valued in the physics community and in physics education, which have a culture of "Fermi estimates", "back of the envelope" calculations, and "order of magnitude" estimates. For example, *The Physics Teacher* publishes a "Fermi Question" in each issue (Weinstein 2018), and several universities have undergraduate courses in estimation (Phinney, Chiang).

We chose to investigate estimation because performing estimates generally requires students to think about their everyday experience and find methods of quantifying it, often while building equations that multiply various such terms together. Thus, it forces students to use intuition and a formal understanding of mathematics simultaneously.

A case study by Modir et al. (2014) established an estimation epistemic game involving six moves,

1. Problematize
2. Propose method
3. What to remember

4. See if parts are enough
5. Pure calculations
6. Evaluation

and documented how a student estimated the energy in a hurricane by going rapidly forward and backward between these moves.

In one of our interviews, a group of four students, Amelia, Zane, Jean, and Chris, attempt to estimate the time it would take a submersible submarine to sink to the bottom of the ocean. The group agreed to assume the ocean was 1000 m deep, and Jean calculated a descent time of about 14 s by assuming the sphere fell with ordinary gravitational acceleration. Several group members challenged the notion that the submersible would accelerate during its descent and proposed it would instead fall at terminal velocity, but never reached consensus before the following exchange

1. Amelia: Well if you think about it based on the previous situation that we said, we said it was at a thousand meters (Jean: mmhmm) the force was two thousand newtons. Fourteen seconds technically could be legible just because a thousand meters isn't really a lot. We have a really heavy (Zane: that's true) like submersible, so it kind of makes sense in that situation.
2. Zane: let's go with it.
3. Jean: go with the...
4. Zane: fourteen seconds, yeah.
5. Amelia: It all depends on like, all these variables. With these variables it would make sense that it would be dropping that fast.
6. Jean: And we're assuming there's no um, buoyant force, no viscous force.

Although Zane called on counterintuitions several minutes before this exchange ("it's not going to hit, you know, a hundred thousand miles per hour at the bottom."), and repeatedly argued against the constant acceleration approach, the group decided that their calculation "kind of makes sense", ultimately accepting a highly unreasonable answer. Despite their incorrect conclusion, we see in this passage group members calling on a sense of whether numbers are reasonable for a given physical situation, questioning the relation between unknown parameters and quantities of interest, and examining the simplifying physical assumptions that go into their reasoning. At the conclusion of the interview, the interviewer mentioned that their conclusion had the submersible reaching the ocean floor at roughly 300 miles per hour, and the group burst out laughing. It may be that the group's considerable efforts at sense-making failed largely due to an unfamiliarity with the relevant units, as well as neglecting to convert them into more everyday terms.

In this incident, we see a group negotiating what physical effects to model mathematically and what to ignore. This skill is essential to all physical modeling. For example, in introductory physics we often model the flight of a thrown ball using only a uniform gravitational force, giving a parabolic trajectory. In doing so, we ignore aerodynamic drag, other aerodynamic effects (e.g. lift), nonuniformity of the gravitational field, inertial forces due to Earth's rotation, magnetization of the ball in Earth's magnetic field, the Yarkovsky effect (black-body radiation is red/blue shifted in the ball's reference frame due to its rotation, cause a net torque), momentum imparted by sunlight the ball absorbs or reflects, transfer of material in and out of the ball's surface, and many other effects. Some of these can be important or not for a ball, depending on the accuracy we want and the parameters of the situation. Others are effectively never important for a ball thrown on Earth, but are relevant for, e.g. dust particles in space. Physicists often estimate the sizes of such effects to see whether they belong in more complete and explicit model. By improving student estimation skills, we also empower them to build better-informed mathematical models, and to understand the extent of those models' applicability.

6.4 The Nature of Equations

In physics education, there has been considerable effort to understand the different ways that students view equations epistemologically (Airey and Linder 2009), e.g. whether they ought to map closely to phenomena or be treated formally, be accepted as given by authority or derived from fundamental principles, and their relationship to modeling. Here, we are interested in a different type of view of equations: their ontology, or what types of object they're considered to be.

Earlier, using the example of physicists discussing the Yukawa and Coulomb potentials, we suggested that there is a variety of ways that physicists conceive of the equations they're working with. Physicists in different contexts speaking to different audiences sometimes thought of equations as dynamic objects, with one equation transforming into another, and other times thought of them as static, with one equation being a special case of another. Additionally, when equations changed, sometimes it was the speaker or the audience actively making the change, and sometimes the equation changed without a specific agent being identified.

The three problem-solving strategies introduced so far all call on students to think about equations in new ways—to hold them accountable to common sense (estimation) and to check various features of them (dimensions and special cases). We might wonder whether interacting with equations in certain ways changes the conceptualization that students have of equations.

In watching students play epistemic games with mathematics, we saw a diversity of conceptualizations of equations emerge. For example, Alma, in working the ellipse problem, checked the special case $a = b$ with reference to the formula $A = \pi \left(\frac{a+b}{2}\right)^2$

> ...so a plus b squared over two squared times two is four plus b that would be 2 ab. b squared plus yeah. Okay. Yeah. Okay. So then you would have r squared plus two r squared plus r squared which equals pi four r squared over four, so I guess it's a plus b over two cause you're taking the average. Oh, it's like you're turning into a circle. That's cool. Yeah.

When Alma checked the special case, she described the ellipse as "turning into a circle", but she didn't make this reference while working with a geometric object. Instead, she made it while working with an equation. In other words, the ellipse was "turning into a circle" in that it became the formula for the area of a circle when $a = b = r$. This dynamic picture of an equation mirrors the language that a Yukawa potential "becomes the Coulomb" potential in an extreme case. She was working with the formula, but instead of saying that the formula turns into a formula for a circle, she said "you're turning it into a circle", referencing a geometric object (the circle) while working with a non-geometric object (the formula). We suggest that for Alma, in this moment, there was no significant distinction between the formula and the object it describes, which, if correct, shows a very strong example of binding meaning to an equation.

Similarly, Amelia was examining the equation $N(t) = N_0 e^{-t/\tau}$ for the number N of particles remaining when they decay over time t with a time constant τ. (The interview protocol for this series of interviews is not available because the interviewer asked the questions verbally, writing equations out by hand on a whiteboard. Videos of these interviews for scholarly review may be available on request.) In examining the special case where half of the original number of particles remain, Amelia described actively changing equations via procedural language, such as "I divide each side by the initial amount. I el-en [take the natural logarithm of] each side", but she also described changing equations not according to any fixed procedural rules, "I changed the equation, if I'm doing this logic, because I don't remember what the half life equation is off the top of my head. So I rewrote the equation to say that $Q(t)$ is equal to one half times the initial amount times e to the negative t. t referring to just time..."

In both cases, the agency in changing the equation lies in Amelia herself. In the first case, she follows formal manipulations. In the second, she is "doing this by logic", presumably a reference to some mix of common sense, intuition, and mathematical reasoning, as a contrast to memorization. She created an entirely new equation based off a template from the old one, assigning specific physical meaning to each term she created.

Students can take varied stances towards the types of objects that equations are while manipulating, creating, and interpreting them in many contexts, not simply in the context of the strategies we investigated. We believe this menagerie of conceptualizations of equations and interactions with them is especially rich in these epistemic games that play out with these strategies due to their requirements to blend symbology and physical meaning.

6.5   Blending and Sensemaking

In most frameworks to analyze student use of mathematics, there is a step in which the student manipulates the equations. For example, in ACER, this step is Execution of the mathematics, described as

Transforming the math structures (e.g., unevaluated integrals) in the construction component into relevant mathematical expressions (e.g., evaluated integrals) is often necessary to uncover solutions. Each mathematical tool requires a specific set of steps and basic knowledge. For example, executing a Taylor approximation may require knowledge of common expansion templates (e.g., $\sin x \approx x + x^3/3! + \ldots$) and how to adapt these templates to the mathematical model developed previously. Alternatively, one might need to know how to compute derivatives of complex functions. The mathematical procedures performed in this component are not, at least to experts, context free. In addition to employing base mathematical skills, experts maintain awareness of the meaning of each symbol in the expression (e.g., which symbols are constants when taking derivatives).

Although this description indicates that the operations are not purely formal, and that the problem-solver needs to remember the context and meaning of the symbols, the steps on which we understand the equations' emergent meaning and match them to physical understanding are separate steps from the steps of symbolic manipulation under these frameworks.

Research on the manipulation step has mostly focused on the difficulties that students have in making manipulations or on the procedural resources they use while manipulating equations (for example, thinking of physically sliding a variable from the numerator of one side of an equation to the denominator of another) (Wittmann and Black 2015).

Experts use individual mathematical manipulations as sources of physical sense-making. Kustusch et al. (2014) studied physics professors solving a thermodynamics problem that involved taking partial derivatives. There were many choices for which derivatives to take, and experts used physical insight into the derivatives' meaning to guide their choices. In a review of the literature on mathematical sensemaking inside the mathematical manipulation steps of problem-solving, Kuo et al. (2013) found "no studies that focused upon the mathematical processing step in quantitative problem solving or described alternatives to using equations as computational tools." The same authors then contrasted two students, one who describes a kinematic formula in terms of its meaning via a symbolic form, another who saw the formula essentially as a black-box tool, and found that these students performed the mathematical manipulations in a problem using that kinematic concept differently. The student who understood the formula via a symbolic form was able to blend mathematical and physical reasoning to take a shortcut solution to the problem, while the other student was not.

If we value this sort of blended sensemaking, we should find ways to encourage it in students. We believe extreme-case reasoning is one way to do this. In order to use extreme case reasoning, students must think about formulas and physical systems simultaneously, and as a result, they find new and creative ways of conceptualizing and manipulating equations.

For example, Myra, while considering the "springs in series" problem, has written $\frac{T}{k_1} + \frac{T}{k_2} = \Delta L_{\text{total}} = \frac{T_{\text{sum}}}{k_{\text{series}}}$ and below it $\frac{T}{k_1} + \frac{T}{k_2} = \frac{T}{k_{\text{series}}}$ on her whiteboard, saying

I'm thinking that if you apply a constant force, for k-one will give like this amount of length plus k-two will give like this amount of length, then that's like the total amount of length of the series, which equals to k over T-series. And that makes sense to me. I just don't know how you would like not put the T in the equation.

Although the group did not take up her method and she soon abandoned it, Myra's expression was correct, and a short algebra step away from the desired solution. In generating this expression, Myra didn't start with basic definitions and follow a purely formal procedure. Instead, she blended her conceptual understanding of stretching with the mathematical formalism while manipulating mathematical expressions.

Shortly before, Lelia stated, "and both would contribute just like one would contribute like one would have less change than the other. They'd still both probably be a part of the stretch."

Myra's key insight was to translate this "both contributing" intuition into a symbolic form (Sherin 2001), a basic template for an equation, along with a meaning used to understand entire classes of equations that build on that template. Here, Myra uses what Sherin identifies as the "parts of a whole" template, $[\Box + \Box + ...]$.

Myra fits Lelia's idea about both springs contributing stretch onto this template via the heuristic equation $stretch_1 + stretch_2 = stretch_{total}$. Then, using the definition of a spring constant, which contains a variable ΔL for the stretch of the spring, Myra substitutes in the stretch of each spring, making each term physically meaningful as she does, obtaining $\frac{T}{k_1} + \frac{T}{k_2} = \frac{T}{k_{series}}$.

In a separate instance, Bert was working on the half-Atwood problem. His solution had a sign error, $a = \frac{mg}{m-M}$ instead of the correct $a = \frac{mg}{m+M}$, due to an inconsistency in how he set up his coordinate system. The interviewer introduced and scaffolded the extreme and special case game for Bert, who readily took it up, discovering that his solution had the blocks reversing direction based on their mass, which he rejected as intuitively incorrect. Instead of reworking the entire problem from scratch, Bert tried making small modifications to his answer to eliminate the problem, for example introducing an absolute value in the denominator to keep it from changing signs. As he continued introducing and testing new solutions, he looked at $\frac{M-m}{mg}$ as a potential solution, considered the extreme case where $M \gg m$, and said

So then this is super big that's super small. [pauses, draws a minus sign on M in the numerator] Still doesn't make sense. Still not working. Cause one of these [the masses] are big then it's gonna be big acceleration. That's not what should happen. Should be as this one grows [points to M] it gets smaller, so like that has to be in the denominator.

In suggesting that M must go in the denominator, Bert has repurposed the extreme cases game. Instead of evaluating potential solutions, he is placing constraints on what the unknown correct solution must look like. Like Myra, he blends his physical intuition and symbolic forms to achieve this.

The symbol template Bert uses is a division template, along with a conceptual schema about inverse proportionality. It is a schema where as one quantity increases, another decreases, but in the extreme case it shows that as one quantity grows very large, another becomes very small.

In applying this symbolic form, Bert begins with his intuitive understanding that very large, heavy objects are difficult to move and blends in his formal understanding of inverse proportionality to creatively generate a new instance of the extreme case game.

Bert did not wind up solving the problem; he rejected the correct solution on the mistaken grounds that it was symmetric with respect to interchange of m and M, but despite not coming to a complete solution, he generated unique insights as well as a partial solution by renegotiating his relationships to the equation he was searching for while playing the extreme case game.

6.6 Implications for Instruction

It is common to see backsliding in surveys of student epistemologies over the course of introductory physics. For most courses, students on average exit their college physics course with less-favorable beliefs about how to learn physics than they had when they entered (Redish et al. 1998; Adams et al. 2006). As epistemologies are tied to problem solving strategies (Ataide and Greca this volume), it's likely that students' conceptions of the role of mathematics and their approaches toward using it also deteriorate over most year-long introductory sequences. This means that although we observed surprising and expert-like strategies in our problem-solving interviews, we need to be wary of the possibility that our classes lead to students using these strategies less and less with time.

The reward and feedback structures in many introductory courses focus on evaluating whether a student can perform a certain calculation correctly. This includes grades on homework and exams, and in many circumstances, the verbal feedback students receive from instructors, for example that in "initiate-response-evaluate" questioning (Mehan 1979). In most of the episodes we've cited in this chapter, students wouldn't have received positive feedback from such systems. Bert didn't get the correct answer when he found creative new applications of extreme case reasoning. Myra blended her physical intuitions with formal mathematics in a symbolic form to get an expression equivalent to the correct answer for how springs add, but her group didn't take it up, and they left the interview without having reached a consensus on the correct answer. Alma, when checking the special case of a circular ellipse, used a dynamic ontology of the equation to reinforce her understanding of the test she was performing, but wasn't able to distinguish two answers which both passed that test, and she wound up choosing the wrong answer. Each time, the students were displaying expert-like problem-solving behaviors that we might not expect to see in introductory courses, but because they didn't come to the correct final conclusion, in many classrooms they wouldn't have received points on a test, heard their teachers praise, reiterate, extend on, or dive more deeply into the reasoning, or seen their peers enthusiastically take up the same methods. Because the type of feedback students receive can significantly affect their attitude toward learning (Carlone et al. 2014; Russ et al. 2009), this lack of positive feedback when trying expert-like strategies could easily quench students' fledgling attempts at useful, general ways of solving problems and understanding physics.

It isn't surprising that the techniques that work for experts in problem solving are less effective for novices. Learning to use tools takes practice. Riding a bicycle is much faster and more efficient than walking once you know how to do it, but it can be wobbly, frightening, and even dangerous at first. If we want students not only to try out strategies such as testing special cases or blending intuition and formalism through symbolic forms, they need a freedom to fail, encouragement to try out new ways of thinking, and positive reinforcement when they do so. Spike and Finkelstein (2016), studying recitation sections, found that the extent to which TAs do these things depends on their beliefs about the goals of instruction. When instructors expand their goals beyond seeing students perform calculations correctly (whether quantitative or qualitative) and value the growth of new and useful ways of thinking, classrooms environments can take the seeds of expert-like thought we've observed here and nurture them.

In our own courses, these observations have led us to two ways of encouraging new problem-solving behaviors. The first is asking questions which focus on evaluating the meaning of formulas, as opposed to using them as black boxes. For example, a problem from the textbook by Serway and Jewett (2004) reads

> Consider a gas at a temperature of 3500 K whose atoms can occupy only two energy levels separated by 1.5 eV ... Determine the ratio of the number of atoms in the higher energy level to the number in the lower energy level.

The solution involves using the formula for the Boltzmann factor as a black box tool. To encourage different ways of reasoning about the formula, in a class one of us (Redish) taught recently, a quiz question asked

> When a membrane allows one kind of ion to pass through and not another, a concentration difference can lead to an electric potential difference developing across the membrane. For example, if the concentration of NaCl on one side of a membrane is $c_1 = 10mM$ and $c_2 = 2mM$ on the other, letting only Na+ ions through (and not Cl-) will build up a potential difference across the membrane. This is controlled by the equation that says that the electric potential energy, $q\Delta V$, balances the concentration difference effects via the Boltzmann factor thus:

$$\frac{c_1}{c_2} = e^{\frac{-q\Delta V}{k_B T}}$$

> For a given set of concentrations (c_1 and c_2 fixed) would you expect increasing the temperature to increase, decrease, or leave the Nernst potential, ΔV, unaffected?

This question encourages students to reason about the functional form of the Boltzmann factor, perhaps by imagining extreme cases or using symbolic forms. It also encourages students to think of T not as a fixed entity, but as a parameter that can be tuned to change both the physical behavior of a system and the numerical value in an equation.

In addition to asking questions that encourage students to reason about formulas instead of apply them in order to get the right answer, we also ask questions that encourage students to reflect on formulas without the need to extract a final correct or incorrect answer. For example, in one of our recitation exercises, students are asked to construct their own equation to describe when a worm will begin to suffocate as we scale up its size (reducing its surface area to volume ratio) (Redish and Cooke 2013). We then ask students,

> Our analysis in [the previous part] was a modeling analysis. An organism like an earthworm might grow in two ways: by just getting longer or isometrically – by scaling up all its dimensions. What can you say about the growth of an earthworm by these two methods as a result of your analysis in [the previous part]? Does a worm have a maximum size? If so, in what sense? If so, find it.

These more open-ended and reflective questions ask students to use formulas - formulas they have constructed - for interpretation and coming to new inferences, both about physical systems and about the mathematical properties of equations.

Throughout this chapter, we have searched for a number of creative ways students approach problems, including thinking about the extreme cases, conceptualizing parameters in different ways, and using equations for estimation. In interviews, students do all these things, but they can easily lead the student seemingly nowhere—no correct answer to a question, no encouragement from an instructor, no adoption by peers. To encourage students to try out useful but difficult-to-master new strategies, we continue refining the way we ask questions and attend to student thinking during instruction.

References

Adams, W. K., Perkins, K. K., Podolefsky, N. S., Dubson, M., Finkelstein, N. D., & Wieman, C. E. (2006). New instrument for measuring student beliefs about physics and learning physics: The Colorado learning attitudes about science survey. *Physical Review Special Topicsphysics Education Research, 2*(1), 010101.

Airey, J., & Linder, C. (2009). A disciplinary discourse perspective on university science learning: Achieving fluency in a critical constellation of modes. *Journal of Research in Science Teaching, 46*(1), 27–49.

Ataide, R., & Greca, I. (this volume). Pre-service physics teachers' theorems-in-action about problem solving and its relation with epistemic views on the relationship between physics and mathematics in understanding physics. In G. Pospiech (Ed.), *Mathematics in physics education research*. Cham: Springer.

Bender, C. M., & Orszag, S. A. (1999). *Advanced mathematical methods for scientists and engineers I*. New York: Springer.

Bing, T. J., & Redish, E. F. (2009). Analyzing problem solving using math in physics: Epistemological framing via warrants. *Physical Review Special Topics-Physics Education Research, 5*(2), 020108.

Bing, T. J., & Redish, E. F. (2012). Epistemic complexity and the journeyman-expert transition. *Physical Review Special Topics-Physics Education Research, 8*(1), 010105.

Bridgman, P. W. (1922). *Dimensional analysis*. New Haven: Yale University Press.

Carlone, H. B., Scott, C. M., & Lowder, C. (2014). Becoming (less) scientific: A longitudinal study of students' identity work from elementary to middle school science. *Journal of Research in Science Teaching, 51*(7), 836–869.

Chiang, E. *Astronomy 250: Order-of-magnitude physics.* http://w.astro.berkeley.edu/~echiang/oom/oom.html. Accessed: 2017-05-04.

Clement, J. J., & Stephens, L. (2009). *Extreme case reasoning and model based learning experts and students.* In Proceedings of the 2009 Annual Meeting of the National Association for Research in Science Learning.

Collins, A., & Ferguson, W. (1993). Epistemic forms and epistemic games: Structures and strategies to guide inquiry. *Educational Psychologist, 28*(1), 25–42.

Giancoli, D. C. (2000). *Physics for scientists and engineers* (Vol. 3). Upper Saddle River: Prentice Hall.

Goffman, E. (1974). *Frame analysis: An essay on the organization of experience.* Cambridge, MA: Harvard University Press.

Gupta, A., & Elby, A. (2011). Beyond epistemological deficits: Dynamic explanations of engineering students' difficulties with mathematical sense-making. *International Journal of Science Education, 33*(18), 2463–2488.

Hammer, D. (2000, July). Student resources for learning introductory physics. *American Journal of Physics, 68*(S1):S52. ISSN: 00029505. doi: https://doi.org/10.1119/1.19520. http://link.aip.org/link/?AJP/68/S52/1&Agg=doi.

Hammer, D., & Elby, A. (2003). Tapping epistemological resources for learning physics. *The Journal of the Learning Sciences, 12*(1), 53–90.

Hassani, S. (2013). *Mathematical physics: A modern introduction to its foundations.* Cham: Springer.

Heile, F. (2015). *Why is the force between two charged particles eerily similar to the force between two large masses?* https://www.quora.com/Why-is-the-force-between-two-charged-particles-eerily-similar-to-the-force-between-two-large-ma answer/Frank-Heile. Online. Accessed 14 December 2015.

Hsu, L., Brewe, E., Foster, T. M., & Harper, K. A. (2004). Resource letter rps-1: Research in problem solving. *American Journal of Physics, 72*(9), 1147–1156.

Kuo, E., Hull, M. M., Gupta, A., & Elby, A. (2013). How students blend conceptual and formal mathematical reasoning in solving physics problems. *Science Education, 97*(1), 32–57.

Kustusch, M. B., Roundy, D., Dray, T., & Manogue, C. A. (2014). Partial derivative games in thermodynamics: A cognitive task analysis. *Physical Review Special Topics – Physics Education Research, 10*(1), 1–16. ISSN: 15549178. 10.1103/PhysRevSTPER.10.010101.

Mehan, H. (1979). *Learning lessons.* Cambridge, MA: Harvard University Press.

Meltzer, D. E., & Otero, V. K. (2015). A brief history of physics education in the United States. *American Journal of Physics, 83*(5), 447–458.

Modir, B., Irving, P. W., Wolf, S. F., & Sayre, E. C. (2014). *Learning about the energy of a hurricane system through an estimation epistemic game.* In Proceedings of the 2014 physics education research conference, pp. 3–6. https://doi.org/10.1119/perc.2014.pr.044.

Morin, D. (2008). *Introduction to classical mechanics: With problems and solutions.* Cambridge: Cambridge University Press.

Nave, R. (2017). *Coulomb's law.* http://hyperphysics.phy-astr.gsu.edu/hbase/electric/elefor.html. Online. Accessed 10 May 2017.

Nearing, J. C. (2003). *Mathematical tools for physics.* New York: Dover Publications.

Phinney, S. *Ph 101 order of magnitude physics.* https://www.its.caltech.edu/oom/. Accessed: 2017-05-04.

Redish, E. F. (2004). *A theoretical framework for physics education research: Modeling student thinking.* arXiv preprint physics/0411149.

Redish, E. F., & Cooke, T. J. (2013). Learning each other's ropes: Negotiating interdisciplinary authenticity. *CBE-Life Sciences Education, 12*(2), 175–186.

Redish, E. F., & Kuo, E. (2015). Language of physics, language of math: Disciplinary culture and dynamic epistemology. *Science & Education, 24*(5–6), 561–590.

Redish, E. F., Saul, J. M., & Steinberg, R. N. (1998). Student expectations in introductory physics. *American Journal of Physics, 66*(3), 212–224.

Redish, E. F., Singh, C., Sabella, M. , & Rebello, S. (2010). *Introducing students to the culture of physics: Explicating elements of the hidden curriculum.* In AIP conference proceedings (Vol. 1289, pp. 49–52). AIP.

Redish, E. F., Bauer, C., Carleton, K. L., Cooke, T. J., Cooper, M., Crouch, C. H., Dreyfus, B. W., Geller, B. D., Giannini, J., Svoboda Gouvea, J., et al. (2014). Nexus/physics: An interdisciplinary repurposing of physics for biologists. *American Journal of Physics, 82*(5), 368–377.

Robinett, R. W. (2015). Dimensional analysis as the other language of physics. *American Journal of Physics, 83*(4), 353–361.

Russ, R. S., Coffey, J. E., Hammer, D., & Hutchison, P. (2009). Making classroom assessment more accountable to scientific reasoning: A case for attending to mechanistic thinking. *Science Education, 93*(5), 875–891.

Schoenfeld, A. H., & Sloane, A. H. (2016). *Mathematical thinking and problem solving.* London: Routledge.

Serway, R. A., & Jewett, J. W. (2004). *Physics for scientists and engineers.* ThomsonBrooks/Cole, 6th edn. ISBN: 9780534408428, 0495142425, 0534408427, 9780495142423.

Sfard, A. (1991). On the dual nature of mathematical conceptions: Reflections on processes and objects as different sides of the same coin. *Educational Studies in Mathematics, 22*(1), 1–36.

Sherin, B. L. (2001). How students understand physics equations. *Cognition and Instruction, 19*(4), 479–541.

Spike, B. T., & Finkelstein, N. D. (2016). Design and application of a framework for examining the beliefs and practices of physics teaching assistants. *Physical Review Physics Education Research, 12*(1), 010114.

Tipler, P. A., & Mosca, G. (2007). *Physics for scientists and engineers.* London: Macmillan.

Townsend, J. S. (2000). *A modern approach to quantum mechanics.* Sausalito: University Science Books.

Tuminaro, J., & Redish, E. F. (2007). Elements of a cognitive model of physics problem solving: Epistemic games. *Physical Review Special Topics-Physics Education Research, 3*(2), 020101.

Weinberg, S. (1967). A model of leptons. *Physical Review Letters, 19*(21), 1264.

Weinstein, L. (2018). *Fermi questions.* URL https://aapt.scitation.org/topic/collections/fermi-questions. Online. Accessed 20 May 2018.

Wikipedia. (2016). *Yukawa potential — wikipedia, the free encyclopedia.*https://en.wikipedia.org/w/index.php?title=Yukawa_potential&oldid=715463242. Online. Accessed 3 October 2016.

Wilcox, B. R., Caballero, M. D., Rehn, D. A., & Pollock, S. J. (2013). Analytic framework for students' use of mathematics in upper-division physics. *Physical Review Special Topics Physics Education Research, 9*(2). ISSN: 15549178. doi: https://doi.org/10.1103/PhysRevSTPER.9.020119.

Wittmann, M. C., & Black, K. E. (2015). Mathematical actions as procedural resources: An example from the separation of variables. *Physical Review Special Topics – Physics Education Research, 11*(2):020114. ISSN: 1554-9178. https://doi.org/10.1103/PhysRevSTPER.11.020114. http://link.aps.org/doi/10.1103/PhysRevSTPER.11.020114.

Zee, A. (2010). *Quantum field theory in a nutshell.* Princeton: Princeton University Press.

Chapter 7
Theorems-in-Action for Problem-Solving and Epistemic Views on the Relationship Between Physics and Mathematics Among Preservice Physics Teachers

Ileana M. Greca and Ana Raquel Pereira de Ataíde

7.1 Introduction

The most frequent complaints voiced by physics teachers, at all levels, are that their students miscomprehend concepts in physics because of weak mathematical skills (Pietrocola 2010; Redish and Kuo 2015), which is seen as one of the principal causes for academic failure. Nevertheless, although mathematical skills are necessary for a full understanding of physics, they are not enough in themselves to guarantee success at physics (Hudson and McIntiry 1977). Moreover, although in some cases students do not in fact master the necessary skills at mathematics, recent research in physics teaching has revealed other possible diagnoses. For example, Romer (1993), Lozano and Cárdenas (2002), Martinez Torregrosa et al. (2006), and Redish and Kuo (2015) all discussed the need to teach students how to read mathematical symbols and equations in physics contexts and to interpret them because, as stressed by Redish (2005), even though mathematics may be the language of science, "maths-in-physics is a distinct dialect of that language." As discussed in this book by Karam, Uhden, and Höttecke using their historical genesis, the way physicists make use of some basic mathematical concepts is specific to physics.

Other problems are to do with the ideas students have of the relationship between physics and mathematics, an interplay that is not presented with sufficient clarity for students. These ideas lead to misunderstandings such as seeing mathematics only as a mere instrument for physics, as the teachers themselves may describe it. This type of thoughts make students believe that they need to know no more than

I. M. Greca (✉)
Department of Specifics Didactics, Faculty of Education, Universidad de Burgos, Burgos, Spain
e-mail: imgreca@ubu.es

A. R. P. de Ataíde
Department of Physics, Universidade Estadual da Paraíba, Paraíba, Brazil

© Springer Nature Switzerland AG 2019
G. Pospiech et al. (eds.), *Mathematics in Physics Education*,
https://doi.org/10.1007/978-3-030-04627-9_7

the equation and its solution, to solve problems in physics, or that they can use equations without any direct association with the principles of physics; i.e., there is no need to understand equations in the context of physics before they are applied (Redish et al. 1998; Adams et al. 2006; Sherin 2006; Mason and Singh 2010). Pietrocola (2002, 2010), based on philosophy and history of science, discussed how the relationship between mathematics and physics might influence its teaching and learning. Dormert et al. (2007) described the epistemological components of students' mind-sets in their understanding of physics equations, finding that the most recurrent mind-sets were how to use an equation to solve problems and how to recognize what the symbols in an equation represent. These results are similar to those found by Redish et al. (1998) on the expectations of introductory physics students regarding equations, who merely used mathematics, it appeared, as a way of calculating numbers.

Using the historical framework provided by Michel Paty for the relationship between mathematics and physics, we studied how the epistemological views of students influenced the way they solved physics problems (Ataide and Greca 2013, 2017). In a case study with undergraduate physics students, we were able to identify (adapting a classification proposed by Karam 2007) three categories related with student views on the role of mathematics in the construction of a physical theory (tool, language, and structure) that correlate with their salient feature in the strategies they used in problem-solving (operational mathematics, conceptualization, and mathematical reasoning). So, students who see mathematics as a mere tool for physics use mathematics as a technique and tend to solve problems by trial and error. Their epistemic views also appear to influence their learning and understanding of physics concepts, because problem-solving is the main activity in the physics classroom.

In this chapter, we present our attempt to understand and to detect the mental "rules" that seem to be behind the way students solve physics problems. With these rules in mind, we prepared a questionnaire with 15 (Likert-type scale) items that may help instructors to detect the behavior of students in relation to problem-solving before their instruction, knowledge that may be useful for designing their classes.

7.2   Theoretical Framework

In this section, we present the theoretical framework used to determine the cognitive aspects involved in the interrelationship between mathematics and physics when students try to understand physics concepts and, in particular, when they solve problems in physics. Our framework is based on the model proposed by Greca and Moreira (2002) that includes Johnson-Laird's mental model theory and Vergnaud's theory of conceptual fields.

In Johnson-Laird's theory of mental models (1983), the mental models are determined and concrete idiosyncratic cognitive structures operating in the working memory of the individual who wishes to understand, to explain, and to predict a specific situation or process, and that act as structural analogues of this situation or process. We can understand them as "simulations" of such situations, similar to analogical computers simulating a physical system (Greca and Moreira 2000). They are characterized as dynamic structures generated to solve particular (often new) situations and incomplete ones. They are recursively modified or updated whenever the individual users detect a mismatch between the predictions generated by the models and external events or detect new information to input into the model, depending on the use they wish to make of it. Hence, these mental models are called "working models" – "disposable" representations the main role of which is their functionality.

As these mental models are "disposable" representations, there is a need to find some way to define the more stable knowledge in the mind of the individual. To do so, we adopt the model proposed by Greca and Moreira (2002), which articulates mental models with the theory of conceptual fields of Gérard Vergnaud. According to the theory of conceptual fields, the knowledge that a subject has is organized into conceptual fields, mastery over which is only achieved over a long period of time, and comes from experience, maturity, and learning. A conceptual field can be defined as an informal and heterogeneous set of problems, situations, concepts, relations, structures, content, and thought operations connected to each other and probably intertwined during the acquisition process (Vergnaud 1982). The conceptual field is therefore the way people organize their knowledge over a set of situations that they consider to be related, through the competencies and conceptions they acquire when they face those situations. So, in a certain conceptual field, there may be a wide variety of situations, and individual knowledge is shaped by situations that are repeatedly encountered and gradually mastered. A learning process that is especially true in the initial situations where individuals accord meaning to concepts and procedures that they wish to learn (Vergnaud 1990). According to Vergnaud, in a problem situation, a concept never appears in isolation, but is articulated with others that, according to the individual, belong in a particular conceptual field. Thus, learning a concept is understood as a broad cognitive process that is related to the number of problem situations that the individual experiences, through which sense is attributed to that concept.

Although mastery of these situations will shape individual knowledge, the sense attributed to a situation by the individual is not in the situation itself, but in the relation between the situation and the internal representation in the mind of the individual. The internal representation for Vergnaud depended on the schemes that the individual forms. These schemes (see Fig. 7.1) are the invariant organization of behavior in a given class of situations (Greca and Moreira 2002). The cognitive elements that lead to individual action should be researched in relation with the schemes. A scheme is a universal construct, an efficient means with which to address a whole range of situations. It can generate different sequences of action, information collection, and control, depending on the characteristics of each situation. So, it is not the behavior in similar situations that is invariant or universal, but the organization of that behavior (1998, p. 172).

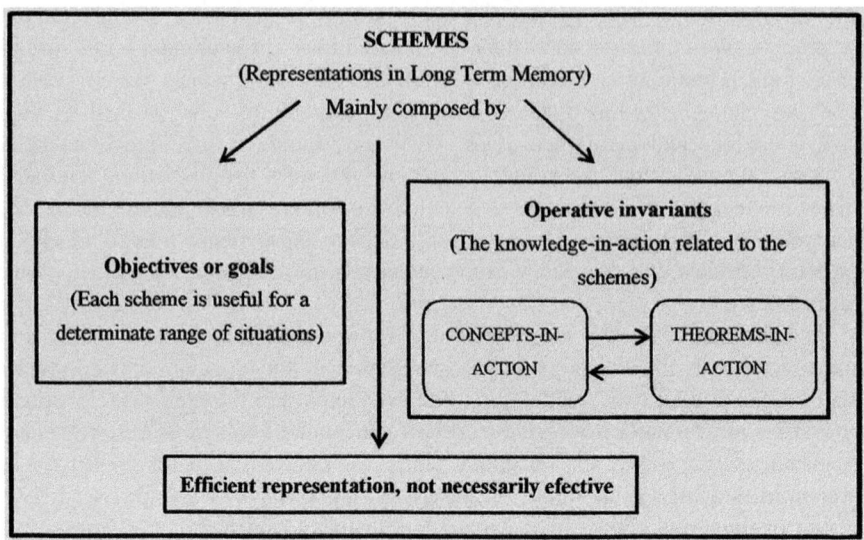

Fig. 7.1 Composition of schemes

The knowledge-in-action of an individual is contained in the operative invariants (theorems-in-action and concepts-in-action) of the schemes. Concepts-in-action allow the subject to organize the world in categories and to pick up from this world the relevant information according with the situation and the schemes he has. Theorems-in-action are propositions that may be true or false, concerning those concepts-in-action.

These operative invariants form the essential link between theory and practice, because the system of concepts-in-action and the underlying theorems-in-action determined the perception, the search, and the selection of information that is available to the individual. In-action means that they are useful for the individual to deal with situations and are therefore mainly implicit. Unless they become explicit, the operational invariants are conceptualized in the hidden part of the iceberg of conceptualization. Nonetheless, they can gradually become real scientific concepts and theorems, and the educational activities are fundamentally to help students to build explicit and scientifically accepted concepts and theorems from implicit knowledge.

There is a dialectical relation between concepts-in-action and theorems-in-action, since concepts are ingredients of theorems and theorems are properties that give concepts their contents. For example, let us think of the situation of moving an object over a surface. An individual without any physics training would not identify all the forces involved in the situation and would also say that the object will stop when the application of the force ceases, because in the schemes of that individual for this class of situations, neither are the concepts-in-action of friction and acceleration present, nor is an appropriate theorem-in-action for the relation between force and speed. So, he cannot "see" these elements in the situation. In contrast, we expect that after following a subject about mechanics, students should recognize the different forces that appear in the situation and solve problems using certain laws. After instruction, the students should begin to incorporate some "ideas" about the different forces and their relation with acceleration, speed, etc. in the schemes that they form around these classes of situations. Nevertheless, the individual may not be able to define properly the physics concepts that are involved, because a concept-in-action is not a scientific concept, just as a theorem-in-action is not a true theorem.

Summing up, in this theory, the knowledge people have for doing things, their procedural knowledge, is organized into schemes, mainly composed of concepts-in-action and theorems-in-action. Concepts-in-action guide our perception and allow us to identify the objects that are considered relevant for acting in the set of situations we face, and the theorems-in-action are the rules that allow our mind to act on those objects. If considered similar situations, then the individual will apply the same schemes, with the same concepts-in-action and theorems-in-action. For example, for solving problems in kinematics, students may have a scheme in which there are some concepts-in-action related to speed, distance, and time that at the same time are the "objects" the student will see in the situation. Also, the scheme has to include some theorems-in-action (generally rules for solving these kinds of problems and also formulas). If the students identify that a problem is a "kinematic" problem, they will apply this scheme. But, students will not apply the same scheme, if identifying the problem as belonging to another class of situation that is not "kinematic." That difference can explain why students very often fail to use a similar "process" to solve problems that are, from the point of view of physics, identical – for example, a block in free fall, blocks connected by a rope, and blocks on a ramp. For many students, they belong to different schemes and different concepts, and theorems-in-action are applied. A similar process happens with the relationship between mathematics and physics. For students, elements of mathematics (knowledge of how to solve equations, to understand a graph, etc.) are very often included in different schemes than those needed to solve physics problems. For example, Planinic et al. show in this book the competences and difficulties of high school students at transferring their knowledge of graphs when presented with similar situations but framed within different contexts, math context and physics context. The same happens with many other mathematical concepts. As teachers, we very often take it for granted that if students have learnt to solve a differential equation, they can solve it anywhere. Nevertheless, students will not share this view, because they appear to have different schemes for differential equations for math or for physics. It is also worth stressing, as Bing and Redish (2007) pointed out, that in general the mathematical elements cannot simply be transported into physics but have to be framed by the physical concepts, an aspect that is not usually stressed in physics classes.

As individuals face different situations, the schemes they hold can become more complex or may be necessary to create new ones, if they fail to understand those situations successfully or to resolve them. So, schemes may be little pieces or big pieces depending on the kind of situations we face. Experts, for example, may have more comprehensive schemes. But also in this case, their schemes depend on the situations they have faced, generally related to their area of expertise. For example, a physicist working with electromagnetism in situations involving electrons would apply different knowledge-in-action than another physicist working with quantum mechanics.

As previously mentioned, the knowledge contained in the schemes is mainly implicit. The knowledge they contain must progressively become explicit, so as to be able to discuss it and thereby to convert this knowledge for action into knowledge for thinking. And in this process, the individual begins to conform true concepts. According to Vergnaud, a concept gradually acquires implicit meaning for an individual through situations and problems, because it is from these situations and problems that the properties constituting the concepts-in-action and the theorems-in-action of an individual will be abstracted. However, these operative invariants have to be explicitly expressed, before the individual can reason with them, through their different representations, and in this way, these invariants start to form true concepts. In fact, modeling, as described by Pospiech in the framing chapter, can be understood as one of the possible mechanisms of explaining concepts and theorems-in-action, an indispensable stage for the acquisition of true scientific concepts. In modeling a situation, students should choose the entities that will be part of the model and establish the relationships between them, which will serve as the basis for the equations that describe it. In this process, they necessarily have to explain the concepts and theorems-in-action that they possess.

Now, the question is how to modify the existing schemes and how to create new ones. For these processes, we used the articulation of mental models and Vergnaud schemes proposed by Greca and Moreira (2002). Accordingly, while the schemes remain as structures in the long-term memory, with their theorems-in-action and concepts-in-action, whenever new situations are encountered, the individual will generate mental models, representations in the short-term memory, that are useful at solving certain tasks.

The relation between mental models and schemes is dialogic. If the situation is known, the individual will not have to construct mental models, once the concepts and theorems-in-action are activated in the schemes of the long-term memory for solving it. But if the situation is perceived as unknown, the individual will need to construct mental models to make inferences and predictions, so as to understand and to master the situation, because the solution strategies are not automated. And these mental models are constructed on the basis of the knowledge the individual already has in other schemes. But, the comparison between the results of the mental models and the actual result of the situation can lead to changes in the invariants of the individual. These changes occur when the individual systematically seeks consistency between internal thinking and data from the outside world (Greca and Moreira 2002). This process is exemplified in Fig. 7.2.

When students are challenged with situations that they understand as new ones, they create mental models. Their mental models may be not correct, and, as teachers, we can help them to change it. But this change in a mental model will not necessarily

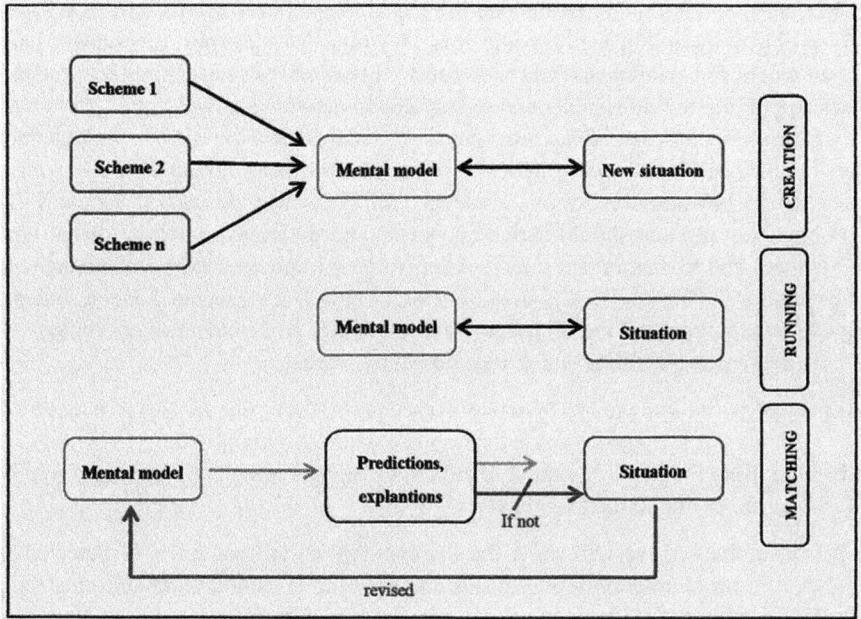

Fig. 7.2 Relation between schemes and mental models when encountering new situation

mean that the student will automatically apply it or, equally, that the student will have changed schemes. If that change is to happen, it is necessary to have encountered several new situations and to have applied the approach (that may not be correct but is useful) repeatedly and successfully. This can explain why students need time for adopting new techniques for solving problems (Roorda et al. 2015). Another possibility is the use of the strategies proposed by Eichenlaub and Redish in their chapter and detected by Lehavi et al. (2017) analyzing teaching patterns in expert high school physics teachers, such as "examine the extreme cases" and "think about the dimensions". These strategies demand that students move systematically from their known schemes and explicitly address the possible mismatches between the solutions that their own mental models produce and the way the world behaves. These strategies appear to be useful especially for the development of appropriate schemes in relation to the physical understanding of mathematical expressions.

Thus, the relation between mental models and schemes is on a continuum, because, in new situations, the individual will construct mental models and in situations that have similarities with those already known, the individual will use stable schemes built from unstable mental models.

It is worth noting that the notion of scheme has some points in common with the resource framework, proposed by Hammer (2000), although in Vergnaud's theory, which also explains concept formation, schemes do not need to be small. Also, for our framework, epistemological resources are not a different kind of representation.

They are theorems-in-action included in the schemes that students have developed for problem-solving in each specific area. Of course, as some physics contents have been taught for many years, i.e., mechanics, it is easier and saves time for students when applying the rules generated in that area to new physics contents.

During the process of learning physics, students develop a series of schemes on both physical and mathematical content. As problem situations are the main activity in physics classrooms, students will have also developed schemes for problem-solving that should include concepts and theorems-in-action related both to physics and to mathematics, as problem-solving in physics applies mathematical formulations. This knowledge-in-action would have been generated as the students gradually appropriated the different conceptual fields linked to areas of physics.

The research questions that guided our study were:

(a) What is the knowledge-in-action – and, particularly, the theorems-in-action – that preservice physics students mobilize when solving problems in physics?
(b) Are these theorems-in-action related to the epistemological views they hold of the role of mathematics in physics?

In this chapter, we will show the theorems-in-action that we have detected in relation to problem-solving in physics and how those theorems relate to the most striking features of students in solving physics-based problems. In the next section, we will present the methodology used to do so.

7.3 Methodology

We conducted a qualitative research study, with two groups of students on the degree course for training high-school school physics teachers at the State University of Paraíba. This degree has a duration of eight semesters. The first group consisted of eight students following the final year degree (veteran students, VS) during the second semester of 2010. The second group was composed of 17 students, enrolled in the second year, during the first semester of 2016 (new students, S). Data collection was done during individual problem-solving sessions in which we used verbal protocols for the identification of mental models.

In verbal protocols, the individuals are expected to express their thoughts out aloud that pass through their minds in the course of the task, and the data obtained is used, in conjunction with theoretical assumptions to generate hypotheses and to draw conclusions on the cognitive products and processes. In our case, the task that the students had to perform was problem-solving, and within our framework, we drew inferences on possible mental models and, subsequently, inferences on concepts and theorems-in-action, as indicated below.

The sessions were recorded on video and had the mediation of the researcher, in order to stimulate student comments, so as to facilitate the identification of the operative invariants (theorems-in-action and concepts-in-action) used during the activity. The problems used during the interviews were similar to those at the end of each chapter of a standard physics textbook.

Data analysis was done through detailed scrutiny of both the written and the verbal documentation of each student, obtained during the problem resolution sessions. We conducted a separate study for each student. First, we sought to determine the mental models used for each problem. To do so, we separated the different parts of the documentation, highlighting key concepts and phrases that appeared to show the way that each student had understood the problem and approached its solution. With this information, we composed an initial sketch of the mental model. Then, the sketch was compared with drawings and writings and with the answers the student had given in the interview. This process gave us the possibility to better define the working model that the student was using for solving each problem.

After this process, we observed whether and how the working models detected were modified for each problem. In other words, as stated in the theoretical framework, the individual has to construct specific mental models to solve a new situation (each problem). But, as the students move from problem to problem, it is possible to identify ideas and propositions that remain intact even when the models are modified. Hence, we inferred that such propositions were theorems-in-action, once they had great stability in the cognitive structure of each student and, supposedly, guided the construction and use of mental models.

The theorems-in-action related to problem-solving that were identified for the group of students in the first phase of the study through the problem-solving sessions were related with the views of the role of mathematics in physics presented by the students, which were identified through direct questioning and have been presented in a previous study (Ataíde and Greca 2013). These views, adapted from those of Karam (2007), were:

- *Tool*: Mathematics is used by physics as a facilitator of the numerical calculations.
- *Language*: Mathematics is a translator of physical thought to the world, a mere manifestation of physics, with the task of representing it in an understandable way.
- *Structure*: Mathematics appears as a physical structuring of thought itself.

With the group of students that was the focus of the second phase of the study, in addition to the identification of theorems-in-action, we also used a questionnaire (Likert scale), constructed with statements relating to some of the theorems-in-action that were detected, described in more detail later on. The results of the problem sessions and questionnaire were later compared, to verify the validity of the questionnaire. The questionnaire was then administered to a broader survey of 80 students.

7.4 Results and Discussion

7.4.1 Inferring Theorems-in-Action During Problem-Solving

As indicated above, theorems-in-action for problem-solving are related to strategies and attitudes that guide this process. We identified these theorems-in-action for each of the students. Before proceeding, we will recall the categories that we used to characterize the most significant strategies that our students have applied when solving problems in previous studies and that we suppose are associated with theorems-in-action (Ataíde and Greca 2013).

- *Operational mathematics* (OM) – Students who use mathematics as a technique and who tend to solve problems by trial and error
- *Conceptualization* (C) – Students who favor conceptual understanding and try, not always successfully, to form a link between the concepts and the maths to be used in problem-solving
- *Mathematical reasoning* (MR) – Students who use mathematical reasoning that is coherent with the situation outlined in the resolution of problems, although they may not work properly with the mathematical techniques

From the analysis, it was observed that eight students applied, in general, the following theorems-in-action in their attempts to find a solution:

- The resolution begins with the explanation of the problem.
- Solving the problem requires a search for meaning between the equations and concepts.
- The result obtained by following a structured sequence is unquestionable.

The students who presented these theorems-in-action always detailed the problem in greater depth before they began to solve it. Their main characteristic concerning how to solve problems can be described as mathematical reasoning. Regarding their epistemological view, all except one appeared to have a view of mathematics as a framework for physics, and, in fact, their theorems-in-action appeared to be based on a structure that expressed a junction between theoretical definitions and mathematical equations.

In what follows, we present some replies from these students in response to questions concerning the procedures used to solve the problems, which contributed to this interpretation.

> To solve a problem I like to know the behavior of each variable and how it appears in the equation that represents the mathematical model to explain the phenomenon ... (S54)
>
> The concepts involved in the problem need to be understood and have a meaning within the equation, if not, we do not understand the physics involved in the problem ... (S60)

For nine other students, the theorems-in-action detected in their process of problem-solving appear as a guide toward a detailed conceptual understanding of the problem before using the equations to find a result. The conceptualizations of these students are their most striking feature during problem-solving, most of whom were categorized as having a view of mathematics as either translation or a language of physics. It should be noted that a small number of these students (two students) tended to leave the mathematical treatment in the background and, in some cases, were not always confident with the mathematical techniques.

The main theorems-in-action detected in the attempts of these students to find a solution can be expressed in the following form:

- The resolution begins with a detailed reading and pictorial representation and explanation of the problem.
- The resolution is facilitated by figures and graphs.
- The resolution requires identification and characterization of the variables involved.
- The resolution requires the placement of equations and its explanation of use.
- The resolution requires manipulation of equations.

> I always start the resolution reading and explaining the problem .... (S11)
>
> Solving conceptual issues is more difficult than questions with direct numerical data, but I understand the concepts better when I see it that way .... (S17)

Another eight students applied theorems-in-action in which mathematical operationality was evident. They have to know the equations, manipulate them, and reach an end result, if they are to solve the problems. Overall, the members of this group have many conceptual shortcomings, both in the areas of physical and mathematical. Most of these students have a view of mathematics as a tool at the service of physics, and the main feature of their problem-solving is the operational application of mathematical formulas.

The main theorems-in-action detected in the attempts to find a solution were:

- The resolution begins with the equations.
- The resolution is facilitated with charts and graphs.
- Solving problems requires manipulation of equations appropriate to the available data.

> I always try to fit the data to the equations, I also know that when we master the mathematical techniques we solve the problem faster .... (S59)
>
> When I master the maths of the problem it becomes easier. If I don't quite understand the problem, I replace the values and solve it.... (S61)

Although included in this group because of how they address problem-solving, two veteran students appeared to act in this way for contrary reasons to those of the other students. Their confidence with the mathematical techniques stands out when asked to solve problems. It should be noted that even when there are conceptual errors, their strong mathematical knowledge can help them to overcome these misconceptions. For example, one of them (VS19), when trying to explain the meaning of the heat generated by a temperature difference, referred to the inexact differential associated with heat in the formalization of the first law and stated that this differential indicates that the heat is characterized by its existence in a process and not in a body.

In Table 7.1, we present a summary of the results.

Table 7.1 Summarization of the results obtained from the problem-solving sessions

Students	View of the role of mathematics in physics	Most striking feature in problem-solving	Theorems-in-action
VS15, S4, S9, S16, S18, S54, S60	Structure	Mathematical reasoning	The attempt to find a solution begins with the explanation of the problem
			Solving the problem requires searching for meaning between the equations and concepts
			The result obtained by following a structured sequence is unquestionable
VS18, S5, S11, S13, S17, S55, S62	Translator	Conceptualization	The attempt to find a solution begins with a detailed reading and pictorial representation and explanation of the problem
			The solution is facilitated by charts and graphs
			The solution requires identification and characterization of the variables involved
			The solution requires the placement of equations and its explanation of use
			The solution requires manipulation of equations
VS17, VS20, VS22, S12, S59, S61	Tool	Operational mathematics	The resolution begins with the equations
			The resolution is facilitated with graphic figures
			Solving the problem requires manipulation of the appropriate equations in relation to the available data
VS16, VS19	Structure	Operational mathematics	The attempt to find a solution begins with the equations
			The solution requires manipulation of the equations
			The result obtained by following a structured sequence is unquestionable

(continued)

Table 7.1 (continued)

Students	View of the role of mathematics in physics	Most striking feature in problem-solving	Theorems-in-action
S15	Tool	Mathematical reasoning	The attempt to find a solution begins with the explanation of the problem
			Solving problems requires manipulation of equations appropriate to the available data
			The solution is facilitated with charts and graphs
			The solution requires identification and characterization of the variables that are involved
			Solving the problem involves a search for meaning between the equations and the concepts
VS13, S29	Tool	Conceptualization	The attempt to find a solution begins with the equations
			The solution is facilitated with charts and graphs
			Solving the problems requires the manipulation of appropriate equations in relation to the available data
			The solution begins with the explanation of the problem

From this sample, only 32% of the students appear to have theorems-in-action that manage to address an appropriate relationship between mathematics and physics. Most of the veteran students appear to have theorems-in-action that guide an instrumental use of mathematics. It is worth stressing that these students were studying thermodynamics and the math required to solve the problems proposed is more complex than the math needed for the other students, who were studying optics. In fact, these veteran students showed that although they had mastered the mathematical techniques of differential calculus (Ataide and Greca 2012), their use reflected an instrumental understanding (Uhden and Pospiech 2013).

The theorems-in-action identified during the problem-solving coincided with the features of epistemic games[1] detected by Tuminaro and Redish (2007):

---

[1] It is worth stressing the similarities between epistemic games and theorems-in-action. Nevertheless, it must be noted that theorems-in-action can be used to understand many other cognitive behaviors, not only the ones related to problem-solving, which appears to be the case of the idea of epistemic games.

- The theorems-in-action that best characterize mathematical reasoning are very similar to the epistemic game "mapping meaning to mathematics."
- Those relating to conceptualization are very similar with the epistemic games "physical mechanism" and "pictorial analysis."
- Those related to operational mathematics are very similar with the epistemic recursive game "plug-and-chug."

The epistemological views of students on the role of mathematics in physics seem to be, for most of the students, related to theorems-in-action consistent with those views: the students who consider mathematics as a structure for physics tend to be guided, in solving problems, by theorems-in-action with solutions that can be characterized as mathematical reasoning; those who merely consider mathematics as a language for the phenomena and laws tend to use theorems-in-action that can characterize the resolution of problems as conceptualization; and, lastly, those who understand that mathematics is only a tool use theorems-in-action that identify a mathematical operationality of action. Thus, in 20 of the 25 students (80%), their views about the role of mathematics in physics seem to "materialize" into theorems-in-action that guide these students in their problem-solving techniques.

So, these results point toward epistemological views behind the ways students solve problems, which relate to the role of mathematics in physics that correlate with the abovementioned strategies. It is very likely that ideas about the relationship between mathematics and physics and theorems-in-action for problem-solving have been developed at the same time in the learning process of students and mutually reinforced. It is worth stressing that the theorems-in-action for problem-solving and the epistemic views appear to be independent of the course or the complexity of the physical and mathematical concepts that are involved.

Finally, related to the theoretical framework, we realized every solution of a problem situation involved a change in mental models, even when the theorems-in-action remained unchanged. That finding is in agreement with the theories that underlie our proposal, the idea that mental models are unstable elements in short-term memory, while theorems-in-action are part of schemes in the long-term memory (Greca and Moreira 2002). So, several specific activities explicitly addressing the relationship between physics and mathematics have to be done, in order to achieve lasting change in the theorems-in-action of students.

7.5 A Questionnaire to Detect the View of the Role of Mathematics in Physics and the Most Striking Features of Problem-Solving

Detection of theorems-in-action is a complicated process. However, as shown in the previous section, the process of detection yields information for teachers that is relevant to the design of appropriate teaching strategies for students. So, we attempted to develop a user-friendly instrument to administer, a Likert-type questionnaire, from the theorems-in-action that were inferred. The construction of this instrument will be described as well as the validation process.

The questionnaire consists of a set of 15 positive statements (the final version appears in Appendix), related to three specific topics that respond to our goal: (a) the role of mathematics in physics (three statements); the strategies used in solving physics-based problems (nine statements); and (c) about the performance and difficulties in curriculum components of physics.

The nine items of Part B of the questionnaire were generated from the theorems-in-action detected for solving problems and indicated in the previous session. For each of the categories (mathematical reasoning, conceptualizing, and operational mathematics), we chose what in our opinion are the three most characteristic theorems-in-action.

The questionnaire was scored so that, for the 4 to 15 affirmations, the value is (5) for the option "Strongly agree," (4) to "Agree," (3) to "Neither agree nor Disagree," (2) to "Disagree," and (1) to "Strongly disagree." The students were expected to choose only one option for the first three statements.

A trial questionnaire (to verify respondent understanding of the meaning of the words, to avoid misinterpretation, and to employ affirmative language) was administered to five high school physics teachers. They pinpointed the difficulties that high school students would have when responding to the questionnaire. It was also tested with ten students of the Degree in Physics, who did not participate in the survey but who answered the questionnaire.

With the suggested changes, the final version of the questionnaire was administered to 80 students following the degree course for high-school school physics teachers at the State University of Paraíba in the first semester of 2016, among whom the 17 students mentioned in the previous section.

After the questionnaire had been administered, the data were coded and passed by a statistical analysis. The internal consistency analysis (Cronbach's alpha) yielded a value of 0.65, which is considered reasonable in social science research.

Regarding the coding of results, responses were grouped, creating new variables that allow us to analyze the obtained data. Statements 4–12 related to the strategies used to solve physics problems generated by the following subscale.

- The students who had the highest sum in statements 4, 5, and 6 were categorized as "operational mathematics" problem solvers (value 1).
- Those with the highest result of the sum in statements 7, 8, and 9 were characterized as "conceptualizing" problem solvers (value 3).
- Those with the highest sum value to statements 10, 11, and 12 were characterized as "mathematical reasoning" problem solvers (value 5).

The values 2 and 4 were used for those cases where there was overlap between the sums, considering that the students shared characteristics of both categories. Thus, where they coincide with the highest values, the sum of items 4, 5, and 6 with the sum of items 7, 8, and 9, the student was assigned a value of 2, and if the sum of items 7, 8, and 9 overlapped with 10, 11, and 12, they obtained a value of 4. If they overlapped three sums, they were left without a category.

Table 7.2 Total number of matches (bold letter) of the most striking feature in resolving the problems detected in problem-solving sessions and the answers to the questionnaire.

Most striking feature in problem-solving (identified in problem-solving sessions) versus most striking feature in problem-solving (identified in questionnaire)	MR	C	OM
MR (value 5)	**2**	0	0
MR/C (value 4)	1	**2**	0
C (value 3)	3	**5**	0
OM/C (value 2)	0	0	**1**
OM (value 1)	1	0	**2**

MR mathematical reasoning, *C* conceptualization, *OM* operational mathematics

Table 7.3 Total number of matches (bold letter) between views of the role of mathematics in physics detected in problem-solving sessions and the answers to the questionnaire

View of the role of mathematics in physics (identified in problem-solving sessions) versus view of the role of mathematics in physics (identified in questionnaire)	Structure	Translator	Tool
Structure	**6**	1	2
Translator	0	**5**	0
Tool	0	0	**3**

We compared the results obtained by the two strategies for the group of 17 students (2nd phase), in order to determine the validity of the questionnaire. For the most striking feature in solving problems, we realized that the responses obtained with the two strategies (resolution sessions and responses to the Likert scale) for the group of students under analysis were quite consistent, as they coincided almost exactly for the 13 students in a universe of 17 students, i.e., a 76% match (see Table 7.2). In relation to epistemological view, where the students were asked to choose only one of the three statements, we obtained an 82% level of coincidence (see Table 7.3).

This degree of coincidence appears to indicate that the questionnaire may be a resource that could be used to identify the views of students regarding the role of mathematics in physics and to characterize how they solve problems with a high degree of reliability. This fact is very important for physics teaching as, if such features may be identified before proposing problem-solving situations, the teacher may refer the students to situations that lead them to rethink their positions. Nevertheless, other samples should be used to check the results of this study.

7.6 Conclusions and Implications for Physics Teaching

In physics teaching a very important point and the focus of this study is the relation between the solution of problems and the viewpoint that students have of the role of mathematics in the construction of physical knowledge. Although these epistemological views or their influence on learning physics are not widely discussed in science education research, they seem to influence the way students approach learning in physics and specifically problem-solving activities. This idea is reinforced in this study.

In relation specifically to problem-solving, the epistemological views of students on the role of mathematics in physics seem to be, for most of them, associated with theorems-in-action consistent with those views, i.e., these views appear to "materialize" into theorems-in-action that guide how these students face problem-solving (Ataide 2013). It is striking that almost 25% of the students of this sample, who are studying to become physics teachers, see mathematics just as a tool for physics and use it only as a problem-solving technique, most of the time solving problems by trial and error. As Hansson et al. argue in their chapter, the approach that merely makes use of mathematics in an instrumental way is reinforced by textbooks, and the instrumental use of mathematics may even be strengthened by future teachers, such as the ones shown in this research who use it in that way.

In this work we have also presented a questionnaire to measure the relations between the epistemological views of students on the Degree in Physics (teacher training) and strategies (theorems-in-action) that they use to solve physics problems with reasonable internal consistency. The questionnaire results have also been validated by comparing them with the results of other analyses. Although the questionnaire should be improved and validated with larger groups, its most important aspect is that it could be used both in research and in physics teaching, for teachers who wish to improve their understanding of their students.

It is important to note that the participants of this study were following a teacher training degree and the views found in this study are likely to be consolidated. Following the PCK model proposed in the chapter by Pospiech et al., these views may influence their orientation toward teaching. They may therefore pass them on, either implicitly or explicitly, to their high school students. It therefore appears necessary not only to work on appropriate techniques to solve problems but also in an explicit way to discuss the epistemological views that students hold of the role of mathematics in the construction of physical knowledge, in order to change those views on the relationship between physics and mathematics that appear to be unproductive views. Some of the strategies proposed in this book could be useful in this sense. For example, the explicit teaching of modeling techniques, as proposed by Pospiech, or the use of history, as in the examples advanced by Karam, Uhden, and Höttecke, show that there is no direct and straightforward application of mathematics simply as a tool in physics.

We should, nevertheless, warn of the complexity involved in understanding the relations between physics and mathematics and more specifically their influence on physics teaching and our understanding of student thinking processes when performing tasks such as problem-solving. Those aspects may incite new research initiatives and are a rich topic for science education and, in particular, for physics teaching.

A.1   Appendix: Questionnaire (Likert Scale)

This questionnaire is a data collection instrument from a survey that aims to identify relations between epistemological views on the role of mathematics in physics and strategies used in problem-solving.

A.1.1   Identification

Name:_____

Registration   number: _____   Semester   that   you   are   attending: _____

A.1.2   About the Role of Mathematics in Physics

(Select one option)

1. Mathematics is an instrument that is used in physics to solve problems.
2. Math works for physics as a language that helps describe and translate the problems.
3. Math is a structure for physics; both are interrelated in such a way that mathematics is crucial in the construction of the concepts and theories of physics.

A.1.3   The Strategies Used in Solving Physics Problems

4. Equations are the first step in solving a physics problem.

   ( ) Strongly agree ( ) Agree ( ) Neither agree nor disagree ( ) Disagree ( ) Strongly disagree

5. The manipulation of equations is essential to solve a physics problem.

   ( ) Strongly agree ( ) Agree ( ) Neither agree nor disagree ( ) Disagree ( ) Strongly disagree

6. Adjust the equations to available data is essential to solving physics problems.

   ( ) Strongly agree ( ) Agree ( ) Neither agree nor disagree ( ) Disagree ( ) Strongly disagree

7. The use of charts and graphs facilitates the resolution of a problem.

   ( ) Strongly agree ( ) Agree ( ) Neither agree nor disagree ( ) Disagree ( ) Strongly disagree

8. Before solving a physics problem, you need to know how to explain it.

    ( ) Strongly agree ( ) Agree ( ) Neither agree nor disagree ( ) Disagree ( )
    Strongly disagree

9. A detailed reading, with representation through charts and problem explanation,
    is essential to its resolution.

    ( ) Strongly agree ( ) Agree ( ) Neither agree nor disagree ( ) Disagree ( )
    Strongly disagree

10. The identification and characterization of the variables involved in a problem
    are of fundamental importance for its resolution.

    ( ) Strongly agree ( ) Agree ( ) Neither agree nor disagree ( ) Disagree ( )
    Strongly disagree

11. Solving a problem requires the placement of equations and an explanation of
    its use.

    ( ) Strongly agree ( ) Agree ( ) Neither agree nor disagree ( ) Disagree ( )
    Strongly disagree

12. It is of fundamental importance in solving a physics problem to build meaning
    between equations and concepts, so as to understand how they are related.

    ( ) Strongly agree ( ) Agree ( ) Neither agree nor disagree ( ) Disagree ( )
    Strongly disagree

A.1.4 About the Performance and Difficulties in Curriculum Components of Physics

13. I have no difficulty with physics, and my performance in this curriculum
    component is very good.

    ( ) Strongly agree ( ) Agree ( ) Neither agree nor disagree ( ) Disagree ( )
    Strongly disagree

14. I have difficulty with the understanding of the concepts of physics, and my
    performance in this curriculum component is weak.

    ( ) Strongly agree ( ) Agree ( ) Neither agree nor disagree ( ) Disagree ( )
    Strongly disagree

15. I have difficulty with math, and my performance in curriculum components of
    physics is weak.

    ( ) Strongly agree ( ) Agree ( ) Neither agree nor disagree ( ) Disagree ( )
    Strongly disagree

References

Adams, W., Perkins, K., Podolefsky, N., Dubson, M., Finkelstein, N., & Wieman, C. (2006). New instrument for measuring student beliefs about physics and learning physics: The Colorado learning attitudes about science survey. *Physical Review Special Topics Physics Education Research, 2*(1), 1–14. https://doi.org/10.1103/PhysRevSTPER.2.010101.

Ataide, A. R. P. (2013). *O papel das matemáticas na compreensão de conceitos da termodinâmica* (Tese de doutorado). Brasil: Universidade Federal da Bahia/Universidade Estadual de Feira de Santana

Ataíde, A. R. P., & Greca, I. M. (2012). Epistemic views of the relationship between physics and mathematics: Its influence on the approach of undergraduate students to problem solving. *Science & Education, 22*(6), 1405–1421. https://doi.org/10.1007/s11191-012-9492-2.

Ataíde, A. R. P., & Greca, I. M. (2013). Estudo exploratório sobre as relações entre conhecimento conceitual, domínio de técnicas matemáticas e resolução de problemas em estudantes de licenciatura em Física. *Revista Electrónica de Enseñanza de las Ciencias, 12*(1), 209–233.

Bing, T. J., & Redish, E. F. (2007). The cognitive blending of mathematics and physics knowledge. *AIP Conference Proceedings, 883*(1), 26–29. https://doi.org/10.1063/1.2508683.

Domert, D., Airey, J., Linder, C., & Kung, R. L. (2007). An exploration of university physics students' epistemological mindsets towards the understanding of physics equations. *NorDiNa Nordic Studies in Science Education, 3*(1), 15–28. https://doi.org/10.5617/nordina.389.

Eichenlaub, M., & Redish, E. F. (2018). *Blending physical knowledge with mathematical form in physics problem solving.* arXiv preprint arXiv:1804.01639.

Greca, I. M., & Ataíde, A. R. P. (2017). The influence of epistemic views about the relationship between physics and mathematics in understanding physics concepts and problem solving. In T. Greczyło & E. Debowska (Eds.), *Key competences in physics teaching and learning* (pp. 55–64). Cham: Springer. https://doi.org/10.1007/978-3-319-44887-9_5.

Greca, I. M., & Moreira, M. A. (2000). Mental models, conceptual models and modelling. *International Journal of Science Education, 22*(1), 1–11. https://doi.org/10.1080/095006900289976.

Greca, I. M., & Moreira, M. A. (2002). Além da detecção de modelos mentais dos estudantes: uma proposta representacional integradora. *Investigações em Ensino de Ciências, 7*(1), 30–45.

Hammer, D. (2000). Student resources for learning introductory physics. *American Journal of Physics, 68*(S1), 52–59. https://doi.org/10.1119/1.19520.

Hudson, H. T., & Mcintiry, W. R. (1977). Correlation between mathematical skills and success in physics. *American Journal of Physics, 45*(5), 470–471. https://doi.org/10.1119/1.10823.

Johnson-Laird, P. (1983). *Mental models: Towards a cognitive science of language, inference and consciousness.* Cambridge, MA: Harvard University Press.

Karam, R. A. S. (2007). Matemática como estruturante e física como motivação: Uma análise de concepções sobre as relações entre matemática e física. In E. Fleury (Organizer) (Ed.), *VI Encontro Nacional de Pesquisa em Educação em Ciências.* Florianópolis: Conference held by Abrapec.

Karam, Uhden & Hottecke (this book). The math as "prerequisite" illusion: Historical considerations and implications for physics teaching.

Lehavi, Y., Bagno, E., Eylon, B. S., Mualem, R., Pospiech, G., Böhm, U., Krey, O., & Karam, R. (2017). Classroom evidence of teachers' PCK of the interplay of physics and mathematics. In T. Greczyło & E. Dębowska (Eds.), *Key competences in physics teaching and learning* (pp. 95–104). Wrocław/Cham: Springer.

Lozano, S. R., & Cárdenas, S. (2002). Some learning problems concerning the use of symbolic language in physics. *Science & Education, 11*(6), 589–599. https://doi.org/10.1023/A:1019643420896.

Martínez-Torregrosa, J., López-Gay, R., & Gras-Martí, A. (2006). Mathematics in physics education: Scanning the historical evolution of the differential to find a more appropriate model for teaching differential calculus in physics. *Science & Education, 15*(5), 447–462. https://doi.org/10.1007/s11191-005-0258-y.

Mason, A., & Singh, C. (2010). Surveying graduate students' attitudes and approaches to problem solving. *Physical Review Special Topics—Physics Education Research, 6*(2), 1–16. https://doi.org/10.1103/PhysRevSTPER.6.020124.

Pietrocola, M. A. (2002). Matemática como estruturante do conhecimento físico. *Caderno Brasileiro De Ensino De Física, 19*(1), 89–109.

Pietrocola, M. (2010). Mathematics structural language of physics thought. In M. Vicentini & E. Sassi (Eds.), *Connecting research in physics education with teacher education* (pp. 35–48). New Delhi: Angus & Grapher Publishers.

Planinic et al. (this book). Student understanding of graphs in physics and mathematics.

Pospiech, G. (this book). Mathematics and physics their interplay and its relevance for teaching.

Redish, E. (2005). *Problem solving and the use of math in physics courses.* Invited talk presented at the conference, world view on physics education in 2005: focusing on change, Delhi. http://www.physics.umd.edu/perg/papers/redish/IndiaMath.pdf

Redish, E. F., & Kuo, E. (2015). Language of physics, language of math: Disciplinary culture and dynamic epistemology. *Science & Education, 24*(5–6), 561–590. https://doi.org/10.1007/s11191-015-9749-7.

Redish, E. F., Saul, J. M., & Steinberg, R. N. (1998). Student expectations in introductory physics. *American Journal of Physics, 66*(3), 212–224. https://doi.org/10.1119/1.18847.

Romer, R. H. (1993). Reading the equations and confronting the phenomena: The delights and dilemmas of physics teaching. *American Journal of Physics, 61*(2), 128–142. https://doi.org/10.1119/1.17327.

Roorda, G., Vos, P., & Goedhart, M. J. (2015). An actor-oriented transfer perspective on high school students' development of the use of procedures to solve problems on rate of change. *International Journal of Science and Mathematics Education, 13*(4), 863–889. https://doi.org/10.1007/s10763-013-9501-1.

Sherin, B. (2006). Common sense clarified: The role of intuitive knowledge in physics problem solving. *Journal of Research in Science Teaching, 43*(6), 535–555. https://doi.org/10.1002/tea.20136.

Tuminaro, J., & Redish, E. F. (2007). Elements of a cognitive model of physics problem solving: Epistemic games. *Physical Review Special Topics-Physics Education Research, 3*(2), 1–22. https://doi.org/10.1103/PhysRevSTPER.3.020101.

Uhden, O., & Pospiech, G. (2013). Die physikalische Bedeutung der mathematischen Beschreibung – Anregungen und Aufgaben fur einen neuen Umgang mit der Mathematik. *Praxis der Naturwissenschaften – Physik in der Schule, 62*(2), 13–18.

Vergnaud, G. (1982). A classification of cognitive tasks and operations of thought involved in addition and subtraction problems. In T. Carpenter, J. Moser, & T. Romberg (Eds.), *Addition and subtraction. A cognitive perspective* (pp. 39–59). Hillsdale, N.J: Lawrence Erlbaum.

Vergnaud, G. (1990). La théorie des champs conceptuels. *Récherches en Didactique dês Mathématiques, 10*(2–3), 133–170.

Chapter 8
Learning to Use Formulas and Variables for Constructing Computer Models in Lower Secondary Physics Education

Onne van Buuren and André Heck

8.1 Introduction

The main goal of physics is to understand physical processes, that is, the behaviour of physical systems. An important aspect of physical processes is the covariation between the involved quantities. The mathematical counterparts of these physical quantities are variables. The laws of nature that govern the behaviour of these quantities are represented by mathematical formulas. Therefore, the mathematical concepts of formula and variable, and skills involving these two concepts, are important in physics education.

The mathematical skills of young secondary students are limited, however. Even solving simple equations, in which all but one of the variables has been replaced by numbers, can be cumbersome for them. It is even more difficult for students to solve physics problems in generic ways, in which the variables represent truly varying quantities: it often exceeds their mathematical capabilities. Therefore, it is not surprising that in school practice, most attention is paid to the solving of problems in which the main task is to calculate only one constant value, or one mean value, or one momentary value of some physical quantity. To reach this goal, it usually suffices to replace as many symbols in the formulas as possible by numbers and then solve the resulting equations. In this way, the idea is strengthened that each symbol in the formula represents only one, known or unknown, number. It

O. van Buuren
Faculty of Behavioural and Movement Sciences, Vrije Universiteit Amsterdam, Amsterdam, The Netherlands
e-mail: o.p.m.van.buuren@vu.nl

A. Heck
Korteweg-de Vries Institute for Mathematics, University of Amsterdam, Amsterdam, The Netherlands
e-mail: a.j.p.heck@uva.nl

© Springer Nature Switzerland AG 2019
G. Pospiech et al. (eds.), *Mathematics in Physics Education*,
https://doi.org/10.1007/978-3-030-04627-9_8

may also contribute to the well-known tendency of students to view a mathematical expression not as a proper answer to a question (see, e.g. Booth 1984, 1988). For students, there is little reason to consider variations of quantities and the way in which such variations are related.

Usually, only situations are considered in which there is some sort of equilibrium and all quantities are constant or, when there are varying quantities, in which varying quantities can be replaced by one single number. An example of a situation in which most quantities have become constant is the case of an object that has been falling for a period of time, long enough for an equilibrium of gravity and air friction to have set in; as a consequence, the velocity has become constant. In school practice and in textbooks, little attention is paid to the initial phases of this movement, in which there is no equilibrium and in which air friction and velocity are changing.

A typical example of the reduction of a varying quantity into one average number in school practice is the problem of a car, accelerating from some initial to some final velocity in some interval of time. When asked to calculate the distance travelled, many young students do not use the average velocity, but only the final velocity or initial velocity in their calculations. Apparently, young students are not sufficiently aware of the changing nature of the velocity in this situation. In interviews that we held with young students, they often stated that they did not understand the term "average velocity" in such situations. After an explanation that this single and constant velocity replaces the changing velocity in the real situation, they understood it better but also made clear that they would then prefer the term "replacement velocity".

Part of the mathematical limitations of students can be overcome by using computers for doing the calculations. This is the case with computational modelling and with analysing experimental data with the help of ICT. By doing the calculations, the computer facilitates the study of subjects that have a complexity surpassing the mathematical capabilities of the students. Variations of variables come to the fore in this approach, and it is no longer necessary to reduce a situation into one equation in which each symbol stands for only one number. Results of the computations are not restricted to some numeric values but can be whole functions, presented as graphs. Because the computer does the calculations, the students' focus can be less directed towards the process character of the formula (in the mathematical sense of the word) and more on the object character. As a consequence, students get an overview over the whole problem situation and how this situation changes in time. Questions about how a process depends on initial values and on values of the involved parameters become more important. This especially holds when students are involved in the construction of the computer models.

With this in mind, Van Buuren (2014) started a design study with the goal to develop and test a learning path on modelling that is (1) integrated into the Dutch physics curriculum (currently for the first 2 1/2 years), (2) fits in the standard curriculum as much as possible, and (3) starts from the initial phases of physics education, grade 8 (age, 13–14 years). The three most important pedagogical goals for the modelling learning path were that students

- learn how to use and create models themselves;
- get a better understanding of the relation between physics and the real world;
- can get some understanding via computer-based modelling of the physics of more complex and more dynamic situations than they can obtain by the traditional pencil-and-paper-based methods.

For becoming proficient in the modelling process, many (sub)competencies are required, including, for instance, competencies to specify a systemic structure as a set of related objects for a given problem situation, competencies to set up a physics-based mathematical model from the real model, representational fluency, and critical thinking with regard to interpretation and validation of model results. In this chapter we only discuss the demands that modelling puts on students' understanding of the mathematical notion of variable and formula. Thereto, we first explain the modelling approach we have used and subsequently discuss the pros and cons of this approach and how it affected the students' understanding of variable and formula.

8.2 Graphical Modelling as a Modelling Approach

We chose the graphical system dynamics approach developed by Forrester (1961) as a modelling approach, because of its wide range of applications to phenomena that can be modelled as systems whose states change over time and because it was considered by an advisory committee (Savelsbergh 2008) an appropriate candidate for a modelling approach in new Dutch secondary mathematics and science curricula. Henceforth, we refer to this approach as graphical modelling. With the term modelling, the entire modelling process outlined in Fig. 8.1 is addressed. This picture was derived from proposals for the renewal of Dutch science curricula, stating that

> the student must be able to analyse a situation in a realistic context and reduce it to a manageable problem, translate this into a model, generate outcomes, interpret the outcomes, and test and evaluate the model.

Several computer environments exist in which this graphical approach can be implemented. Examples are Modus (Klieme and Maichle 1991), STELLA (Chonacki 2004; Steed 1992), Co-Lab (Van Joolingen et al. 2005), and Coach 6 (Heck et al. 2009). These systems all adhere to the same basic principle: the students specify models drawn as graphical structures that can be executed (simulated). Research (see, e.g. Angell et al. 2008; Sander et al. 2002; Zwickl et al. 2015) indicates that combining experimentation and modelling helps students make connections between the concrete realistic situations that are modelled and the more abstract models and model outputs. With this in mind, we selected the Coach environment because it is one of the few integrated computer learning and multimedia authoring

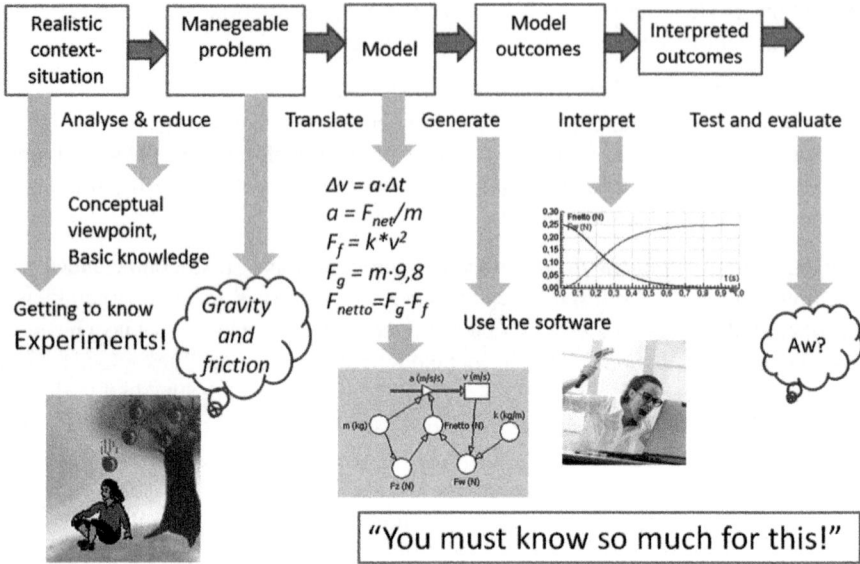

Fig. 8.1 The model of the ICT-supported physical-mathematical modelling process used by Van Buuren (2014) and annotated to point at the main relevant aspects. The statement in the lower-right corner is a quotation of a student about modelling

environments in which computer modelling can be combined with measurements through video and/or sensors and with model-based computer animation. A practical advantage is that Coach is available at the majority of Dutch secondary schools.

In a graphical model, the variables and relationships between variables are visually represented as a system of icons in a diagram. Five types of variables can be distinguished: (1) stock (or state) variables, (2) flow variables, (3) the independent variable (which by default is time in most modelling tools), (4) auxiliary variables, and (5) constants. The icons used in the Coach environment, except the one for the independent variable, are shown in Fig. 8.2. The type of a variable is not an intrinsic property, but follows from its role in the equations and can actually be manifold. A graphical model can be built on the computer by adding the suitable icons. Flow variables, auxiliary variables, and constants must be defined by entering adequate formulas and values. The difference equations (or differential equations) by which stock variables are defined are not entered as formulas; however, they follow directly from the flow icons that are connected to the stock icons. Only the initial value for the stock variable must be entered.

Several researchers (e.g. Niedderer et al. 1991) have suggested that the visual representations in graphical models provide students with an opportunity to express their own conceptual understanding of physical phenomena and can help to shift the focus from learning and working with mathematical formulas to more qualitative conceptual reasoning. Forrester (1961, p. 81) considered such a diagram as "an intermediate transition between a verbal description and a set of equations". Its main

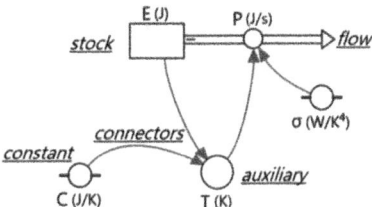

Fig. 8.2 Graphical model in Coach for the radiation of energy by an object with heat capacity C and an area of 1 m^2. The corresponding equations are the difference equation $\Delta E = -P \cdot \Delta t$ and the relations $T = E/C$ and $P = \sigma \cdot T^4$. In graphical models, a stock variable (E) is represented by a rectangle and a flow variable (P) by a circle on a thick arrow. Each stock variable needs an initial value. Variables and constants that are not part of a difference equation are referred to as auxiliary variables (T) and constants (C and σ) and are represented by two types of circles. Connectors (thin arrows) indicate which variables are used to define a particular variable. The independent variable, time t, is not visualized in this model by default

goal is to communicate the causal assumptions and the main features of the model, in a way comprehensible to people with less mathematics education (cf., Lane 2008). Understanding of difference equations would not be required. The diagrams representing the difference equations often can be understood metaphorically, as material flows into and out of stocks. Research has shown that students using an environment for graphical modelling can indeed reason qualitatively and intuitively about systems (Doerr 1996) and that graphical modelling seems to be effective for learning to reason with complex structures (Van Borkulo 2009). Schecker (1998) reported that half of his students after a mechanics course with STELLA were able to construct a qualitative causal reasoning chain on a new subject.

However, graphical modelling is not without problems in education practice, and many authors reported on difficulties that students have, especially when designing or adapting graphical models (see, e.g. Lane 2008; Ormel 2010; Van Borkulo 2009; Van Buuren et al. 2011, 2012; Westra 2008). Groesser (2012) pointed out that the information provided by system dynamics models can benefit only students who are familiar with system dynamics methodology and who are thus able to read and interpret graphical models. Van Buuren et al. (2015) found that students can use a given graphical model to reason about a situation to which they are already somewhat familiar but that these students are not able to build a graphical model themselves if they do not sufficiently understand the relation between the icons and the underlying mathematical equations. For novice graphical modellers, the hiding of mathematical details may even be counterproductive. Many a student is not sufficiently aware of the presence of the time step Δt in the difference equations. The fact that for a stock no formula but only the initial value must be entered may contribute to the misconception that a stock is not defined by a formula at all. And even if the use of difference equations may be avoided, students still need the ability to use formulas for variables that are defined by direct relations[1] instead of difference equations.

Van Buuren et al. (2012) found that many grade 9 students had difficulties using direct relations in graphical computer models. These students tried to define a variable in a graphical model by means of numbers or by means of expressions consisting of numbers only, and not by formulas consisting of symbols. They understood the calculation process on a numerical level, but they had problems with the (re)creation and use of formulas in which the variables are symbolized. A concrete example is a modelling exercise in which students must define the variable for the amount of interest A due to a constant interest rate of 2.4% per year and an amount of money S on the savings account that not only changes because of interest but also because of irregular deposits, where the initial value of S is 150 Euros. Students defined A as $0.024 \cdot 150$, which is a constant, instead of $0.024 \cdot S$, which is a variable expression. Another example is discussed in more detail in Sect. 8.5.

We advocate that instead of looking at a graphical model as graphical expression of a mental model that is turned into a quantitative computer model after entering the required relations, one can also consider it as a way of expressing the meaning of equations and of the structures of equations. In this perspective, worked out by Van Buuren (2014) in his modelling learning path, equations form the starting point for model construction, and the stock-flow diagram clarifies the meaning of the equations. In this modelling approach, the relations between the variables, i.e. the difference equations and direct relations, are the dominant building blocks of models, instead of the individual variables. We are of opinion that this fits better to the existing physics curriculum, in which formulas are introduced and used almost from the beginning: it connects to an already existing conceptual network and may enhance the development of mental models. Central to what Van Buuren calls the relation approach are the link between a difference equation and a stock-flow diagram and the interpretation of connectors as indicating structures of direct relations.

8.3   What Is a Formula in the Eyes of Students?

In the previous section, we noted that many young students have limited notions of variable and formula. Variables appear to be seen as having only one value in a certain situation, whether this value is known or yet unknown. Formulas appear to be used as merely providing general rules for doing calculations with such fixed variables. In other words, formulas are mainly used in the operational, process-oriented sense and less in a relational sense, that is, in a perspective where formulas are seen as mathematical objects that can be manipulated. We refer to Heck and Van

---

[1] By a direct relation, we mean a mathematical relationship between symbolized quantities in which at least one quantity can be isolated and written as a closed form expression of the other quantities.

Buuren (this volume) for a brief discussion of the process-object duality and the difficulties that students at all educational levels have in coping with this duality. This confusion has its roots early in the school career. When one asks lower secondary students about variables, formulas, equations, and simple calculations, many of them only have a vague understanding of these notions. As one well-achieving grade 9 student put it in a small-scale interview: "I have never understood what exactly a formula is". As we argued in Sect. 8.1, the uncertainty of students is strengthened by the way formulas and variables are used in exercises in physics education without much discussion of terminology.

To study students' notions, Van Buuren et al. (2011) explored what lower secondary students call a formula. They asked in three grade 9 (age, 14–15 years) physics classes, labelled 9A, 9B, and 9C, the simple question whether $y = 7 \times 8 + 27$ is a formula or not, and they gave five options: "yes", "I think so, but I'm not sure", "I am in doubt", "I don't think so, but I'm not sure", and "no". The results are shown in Table 8.1.

The results seem mixed: the answers in class 9B clearly deviate from the answers given in the other two classes. But this can be explained: the mathematics teacher of class 9B had discussed formula notation and function notation a week before, but such discussion had not taken place in the other classes. What it shows is that discussing such a subject makes a huge difference for students.

In a second study in 2012, the same question was given to 506 grade 9 students (age, 14–15 years) from 19 classes of 11 teachers at 6 schools spread around the Netherlands. The question was also given to 63 grade 10 students (age, 15–16 years) from 4 classes at 2 schools. All students were in the upper general secondary education stream or in the preuniversity stream. Because of the larger amount of data, results of the question are presented graphically in Fig. 8.3. The horizontal axis is used for the different classes. Grades of these classes are labelled "9" and "10", respectively. General secondary stream classes and preuniversity classes are labelled "G" and "U". The negative vertical axis, with lower bars, is used for the percentage of students per class who think that the expression is not a formula (although they may not be sure about it); the positive vertical axis, with upper bars, gives the percentage of students per class who are of opinion that it is a formula (even if they are not sure about it). If in one class both the upper and lower bar are short, this is an indication that there is much insecurity in this class. If in one class the upper and lower bar have approximately equal length, this is an indication that there is little consensus in the class about what a formula is. Figure 8.3 illustrates

Table 8.1 Results of the question "$y = 7 \times 8 + 27$ Is a formula, according to you?"

Class	"Yes" and "I think so, but I'm not sure"	"No" and "I don't think so, but I'm not sure"
9A (28)	68%	21%
9B (28)	28%	64%
9C (28)	67%	13%

The number in brackets is the number of respondents

↑ 'Yes' & 'I think so, but I'm not sure'

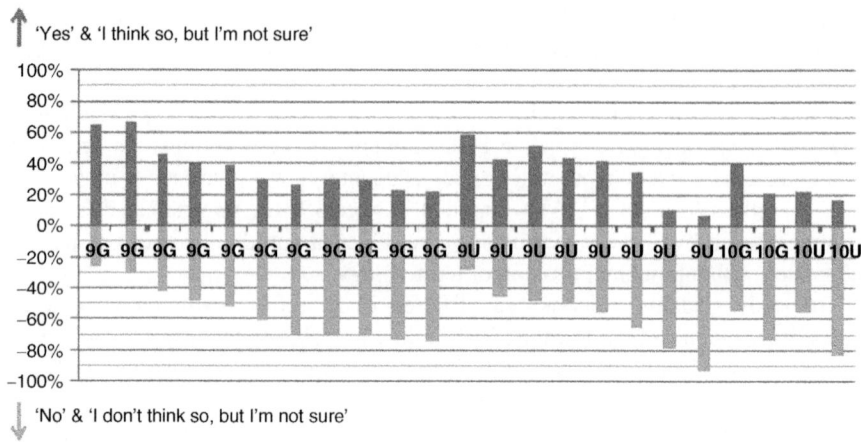

↓ 'No' & 'I don't think so, but I'm not sure'

Fig. 8.3 Percentages of students in 23 classes who consider the expression $y = 7 \times 8 + 27$ to be a a formula (positive vertical axis) or not (negative vertical axis). The classes are labelled by the grade and its type (general or preuniversity, labelled "G" and "U", respectively). This diagram illustrates the lack of consensus about the notion of formula across physics classes in the Netherlands

that the student confusion about the notion of formula exists at many schools, even at upper secondary level.

Van Buuren et al. (2011) also asked students in the aforementioned classes 9A and 9C to classify the mathematical statements shown in Fig. 8.4 as formulas or not. It illustrates that the notion of formula was diverse for the students. The equal sign apparently was not required to consider an expression as a formula: many students called m/V a formula, too. The results may indicate that the understanding of the notion of variable of many a student is on the level of polyvalent name and placeholder: they apparently consider a mathematical statement as a formula whenever one can calculate something with it and at least one variable is present, but not necessarily an equal sign. For example, $a = 17^2$ is considered a formula by approximately 40% of these students. Similar findings about students understanding of formulas and equations have been reported by other researchers (Godfrey and Thomas 2008; Hansson and Grevholm 2003). Student answers on the same questions in the second study, amongst 506 grade 9 students, exhibited similar patterns (as illustrated in Fig. 8.3). Many factors may influence the amount of consensus within one physics class. To mention a few, the interest for physics or for mathematics in a class, the textbooks that are used, the amount of peer influence with a class, and the mathematics teacher. We found differences between different classes that were taught by the same physics teacher. But we also found indications that explicit attention by the teacher to the subject can be effective. One teacher explicitly demanded that his students would always write down the "formula" and subsequently the "calculation". In this way, students apparently became more aware of the difference between a formula and a calculation. In three out of four of his classes, there was relatively much consensus amongst the students. We concluded

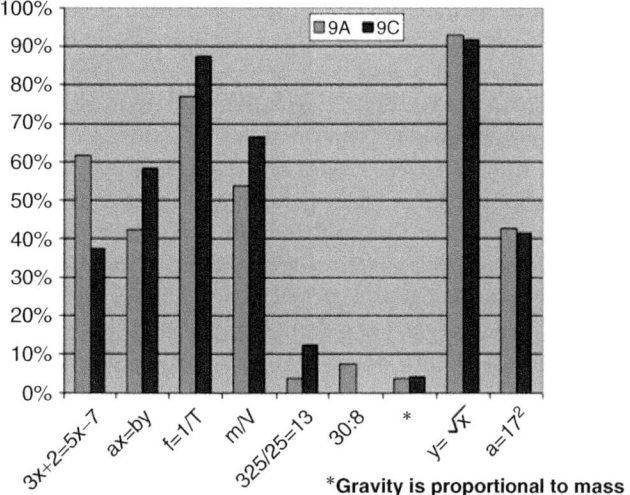

Fig. 8.4 Percentages of students, of two different grade 9 classes from one school (9A and 9C), who consider the expressions on the horizontal axis as a formula. This diagram shows amongst other things that many students consider m/V as a formula, in spite of the lack of an equal sign, and also that many students consider the expression $a = 17^2$ as a formula

that providing students with operational definitions of formula and variable might be effective. We decided to incorporate such definitions in the modelling learning path.

We think that the preference of lower secondary students to consider formulas and equations in a calculation perspective may be related to their gradual process of coming to grips with what is called "generalized arithmetic" in mathematics education. When variables are introduced in school algebra, students mostly learn to work with mathematical expressions like $2x + 1$, and they are told that x is not one specific number here but can represent *any* number (even though it is sometimes restricted to a small set of numbers—therefore the term "polyvalent name"). However, when working with the expression $2x+1$, one thinks of x as some specific but unspecified element of a number set: for example, when the equation $2x + 1 = 0$ must be solved, x really is one specific, albeit yet unknown element. For students, it is often not clear when x can be seen as generic, unspecified, or specific. Yet, this will not be their first and only encounter with ambiguity in mathematical thinking; Byers (2007) gives other mathematical examples where ambiguity plays a central creative role.

Let us look at an example from physics, which we will use again in Sect. 8.6: the vertical free fall of an object of mass m dropped at height y with zero speed and hitting the ground with speed v_{hit}. The law of conservation of energy implies that $mgy + \frac{1}{2}mv^2 = k$, for some constant k and constant of gravity g, whatever the height y of the object and the speed v at this height is. The law holds for *any* situation during the free fall motion, i.e. for *any* pair (y, v). Yet one applies the law in this problem to two specific situations: when the object is released at height h, then $y = h$, $v = 0$ and one gets $mgh = k$; when the object hits the ground $y = 0$, $v = v_{\text{hit}}$ and $\frac{1}{2}mv_{\text{hit}}^2 = k$. Equality gives $mgh = \frac{1}{2}mv_{\text{hit}}^2$, which can be rewritten as $v_{\text{hit}} = \sqrt{2gh}$. During the algebraic manipulation, the symbols stand for specific values of context-related quantities. In this sense, the last formula allows one to compute the speed with which the object hits the ground when it released at the given height h; many a student can hardly wait to plug in numbers. But $v_{\text{hit}} = \sqrt{2gh}$ can also be considered as a functional relationship between two physical quantities. In this case, the height h represents *any* release height. Thus one can think of h in $v_{\text{hit}} = \sqrt{2gh}$ as a specific and general number, and probably it is best to have these interpretations simultaneously in mind during the problem-solving activity.

The reason that we will come back to this physics example in Sect. 8.6 is that in school practice the formula $v = \sqrt{2gh}$ is problematic because the speed v and the height h in this formula are actually connected with different moments in time, namely, when the object hits the ground and when it is released. But students often forget this and think of speed at height h. They also tend to forget that this formula only holds when the object is dropped with zero speed. The assumptions, and thus the binding of the formula with the real context, as well as the link with the law of conservation of energy get lost. Sloppy use of formulas in relation to their physical meanings quickly leads to wrong interpretations; it probably explains mistakes made in exams where this formula occurred. An ICT-supported investigation of a similar context, namely, a bouncing ball, serves in Sect. 8.6 as an example of how ICT can help student keep an eye on the context and the whole motion. It will support our opinion that graphical modelling can help students become more proficient in dealing with the general-specific ambiguity and appreciate the strength of this because variables and formulas are then used in a way in which the varying and therefore general character of the variables comes to the fore.

8.4 Design of the Modelling Learning Path

For computer modelling, students must acquire many (sub)competencies. Van Buuren (2014) categorized these competencies into five intertwined partial learning paths that focus on (1) the modelling software, (2) graphs, (3) variables and formulas, (4) the elements of graphical models, and (5) the evaluation and nature of models. Here, we only describe the partial path on variables and formulas, and we focus on the mathematical notions required for modelling.

In the usual approach to graphical modelling, the five types of variables are considered as the building blocks of a graphical model, and the way they are linked to each other reflects a mental model that must be turned into a quantitative computer model. As we have argued in Sect. 8.2, this approach to graphical modelling leads to serious problems when we want our students to construct models. Therefore, we chose a relation approach to graphical modelling. In this approach, the mathematical relations between the variables are the dominant building blocks. Two types of formulas can be distinguished: difference equations and direct relations. Model conceptualisation starts with establishing the equations. This can be done by (1) selecting already known laws of physics, (2) doing experiments and deriving equations from the results, or (3) constructing the equations. The second, experimental way, requires function-fit abilities from students. They must learn how to determine mathematical functions that can describe experimental data. The third, construction way, demands more abilities and insight with respect to mathematization. We focus on the first and third way of establishing equations.

With respect to the development of proper notions of variables and formulas, we consider the following five goals important for our learning path:

(1) Students must learn to construct simple formulas themselves. Therefore, (word) formulas are never just given to our students but are constructed, starting with comprehensible situations and calculations with simple numbers. These are subsequently generalized into word formulas, by replacing the numbers by words and, subsequently, symbols. In exercises, students are asked to practise with construction of formulas for simple, concrete cases. We expect this approach to support the development of students' level of understanding of the variables in the formula to the level of generalized number. Examples of formulas at this level of understanding are the relation between mass, volume, and density of some substance and Ohm's law relating voltage, resistance, and current. These formulas have a general character, although the variables in these formulas usually are not varying.

(2) Students must learn to distinguish between finite difference equations and direct relations. Thereto, Δ-notation is used from the first difference equation, early on in the learning path. For direct relations, the term "direct formula" is used, and for difference equations, the term "Δ-formula" is introduced. The reason for using terms that both contain the word "formula" is that, otherwise, many students think that one of these two is not a formula at all, as we found in a pilot project (Van Buuren et al. 2011).

(3) For modelling purposes, students must learn to distinguish between a formula (consisting of variables), an equation (in which the remaining variables only have to be determined), and a "simple" calculation. We defined a formula as a relation between at least two symbolized physical quantities. Simultaneously, an equation is introduced as a relation containing only one symbolized quantity of which the value is actually fixed but yet must be determined, by solving the equation. In case this quantity already is isolated, and the other side of the equation only consists of numbers, the relation is called a simple calculation. How these definitions are introduced in the textbook is shown in Fig. 8.5. Students first apply these definitions in exercises and, subsequently, in all modelling tasks on the learning path. Note that these are operational definitions meant to guide the students in their first steps in using algebra in a physics context.

(4) Students must understand the process of numerical integration. Numerical integration can be introduced early on the learning path. At this stage, the

1. *A **formula** provides a (computational) relation between two or more different physical quantities.*
2. *These physical quantities are abbreviated, for surveyability. The abbreviations are called 'symbols'.*
3. *The values of the physical quantities are not all fixed in advance.*
4. *In most computer programs, only one physical quantity can appear at the left side of an equal sign. That must be the physical quantity that is going to be calculated by the computer program.*

*You, as a human being, usually start an exercise by filling in numbers into the formula. For example, $R = 5\Omega$ and $U = 3V$ can be entered into the formula $R = \dfrac{U}{I}$. We get $5 = \dfrac{3}{I}$. This, we will call an **equation**. At this point, the value of I actually is fixed, but this value is still unknown. An equation can be solved, that is: we can search and find the value for the unknown quantity. For example, the above equation can be transformed into $I = \dfrac{3}{5}$. What's left is a **simple calculation** that can be carried out easily.*

Fig. 8.5 This text is used in the instructional materials to provide students with operational definitions of formula, equation and simple calculation (original in Dutch)

integrand is a rate of change that is given in graphical and tabular format as a function of time. This can support students' notion of variable as varying object, because students must explicitly deal with variables of which the values are changing during the process. At a later stage, an integrand is used that for each step of the integration process must be calculated. In order to understand this process, students perform several steps of this process on a numerical level, i.e. by doing calculations by hand.

(5) Students must develop a notion of variable on the level of varying object. A step towards this notion can be made with the help of ICT. Students are asked to elaborate an experiment in which the same calculation must be made many times. For the calculations in the computer learning environment, a direct relation is used. For each calculation, the variables can be considered as isolated numbers, but from one calculation to another, students are dealing with variables that clearly change. As final step towards the notion of varying object, formal definitions of variable, as a symbolized varying quantity, and constant, as a quantity that is not changing, are given to the students. In exercises, students practice distinguishing between variables and constants. In these exercises, attention is paid to the fact that quantities that are constant in one situation can be variable in another. An example of such a quantity is the electrical resistance of a light bulb.

The steps towards these goals were distributed over the entire curriculum over various contexts, as is shown in Table 8.2.

Table 8.2 Overview of the implementation of the partial learning path on variables and formulas, distributed over the entire curriculum

Module title	Content regarding formulas and variables
Second year of secondary education (grade 8; age, 13–14 years)	
Doing research	Introduction of the notion of physical quantity as a measurable property of an object
Density	First word formula
	First direct relation
Velocity	First Δ-formula
	Process of numerical integration for the graph of a varying quantity
	Introduction of the equivalence between an expression consisting of numbers only and its outcome, a single number
Forces and bridges	Two new direct relations
Energy and power	Second Δ-formula
Third year of secondary education (grade 9; age, 14–15 years)	
Resistance and conductance	Two new direct relations
	Operational definitions of formula, equation, and simple calculation
	Step towards the notion of varying quantity: Use of a direct relation in software for doing calculations on data from tabulated values
The vacuum pump	Process of numerical integration on a numerical level
	Introduction of operational definitions of variable and constant
	(Re)construction of simple direct relations
	Formal introduction of process of numerical integration
	First explicit use of a direct relation in a computer model
Sound	Practising with the concepts of variable and constant
	Second use of a direct relation in a computer model
	Introduction of networks of direct relations in graphical models
Force and movement	Construction of a simple Δ-formula for a new realistic situation
	Integration of all elements required for graphical modelling
	Construction of a model based on known equations

8.5 Investigating Student Conceptions and Use of Formula

In order to investigate student notions of formula and their abilities to use formulas and variables for constructing computer models, Van Buuren (2014) did classroom observations, made screen and audio recordings of student dyads and triads working on ICT tasks and modelling tasks, collected computer results, and collected the answers of students on questions of the regular tests that were taken during the curriculum. The classroom observations often led to small-scale in-depth interviews; the test questions were specially developed to investigate students' notions and abilities.

Table 8.3 shows the results of a question from a regular test in which students in the first months of grade 9 were asked to classify expressions as formula, equation, or simple calculation, according to the operational definitions that were given to

Table 8.3 Students' classifications of mathematical expressions in physics contexts (in percentage)

Expression	Simple calculation	Equation	Formula	No answer
Version 1				
$\Delta x = 27.8 \cdot 5.2$	**60**	25	12	4
$12 = \rho \cdot 3.2$	21	**60**	8	12
$P = \Delta E / \Delta t$	15	9	**74**	2
$F_z = m \cdot 9.8$	6	50	**34**	10
Version 2				
$\rho = 23/7$	**76**	12	10	2
$120 = v_{av} \cdot 8.7$	26	**52**	12	10
$\Delta E = P \cdot \Delta t$	2	19	**79**	0
$m = F_z / 9.8$	19	40	**21**	20

Results from the two versions of the regular test after the module *Resistance and Conductance*. In this test, grade 9 students were asked to classify the expressions from the first column of this table as simple calculation, equation, or formula. Version 1 was given to 52 students, and version 2 was given to 42 students. In the columns 2 to 5 is shown how students classified the expressions. Correct classifications (in the operational definitions used) are in bold

them. The majority of students could grasp these notions, although some students had some difficulty distinguishing between simple calculation and equation, and other students probably were troubled by the alternative conception that a formula must consist of symbols only. These results were confirmed in small-scale student interviews.

Results of the tests after a next module showed that the vast majority of grade 9 students could distinguish between variable and constant. Most errors occurred with a quantity that was constant in one situation, but of which the value could be different in an entirely different situation.

These improved notions of students of variable and formula had a positive effect on their performance in modelling tasks, and this performance improved during the year. The first indications of this improved performance were the results from the same test questions as used earlier by Van Buuren et al. (2012) in the context of a student task to model a vacuum pump. The fraction of students that understood the calculation process of this model on a numerical level had hardly changed: this time too, around 60% were able to perform several cycles of this calculation process by hand. But the fraction of the students that was able to (re)construct a formula and use this formula for defining a variable in the computer model increased from 16% to 36%, and the quality of incorrect answers was much better. Almost no students tried to define this variable by means of numbers instead of a formula anymore. Yet, although most students could distinguish constants and variables, and could understand that a formula must contain variables, not all students were able to apply these notions in all circumstances. Many students did not realize that varying quantities in computer models must be defined by formulas in order to be variable. The notions of simple calculation and constant and of formula and variable must yet be coupled. The following typical classroom conversation between a physics teacher (T) and student (S) in the context of a graphical model of a system consisting of a vessel and an air pump illustrated how this coupling can be achieved. What had happened just before was that the student had entered $0.1 * 500$ as definition for the variable N_{pump} in a mathematical model equation instead of the required expression $0.1 * N_{tot}$, where N_{pump} and N_{tot} stand for the number of gas molecules in a pump and in the vessel plus pump, respectively (the number 500 is the first value of N_{tot}).

T: "If you enter $0.1 * 500$, do you have a formula, an equation, or a simple calculation?"
S: "A simple calculation, I think."
T: "A simple calculation. Is its outcome a variable or a constant?"
S: "A constant"
T: "Okay"
S: "and it must not be a constant."
T: "It must not be a constant."
S: "But, I am thinking, how can this be done?"

What followed was a discussion on the use of symbols in order to make the expression more general. What this excerpt shows is not only the confusion of the student but also that an early careful introduction of the notion of simple calculation, constant, and variable helped the teacher to create a conceptual conflict in the student's mind and a need for a formula consisting of variables represented by symbols. Even more important is that the operational definitions of formula and variable provided the students and the teacher with a language to communicate in. At the end of grade 9, almost all students used formulas for defining varying quantities in computer models.

8.6 Long-Term Effects of the Modelling Learning Path

Recently, the lower secondary modelling learning path has been extended into the beginning of the first year of upper secondary education (grade 10) with a module on dynamics. The combination of modelling and ICT-supported experimentation enabled to further shift attention from calculations with formulas to manipulations of graphs of whole functions. Via ICT, students can create models and analyse experimental data. They learn to make function fits of measured data, they use the computer for calculations by defining the variables that must be calculated by formulas, and they let the computer calculate derivatives of whole graphs instead of using tangents to determine momentary values of this derivative only. In this way, functions are gradually changed into objects that can be manipulated and discussed as a whole. Results of the research on this extension of the modelling learning path must still be thoroughly analysed. However, because the researcher is also the teacher of these students, we already have indications of possible long-term effects of the modelling learning path. Here, we give an example of these indications from an activity in grade 11 (age, 16–17 years) on conservation of energy.

In upper secondary education, the law of conservation of mechanical energy is often written as

$$\frac{1}{2}mv_1^2 + mgh_1 = \frac{1}{2}mv_2^2 + mgh_2.$$

In this formula, m is the mass of a moving object, g is the acceleration of gravity, and v and h are the velocity and height of the object. The term $\frac{1}{2}mv^2$ is the kinetic energy of the object and mgh is the potential energy. The indices 1 and 2 are labels that indicate two different moments in time. Written in this way, energy conservation is already transformed from a formula that describes variables v and h that continuously change in time into a relation between two pairs of momentary values. Novice learners often do not realize that two situations are involved and realize even less that the energies are continuous functions of time. Students often forget to use the indices 1 and 2 or forget the meaning of these indices, and regularly they erroneously mix up moments. Video measurements and data analysis in physics software enable students to work with all quantities and relations between these quantities, including the energies, as functions of time. The task for the grade 11 students was to make a video of a bouncing ball and to analyse the resulting movement with respect to all variables: velocity, height, kinetic energy, potential energy, and the sum of kinetic and potential energy. This task used to be very difficult for most students who had not followed the modelling learning path. Especially, they did not understand that the kinetic energy and potential energy are variables and must be defined by means of formulas. Students who followed the modelling learning path clearly needed less assistance from the teacher, even though the learning path that they followed had not yet finished. Figures 8.6 and 8.7 show typical graphs made by students in the Coach environment.

In order to be able to analyse the motion of the ball in this way, students first must understand the concept of derivative, they must be able to define the variable energies by means of adequate formulas in the computer program, and they must understand the effects of measurement noise as a result of minor flaws in the measurements and of noise as a result of digitalization. This digital noise demands some mathematical understanding. But when students are sufficiently able to use the computer for these purposes, they can study

- the relation between the height of the ball and the velocity: explain why the velocity changes suddenly, and investigate whether the movement can be considered as free fall when the ball is not in contact with the floor by determining the acceleration; and
- the energies: explain that the mechanical energy is constant at each moment between a pair of bounces (and learn to deal with the effects of noise), see that the sum of kinetic and potential energy is minimal when the ball hits the floor and explain where this energy has gone, and determine how much energy is dissipated during a hit.

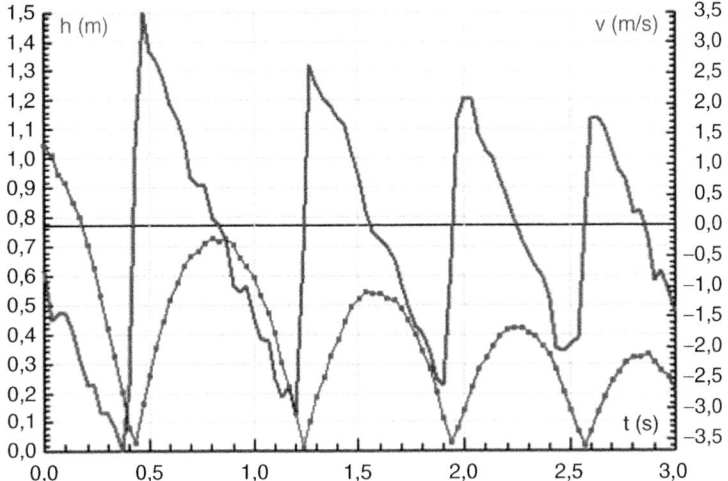

Fig. 8.6 Results of a video measurement by students of a bouncing ball. The quantity on the left axis is the height h of the ball, and the dots in the graph of h are the actual measurements. The symbol v on the right axis represents the computed velocity of the ball. The students have used the option "Derivative" in Coach to compute this velocity

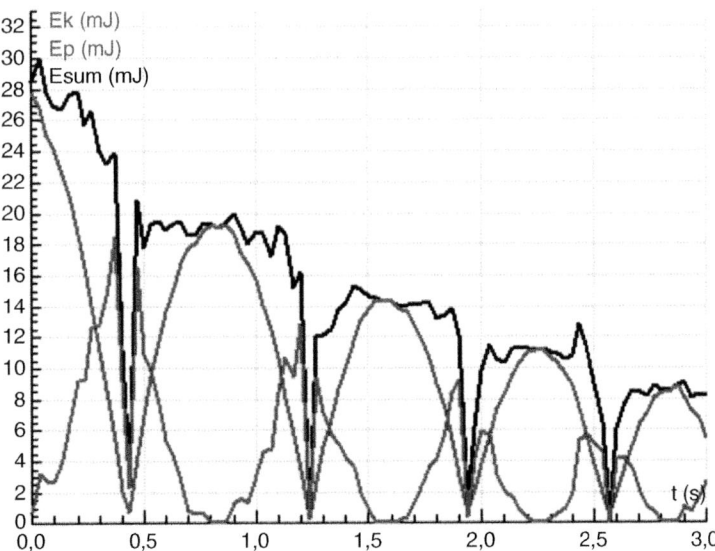

Fig. 8.7 The kinetic energy E_k, potential energy E_p, and the sum E_{sum} of E_k and E_p of the bouncing ball from Fig. 8.5 as functions of time

In other words, computer-based video measurements and data analysis enable students to investigate the whole motion, that is, to study all relations between all variables as functions of time. Computer-based modelling offers similar advantages.

An extra advantage of modelling can be the absence of noise, but a disadvantage is that modelling the loss of energy is rather difficult: it requires embedding of discrete events in the system dynamics-based modelling environment or modelling of the motion during the bounce phase (Heck et al. 2010). In order to investigate whether students really learn from such a task, as a pilot, we did short student interviews in which we asked the students whether they considered this task as instructive and what they had learned from it. Answers of the students were encouraging, like the one quoted below:

> We definitely think that we understand the subject better now. The graphs of the energies provide us with a clear overview of how the energy changes during the bouncing process. Normally, you make calculations for only one point and that does not give you an image of how things are changing. This overview has helped us to understand the subject.

Another student explicitly mentions the role of the formula (in the sense of the law of physics):

> When I do practical work with Coach, I do understand the formula better because I make up everything myself and therefore understand the effects of the formula.

Of course, the effects of graphical modelling and analysing experiments with ICT on student learning and understanding must be investigated more thoroughly, but the answers of these students show us in which direction to look for learning effects.

8.7   Conclusions

Providing students with operational definitions of formula and variable can help students to develop a notion of formula on a more generic level and a notion of variable as varying object. As we have shown, students can understand these definitions, and these definitions provide teacher and student with a useful language for discussing notions of variable and formula.

We found that students' modelling abilities improved by these notions, but we also had clear indications that modelling and analysis of experiments with the help of ICT can contribute to a the development of more general notions of formula and variable and can help students develop a better, more general, and more dynamic understanding of the laws of physics.

We advise teachers to pay more attention to the varying nature of many quantities in physics, at an early stage in the physics curriculum.

References

Angell, A., Kind, P. M., Henriksen, E. K., & Gutterrud, Ø. (2008). An empirical-mathematical modelling approach to upper secondary physics. *Physics Education, 43*(3), 256–263.

Booth, L. R. (1984). *Algebra: Children's strategies and errors*. Windsor: NFR-Nelson.

Booth, L. R. (1988). Children's difficulties in beginning algebra. In A. F. Coxford (Ed.), *The ideas of algebra, K-12* (1988 Yearbook, pp. 20–32). Reston: National Council of Teachers of Mathematics.

Byers, W. (2007). *How mathematicians think: Using ambiguity, contraction an paradox to create mathematics*. Princeton: Princeton University Press.

Chonacki, N. (2004). STELLA: Growing upward, downward, and outward. *Computing in Science and Engineering, 6*(3), 8–15.

Doerr, H. M. (1996). Stella ten years later: A review of the literature. *International Journal of Computers for Mathematical Learning, 1*(2), 201–224.

Forrester, J. (1961). *Industrial dynamics*. Cambridge: MIT Press.

Godfrey, A., & Thomas, M. O. J (2008). Student perspectives on equation: The transition from school to university. *Mathematics Education Research Journal, 20*(2), 71–92.

Groesser, S. N. (2012). Model-based learning with system dynamics. In N. M. Seel (Ed.), *Encyclopedia of the sciences of learning* (pp. 2303–2307). New York: Springer.

Hansson, Ö., & Grevholm, B. (2003). Preservice teachers' conception about $y = x + 5$: Do they see a function? In N. A. Pateman, B. J. Dougherty, & J. Zilliox (Eds.), *Proceedings of the 27th Conference of the International Group for the Psychology of Mathematics Education* (Vol. 3, pp. 25–32). Honolulu: University of Hawaii. Retrieved 25 Feb 2017 from http://files.eric.ed.gov/fulltext/ED500858.pdf

Heck, A., Ellermeijer, T., & Kedzierska, E. (2010). Striking results with bouncing balls. In: C. P. Constantinou & N. Papadouris (Eds.), *Physics curriculum design, development and validation* (pp. 190–208). Selected papers of the GIREP 2008 conference. Nicosia: University of Cyprus.

Heck, A., Kedzierska, E., & Ellermeijer, T. (2009). Design and implementation of an integrated computer working environment for doing mathematics and science. *Journal of Computers in Mathematics and Science Teaching, 28*(2), 147–161.

Klieme, E., & Maichle, U. (1991). Erprobung eines Systems zur Modellbildung und Simulation im Unterricht. In P. Gorny (Ed.), *Informatik und Schule 1991: Wege zur Vielfalt beim Lehren und Lernen* (pp. 251–258). London: Springer.

Lane, D. C. (2008). The emergence and use of diagramming in system dynamics: A critical account. *Systems Research and Behavioral Science, 25*(1), 3–23.

Niedderer, H., Schecker, H., & Bethge, T. (1991). The role of computer-aided modelling in learning physics. *Journal of Computer Assisted Learning, 7*(2), 84–95.

Ormel, B. (2010). *Het natuurwetenschappelijk modelleren van dynamische systemen: Naar een didactiek voor het voortgezet onderwijs* [Scientific modelling of dynamical systems: Towards a pedagogical theory for secondary education] Doctoral dissertation, University of Utrecht. Utrecht: CD-β Press. Retrieved 25 Mar 2017 from https://dspace.library.uu.nl/handle/1874/37371

Sander, F., Schecker, H., & Niedderer, H. (2002). Computer tools in the lab—effects linking theory and experiment. In D. Psillos & H. Niedderer (Eds.), *Teaching and learning in the science laboratory* (pp. 219–230). Dordrecht: Kluwer Academic Publishers.

Savelsbergh, E. (Ed.). (2008). Modelleren en computermodellien in de β-vakken: Advies aan de gezamenlijke β-vernieuwingsacommissies [Modelling and computer models in the sciences: Advice to the joint curriculum innovation committees]. Utrecht: FIsme. Retrieved 23 Mar 2017 from http://www.nieuwenatuurkunde.nl/download/id/40/Modelleren.betavakken.pdf

Schecker, H. (1998). Physik—Modellieren [Physics—Modelling]. Stuttgart: Ernst Klett Verlag. Retrieved 18 June 2018 from https://www.researchgate.net/publication/312097833_Physik_modellieren

Steed, M. (1992). Stella, a simulation construction kit: Cognitive process and educational implications. *Journal of Computers in Mathematics and Science Teaching, 11*(1), 39–52.

Van Borkulo, S. P. (2009). *The assessment of learning outcomes of computer modeling in secondary science education.* Doctoral thesis, Twente University. Twente University. Retrieved 25 Mar 2017 from http://doc.utwente.nl/61674/1/thesis_S_van_Borkulo.pdf

Van Buuren, O. (2014). *Development of a modelling learning path.* Doctoral thesis, University of Amsterdam. Amsterdam: CMA. Retrieved 25 Feb 2017 from http://hdl.handle.net/11245/1.416568

Van Buuren, O., Heck, A., & Ellermeijer, T. (2015). Understanding of relation structures of graphical models by lower secondary students. *Research in Science Education, 46*(5), 663–666.

Van Buuren, O., Uylings, P., & Ellermeijer, T. (2011). A modelling learning path, integrated in the secondary school curriculum, starting from the initial phases of physics education. In W. Kaminski & M. Michelini (Eds.), *Teaching and learning physics today: Challenges? Benefits?* (pp. 609–621). Proceedings of selected papers of the GIREP-ICPE-MPTL conference 2010, Reims. Udine: University of Udine. Retrieved 25 Mar 2017 from http://iupap-icpe.org/publications/proceedings/GIREP-ICPE-MPTL2010_proceedings.pdf

Van Buuren, O., Uylings, P., & Ellermeijer, T. (2012). The use of formulas by lower level secondary school students when building computer models. In A. Lindell, A.-L. Käkönen, & J. Viiri (Eds.), *Physics alive* (pp. 140–149). Proceedings of the GIREP-EPEC Conference 2011, Jyväskylä. Retrieved 25 Mar 2017 from http://urn.fi/URN:ISBN:978-951-39-4801-6

Van Joolingen, W. R., de Jong, T., Lazonder, A. W., Savelsbergh, E. R., & Manlove, S. (2005). Co-Lab: Research and development of an online learning environment for collaborative scientific discovery learning. *Computers in Human Behavior, 21*(4), 671–688.

Westra, R. (2008). *Learning and teaching ecosystem behaviour in secondary education.* Doctoral thesis, University of Utrecht). Utrecht: CD-β Press. Retrieved 25 Mar 2017 from https://dspace.library.uu.nl/handle/1874/26253

Zwickl, B. M., Hu, D., Finkelstein, N., & Lewandowski, H. J. (2015). Model-based reasoning in the physics laboratory: Framework and initial results. *Physical Review Special Topics—Physics Education Research, 11*, 020113. https://doi.org/10.1103/PhysRevSTPER.11.020113

Chapter 9
Graph in Physics Education: From Representation to Conceptual Understanding

Alberto Stefanel

The language of science is an integration of different representation instruments, including words, pictures, equations, and graphs. Graphs have a fundamental role in physics and in physics education. A wide literature in physics education evidenced the difficulties of students in reading, constructing, and interpreting graphs. The use of sensors connected to the computer opened new learning opportunities in that area of concern, aimed at constructing physics concept and developing graphing competencies. Two studies will be discussed regarding the role of graphs acquired in real time for learning. The first study regards students aged 15–16 exploring motion with sonar ranger sensor. Their learning is compared with those of first year university students and a group of prospective middle school teachers. The second study concerns secondary school students learning by analyzing light diffraction pattern acquired with sensors. Students were involved in inquiry-based laboratories following the suggestions of a research-based educational proposal concerning the specific topic considered. Monitoring the students' learning path with tutorials, it was possible to highlight the role of real-time graphs and of the active learning environment: for developing graphing skills and competencies, for connecting the processes underlying the phenomenon observed and the specific features of the graph acquired in real time, for promoting the construction of the capability to attribute physical meaning to the formalism, and for activating conceptual models of processes and phenomena.

A. Stefanel (✉)
Physics Education Research Unit, Department of Chemistry,
Physics and Environment, University of Udine, Udine, Italy
e-mail: alberto.stefanel@uniud.it

© Springer Nature Switzerland AG 2019
G. Pospiech et al. (eds.), *Mathematics in Physics Education*,
https://doi.org/10.1007/978-3-030-04627-9_9

9.1 Introduction

Science language integrates and interlaces different representations: words, diagrams, pictures, graphs, maps, equations, tables, charts, and other forms of visualization and formalization (Lemke 2003). These different representations are a fundamental part of science not only as communicating tools but also because they contribute to define the nature of science itself, as well as to science understanding (Windschitl et al. 2008). The multi-representation plays, in fact, an important epistemic role in the development of science (Fisher et al. 2011; Gilbert 2007; Wainer 1992) and physics (Guttersrud and Angell 2014). Graphical representation is a powerful effective tool for synthetic data rendering and representation of relations between physical quantities (Fan 2015) and for data interpretation (Klein 2001; Tufte 2001) giving access to the "revelation of the complex" (Tufte 2001). Physicists plot a graph to obtain a clear picture of the data, obtaining a synthetic overview of these data, revealing aspects that might not be obvious from a table, as well as regions of interest suggesting further analysis (Deacon 1999), or evidence for review and modify their theory (Glazer 2011).

This role and value of the graphing have always been stressed in the teaching of physics, although it is often assumed that students should develop graphing abilities by osmosis. Since several years already in the teaching of other scientific disciplines at all education levels, more and more importance has been given to the development of skills related to the construction, reading, and interpretation of graphs (Lemke 2003).

Rather than developing such skills for those who will undertake science-based degrees and careers, the focus has been on developing basic skills for the literacy of the twenty-first-century citizen: working with data, organizing it in tables and graphs, making inferences from data, and finding trends and support claims and evaluation (American Association for the Advancement of Science 1993; European Commission 1995; Fan 2015; National Research Council 1996). The use of graphical tools has become the subject of development in the scientific teaching since graphs play a dominant role in the inquiry process and are tools used to analyze and display quantitative relationships. Therefore, graphing competence is a fundamental requirement for doing inquiry and a major component both in science and math education activating inquiry-based learning strategies (Glazer 2011; National Council of Teachers of Mathematics 2000).

Despite the importance of graphic representation, researches on students' learning in physics, in science, and in math highlighted the ineffectiveness of the traditional transmissive teaching in building graphing skills (Beichner 1994; Duit 2009; McDermott et al. 1987, 2000; Shah et al. 2005; Thornton and Sokoloff 1998; Trowbridge and McDermott 1980, 1981) and in the use of the mathematical competencies for such objectives (Curcio 1987; Leinhardt et al. 1990; Meltzer 2002). As McDermott observed "the ability to relate actual motions and their graphical representations does not automatically develop with acquisition of simple graphing skills, such as plotting points, reading coordinates and finding slopes" (McDermott 1993).

The use of Real-Time graphical representation in educational Laboratory (RTL) made possible to develop effective active educative strategies to improve the competence of students in graphing at different levels and in the understanding of the physical processes through graphics (Linn et al. 1987; Mokros and Tinker 1987; Sokoloff et al. 2004; Thornton and Sokoloff 1990, 1998), as well as in modeling these processes (Angell et al. 2008; Glazer 2011; Guttersrud and Angell 2014; Hofstein and Lunetta 2004). Many questions remain open concerning the role of RTL graphs for conceptual learning: in connecting a phenomenon and the representation of the physical quantities describing it (e.g., position in the case of the motion of a body, temperature considering the heating of a system), in connecting data and construction of the phenomenological laws describing processes, and in creating bridges from the descriptive-phenomenological level to that of interpretation.

About this problem area, the present contribution aims to highlight the following research questions:

RQ. 1. In which way could RTL (a) activate conceptual understanding, (b) promote graphing competencies, (c) develop formal thinking, and (d) construct physical meaning of the mathematical representation of physical quantities and relation between them?

RQ. 2. Concerning the process of constructing the law of a phenomenon: (a) what are the students' representations and models activated? (b) What are their difficulties?

Two studies will be discussed regarding the role of RTL graph for learning. In the first, learning of students aged 15–16 exploring motion with sonar ranger sensor (Michelini 2010; Michelini et al. 2002) is compared with those of university students of the first year of the agricultural degrees and a group of prospective middle school teachers. The second study concerns secondary school students learning by analyzing optical diffraction pattern acquired with on-line sensors (Michelini et al. 2014).

The next sessions discuss the research results on graphing, introducing the theoretical framework, the main assumptions, and the choices at the base of the studies presented. The research contexts and instruments will be then briefly presented, stressing the strategies adopted in using real-time graphs. The core of the paper documents the two studies regarding the role of RTL sensors for learning, concluding with the main outcomes and answers to the research questions.

9.2   Theoretical Framework

Graphing may occur in two main contexts: inquiry and reading (Friel et al. 2001). In an inquiry context, individuals engage in empirical investigation of actual data, where they produce and/or analyze data, interpret their own data and results, and report findings and conclusions. In reading contexts, people are data consumers, also when they are requested to make hypotheses or to read data. Many perceptual and conceptual aspects are involved in facing a graph. According to Bertin (1983), three steps occur immediately when reading a graph: (A) external identification, the reader perceives the external factors of the graph (e.g., the title, axis labels, and scales); (B) internal identification, the reader perceives the internal factors, as the bars, lines, or dots representing data; and (C) perception of correspondences, the reader combines the details identified via stages (A) and (B) to capture information displayed in the graph.

Graph comprehension or interpretation (according to the author) is defined as the ability of a reader to derive or obtain meaning from graphs created by themselves or others (Curcio 1987; Glazer 2011). Different reviews (Friel et al. 1997; Glazer 2011) summarized the graph comprehension into three levels, as synthesis of many researchers (Bertin 1983; Brussolo 2010; Costas 2010; Trowbridge and McDermott 1981):

(L1) An elementary level – reading data or extracting immediate information by the graph
(L2) An intermediate level – reading between the data, requiring almost a logical inference to find relationships between data, as, for instance, reducing data categories, finding points with value greater than an assigned one (or than a certain value), confronting slopes, without referring to the specific meaning of the slopes
(L3) Advanced/overall level – to read beyond the data, as, for instance, reducing all data to a single statement/relationship (Bertin 1983; Carswell 1992); synthesis or integration of most or all the data (Carswell 1992); extrapolation from the data; extending, predicting, or inferring from the representation; answering to questions requiring prior knowledge (Curcio 1987; Friel et al. 1997, 2001; Glazer 2011; Wainer 1992); and interpreting relationship between data or determining values of the data conveyed in the graph

Gal (2002) unified the first two levels.

The graph comprehension level can be affected by different factors related both to the graph itself, to the context of the graph, to the reading task, and to the reader. Examples of these factors can be the purpose for using a graph and the context in which the represented data are situated, the perceptual features of the graph, the request of reading variables, performing a computation (sum/differences of values; mean value; comparison between values, derive a relation), and identifying or comparing trends on the basis of qualitative or quantitative information (Friel et al. 1997). The reader's epistemic beliefs in attributing physical meaning to formal entities (Von Korff and Rebello 2013) as well as his prior knowledge strongly affect the comprehension of a graph (Glazer 2011): previous knowledge or difficulties on graphing (Shah et al. 2005); prior theory/beliefs concerning the context related to the graph, in particular dealing with data contradicting hypotheses or with anomalous data (Chinn and Brewer 1993); and prior knowledge of the content displayed in the graphs or explanatory skills. The salience of the different aspects of prior knowledge with respect to graph comprehension seems affected by grade or age differences, younger children showing a greater need for knowledge about the "concrete," visible, explicit aspects of a graph. The most salient prior knowledge is the ensemble of the mathematical concepts needed to read a graph and to extract information from it, independently by age (Aberg-Bengtsson and Ottosson 2006; Curcio 1987; Leinhardt et al. 1990). According to their model of data theory, Chinn and Brewer (1993, 2001) showed that students are more likely to notice problems encountering data inconsistent with their theory than data confirming their theory. This occurs when a student is able to elaborate an alternative conception capable to explain data. If they are unable to do so or are firmly convinced of their conception, they typically refuse anomalous data (Chinn and Brewer 1993), also when the anomalies evidently contradict students' previews (Champagne and Gunstone 1985; Park et al. 2001). Considering familiar data on which they have consistent expectation, students more frequently show a global vision of the graph and tend to describe more frequently the global trend of data. On the contrary analyzing unfamiliar data, they tend to consider local features and aspects of the graph (Shah 2002; Shah and Hoeffner 2002). The familiarity with data and content of a graph can influence both novice and expert interpretation of data, suggesting that the graph interpretation is strongly context dependent in any case (Roth and Bowen 2003).

Researchers highlighted that students have trouble making connections between graphs and other representations, as data sets and tables, algebraic functions, and other types of graphs (Friel et al. 1997; Roth and Lee 2004). Moreover, in general very few students show the transfer of reading graph skills from math to other contexts (Planinic et al. 2013), being line graphs the most difficult for students (McDonald-Ross 1977; Padilla et al. 1986; Shah and Carpenter 1995). To construct a line graph, students need to be able to draw and scale axes, assign variables to axes, and plot points. To interpret a line graph, the skills needed are determination of point coordinates, use of line best fit, interpolation and extrapolation, or stating a relationship between variables. In the specific case of time evolution of physical quantities, students must connect a specific graph to the graphs correlated; as, for instance, from the position vs time (x vs t) graph, construct the velocity vs time (v vs t) graph and acceleration vs time (a vs t) graph (McDermott et al. 1987). Students seem more able to represent correctly the slope of a linear graph than to individuate the correct value of intercept, also when they do not possess previous mathematical instruction (Hattikudur et al. 2012). Moreover, when they can choose, they tend to use more frequently formulas than graph to extract information (Knuth 2000; Planinic et al. 2013). In experimental lab setting analyzing straight-line graphs derived from their own data, students have been able to achieve a considerable development toward a concept of slope, or gradient, and how it relates to the concept of proportionality, but they continue to demonstrate a great resistance to applying their mathematical knowledge to physics (Woolnough 2000). In that process, many young students characterize as proportionality each increasing relation (Pospiech 2015).

McDermott (McDermott et al. 1987), in the context of kinematics, evidenced the difficulties in discriminating between slope and height of a graph or between quantity variation and quantity value, as well in interpreting changes of a quantity and change in slope, and tendencies observed also in other contexts (Trowbridge and McDermott, 1980, 1981). The difficulties in interpreting kinematic graphs seem more related to a lack of understanding or applying physics concepts than a lack in mathematical knowledge (Planinic et al. 2013). Another area of concern regards the connection of graphical representation with the real world. Some aspects evidenced in studies are found in distinguishing real-world trajectory and position-time graph or in representing negative velocity. The difficulties in connecting graph and real-world interlace with difficulties in connecting the graphs of derived quantities, as, for instance, matching the graphs v vs t and a vs t with the behavior of the graph x vs t (McDermott et al. 1987; Suri and Clarke 2009).

New technologies and real-time-based laboratory (RTL) opened new educational opportunities for IBL approaches (Krajcik and Layman 1993; McDermott 1991; Sokoloff et al. 2004) that enable the implementation of active learning-based laboratory to develop physical concepts (Michelini 1988, 2006, 2010; Michelini et al. 2014; Sassi 1996; Thornton and Sokoloff 1990) and mathematical concepts (Educational Studies in Mathematics 2004; Hale 2000). The research outcomes on RTL showed improvement of students' comprehension of graphs (at all level of age) and conceptual understanding of physics (Krajcik and Layman 1993; Stefanel et al. 2002; Svec 1995; Thornton 2004; Thornton and Sokoloff 1998).

RTL is particularly effective to develop reading skills of the graph x vs t (90–100% of cases), of the graph v vs t (80%), and of the graph a vs t (56%), as well as conceptual understanding of kinematical quantities and development of graphics problem-solving competencies (Svec 1995). There is evidence that students interpret graphs more readily than they can read, even when they lack the appropriate interpretative competencies (Michelini et al. 2002; Shah et al. 2005).

Typically, novices and low-level students tend to show a local view of RTL graphs, emphasizing, for example, the presence of experimental irregularities blurring the vision of the global trend (Corni et al. 2005; Testa et al. 2002).

These outcomes show that RTL graphs can become, in appropriate learning environments, powerful resources for learning physical concepts and activating and developing important skills, based on a deep, rich, and generative (if intuitive and sometimes limited) understanding of representation (diSessa and Sherin 2000). A graph provides, in fact, a bridge between more abstract mathematical representation and its physical meaning (Deacon 1999). Extending Vygotsky (1962), we can suppose that the RTL graphs help students to develop the concepts related to the phenomenon explored, as is the case when a child uses words he or she is helped to develop concepts. According to Sokoloff and Thornton (Thornton 2004; Thornton and Sokoloff 1990, 1998), RTL offer the opportunity to design and activate learning environments encouraging students to use and interlace the multiple representations constituting the science language (Jones 1993; Lemke 2003). In addition we can hypothesize that the Prevision-Experiment-Comparison (PEC) strategy (Theodorakakos and Psillos 2010) activates the process to make explicit students' conceptions, through which the inchoate (intuitive or naive) understandings within their (mind/brain) system are made available to the system itself, namely, the verbally mediated, conscious processing through which an individual becomes aware of his or her own beliefs (Karmiloff-Smith 1988).

Finally, we hypothesize that the real-time graph can be a tool that allows to build formal thinking (Bisdikian and Psillos 2002; Michelini 2006), that is, the acquisition of networks which assigns meaning to symbolic elements and which allows students to explore and interpret the world through the formal instruments of physics (Michelini 2010; Michelini et al. 2010). In the following, this important aspect will be investigated analyzing: how students connect the features characterizing a RTL graph to the specific characteristic of the phenomenon studied (e.g., critical points, phases, slope); how they extract information from the graph and use these to construct new quantities (e.g., velocity of the represented quantity change); how they go in-depth in the graph understanding, as well as how they analyze interpolations whose meaning is constructed using an IBL approach; or how they construct a fit of data on the base of theoretical hypotheses.

9.3   Instruments and Methods

9.3.1   The Research Environment

The researches discussed here regard the students' learning paths when they face the conceptual knots related to each of the specific phenomenological context considered, in an inquiry-based educational environment (McDermott 1991, 1993; McDermott et al. 1987, 2000; Michelini 2010). Students explore phenomena and face the related conceptual problems in learning environments named Conceptual Lab for Operative Exploration (CLOE) (Michelini 2006; Michelini et al. 2010; Stefanel et al. 2002), where a researcher drives the interaction with and between students adopting a methodology based on Rogers's reflection interviews (Lumbelli 1996). On the base of this methodology, the questions asked by students become questions asked to them, to which they answer with phenomenological explorations or simple experiments. The concepts introduced by them are re-examined using the words and the ways they used themselves. The researcher follows the students' reasoning and learning path in the construction of concepts. Each step is the base for the further exploration to build a new conceptual step. The students' learning paths are activated and monitored by IBL tutorials (McDermott 1991; McDermott et al. 2000): promoting preview and reasoning on phenomena, according to a PEC strategy (Sokoloff et al. 2004; Theodorakakos and Psillos 2010), and stimulating the passage from the phenomenological level to interpretation, starting from the distinction between the considered phenomenon, the process involved, and its explanation (Michelini 2006). This promotes that students' conceptual gains become explicit through inscriptional practices (Woolnough 2000; Wu and Krajcik 2006) and sharing knowledges by negotiation of meaning (Griffiths and Guile 2003; Von Korff and Rebello 2013; Wellington and Osborne 2001).

The researcher, conducting a CLOE lab, follows the students' suggestion and lines of reasoning having as reference the layout of research-based educational proposals framed in the Duit's model of educational reconstruction (Duit et al. 2005) and designed in a vertical perspective (Costas 2010; Michelini 2010; Suri and Clarke

2009). All these proposals involve students directly in the operative exploration of a specific phenomenology and constructing their conceptual understanding. In that learning process, algebraic and graphical mathematical representation of variables are used for a formalized description of phenomena, as tools for the imaginative reduction of physical concepts (Michelini et al. 2010).

9.3.2 RTL Graphs Exploring Motion

First, a study on students' learning about the motion graphs is discussed, as an example of the role of RTL in the process of construction of physical concepts and in their formal representation. A proposal in vertical perspective on motion was designed (Brussolo 2010; Michelini 2010) and contextualized in the safety perspective (Mossenta et al. 2014) and in sport (Bradamante et al. 2004). Here we consider the following sequence of two steps based on the use of an on-line commercial sensor (Corni et al. 2005; Michelini et al. 2002; Sassi et al. 2005) to analyze two situations: (EA) a person moving in front of the sensor and (EB) a free motion of a toy car launched on a horizontal plane. The analysis of the first situation, after a preliminary informal observation of how the position detected by the sensor is translated in a graph, proposed four situations, involving a person walking: (EA1) away from the sensor, (EA2) at different speeds, (EA3) approaching the sensor, and (EA4) moving first away from the sensor, then stopping, finally approaching the sensor. Every situation has been monitored with an IBL tutorial. Each tutorial first presents a problem situation, asking students to individuate reference frame, trajectory, and type of motion, to make a preview on the expected graph x vs t (in the case of EA1 also v vs t and in the case of EB also a vs t), to perform the experiment, and to report the observed graph, comparing it with the preview graph and discussing analogies and differences. They read the graph, extracting information such as the starting and final positions and times, the displacement, and the mean velocity, to identify the different phases of the motion, to interpolate, or to fit data obtaining the analytical form of the time evolution of the quantity considered, individuating values and physical meanings of the parameters involved.

The focus will be here on the first stage of the construction and analysis of the graphical representation of time evolution of kinematical variables, considering a sample consisting of 134 individuals, of different age, level, and type of school, including a small group of prospective teachers, to individuate parameters of comparison and indications of possible dependence from subjects' age and formation:

- Three groups of students 15–16 years old (grade-10), from three Italian high schools:

  - The first group (mentioned as LM below), consisting of N = 18 students of a high school (scientific lyceum) in Udine in Northern Italy, considered high level by school teachers, with the exception of two students but still positive. The physics teacher had already dealt with the basic concepts of kinematics, adopting a transmissive method without using the laboratory; the math teacher adopted a calculus-based approach and had already dealt with the basic elements of the Cartesian representation of the straight line and parabola.

– The second (LF) consisting of N = 25 students of a scientific lyceum in Crotone in Southern Italy. The math teacher considered the students of middle-low level. Students faced the representation of the straight and the parabola with a formal approach but have no previous knowledge in physics.

– The third (IG) consisting of N = 22 students of a Technical Institute for Commerce in Gemona, a little minor town close to Austria/Slovenian borders. The math/physics teacher considered students of low level, with few cases of sufficient level, and had dealt with the basic concepts of motion without using laboratory and analytical geometry with a transmissive approach.

• Another group composed of two subgroups of 14 and 33 (AG – N = 47) first year university students of the agricultural-food sciences degrees in Udine. From pretest, the level was middle-low in physics, in math, and in graphing skills. Everyone studied the basics of analytical geometry at school and only half had basic knowledge in kinematics.

• The last group included N = 22 prospective teachers of mathematics and sciences in middle school, all graduated in natural science or biology attending a special course for initial preparation (PT group hereafter).

Each group followed the same sequence of experiments EA and EB in 2–3 h of free/not compulsory activity, based on RTL approach described before and using the same tutorials, with few differences specified in the following. In particular, the passage from the graphical representation to the data interpolation was addressed in differentiated ways with the different groups involved in the research, as discussed later. Some changes also will be indicated in the data tables concerning the samples number, because not all attended the full activities proposed on motion. The first group (LM) faced preliminarily a two-question pretest concerning the construction of the equation of the straight line for two points assigned and the equation of the vertical axis parabola for three points assigned, asking values and geometrical meaning of each coefficient involved.

9.3.3 Measuring with On-Line Sensors and Analyzing Light Diffraction Pattern

As a second example of the role of RTL graph for learning, a study will be discussed on how students analyze a single-slit diffraction distribution, acquired by the USB apparatus LUCEGRAFO (Gervasio and Micheliani 2009). Unlike the previous one concerning the time evolution of physical quantities, in this case, the RTL graph represents the light intensity I vs the position x of measurement, transverse with respect to the direction of the light and of the slit. Another interesting difference is the task: students were asked to extract information from the graph, drawing new derived graphs and constructing the formal relationship between the quantities represented.

The analysis presented here is part of the approach developed in other researches on the study of optical diffraction with the use of on-line sensors (Michelini et al. 2014). This approach is based on the analysis of optical diffraction patterns produced by laser light diffracted by a single slit and collected on a screen at a large distance from the slit. First students explore qualitatively the diffraction pattern collected on a white screen, to identify its global characteristics and to predict the associated light intensity distribution. Then they acquire with on-line sensors the distribution I vs x and analyze it quantitatively, by identifying characterizing regularities: (a–b) the linear correlation between minima/maxima positions and order number and (c) the inverse quadratic correlation between angular position X_M and intensity I_M of maxima. Students discuss the inadequacy of the geometric model to interpret the experimental distribution and formulate a wave hypothesis on the nature of light. Students can interpret the experimental pattern fitting experimental data with a model based on the Huygens principle.

Here we focus on the following steps activated by the suggestion and stimuli of the IBL tutorial: (1) predicting the intensity of the light diffracted by a single slit according to the position, (2) experimenting and comparing with the expected graph, and (3) analyzing the experimental distribution, building the relations (a) –(b) –(c) above.

The sample considered here included 168 students of grade 11, aged 16–17, distributed in 8 groups of a scientific lyceum of Treviso, a Northern Italy town. The schoolteachers evaluate students of middle-high level. Students knew the basis of the study of a function and had already addressed qualitatively light interference (Young experiment).

9.3.4 Methodology of Analysis

The students' answers/sentences to the tutorial questions were transcribed and analyzed by keywords and concepts included, according to our research questions. The categories were then defined and identified operatively, a posteriori, representing qualitatively different ways to conceptualize the situation considered (Niedderer 1989), according to the criteria of qualitative research (Denzin and Lincoln 2011; Erickson 1998), and defined operatively through the students' sentences. Students' drawings and formal constructions have been categorized according to the underlying conceptual models, here defined as mental construction about a piece of the physical world or mental representation of the processes producing the phenomenon observed (Nersessian 2007; Perkins and Grotzer 2000; Scott et al. 2007; Windschitl et al. 2008). In some cases, it was possible to specify the category in which to include a response or a graph from the verbal description made by individual students during

Table 9.1 General criteria for the analysis of motion graphs

Graph	Criteria			
x vs t	The position increases from an initial value to a final one	Slope of the graph	Presence of an acceleration phase and a deceleration one	Concavities envisaged in these phases
v vs t	Peak in correspondence of the push phase and gradual decrease for the free motion	Zero speed at the beginning and at the end	Different slopes between push phase and free motion phase	Time correspondence with x vs t graph
a vs t	Positive peak at push stage	Negative constant a in free motion phase		Time correspondence with x/v vs t graphs

the educational labs sessions. Considering how students used the formalism, some connection will be made to the epistemic games of Tuminaro and Redish (2007).

In the following, only line graphs will be considered, representing one variable as function of another one. In the case of motion kinematic graphs, the relative, but distinct, representations of x, v, vs t will be considered. Table 9.1 summarizes the general criteria of analysis for these graphs.

For all the graphs, the following elements were also analyzed:

(a) Presence of sharp points vs emphasis on continuous/smooth trend, to distinguish a mathematical abstract approach to a more physical approach where the changes of the physical quantities are continuous.

(b) Presence or absence of experimental noise and artifacts, in order to identify the emphasis on the iconic/pictorial representation of the observed graphs compared to the predicted ones.

(c) Attention to aspects of the first order describing general trend (i.e., increasing/decreasing graph) or to aspects of the second order, characterizing the specific graph (i.e., different slopes, concavities).

(d) Presence or absence of the units, as an indicator of awareness of what is represented, and of the scales, to distinguish qualitative vs quantitative predictions and comparisons. For v vs t and a vs t graphs, this indicator connects to the time correspondence of x vs t graph and these derived graphs.

In the case of motion data, each category of analysis will be documented with absolute frequency and percentage, even for small groups, to facilitate comparisons (sum 100% because of approximation).

The analysis of the representations of the diffraction light distribution has been made taking into account: the previewed global trend of the distribution (linear, decreasing by law of power, bell shape; presence of maxima and minima), the ratio between the intensity of the central and the lateral one, and any analytical report used to interpolate/fit the data in the case of derivate graphs. Mathematical features have been treated as subordinate characterizing aspects such as the presence of discontinuities.

In discussing the representations of the experimental graphs, we will refer to a relation of direct proportionality when a rectilinear graph passes to the origin and a linear relation when the intercept is other than the zero.

9.4   Analysis of Pretest of the Group LM

As anticipated, a pretest on the basic concepts of analytical geometry was proposed to the group of high school students LM. When asked to find the equation of the straight line for two points assigned [points (1,5; 2) and (−2; −1)], 16/18 (89%) students used an analytical approach, and 2/18 (11%) used a graphic approach. In the analytical approach, two strategies can be identified to determine the coefficients m and q of the equation:

$$y = mx + q. \tag{9.1}$$

In the first, used by ten students (56%), the m-value determined as slope ($\Delta y/\Delta x$) was inserted into Eq. (9.1), obtaining the q value by replacing the coordinates of one of the points. According to the second strategy, six students (33%), having inserted the coordinates of the given points in the expression (9.1), solved the equation system in m and q by substitution method. Only the first strategy corresponded to that used by students analyzing motion graphs. In the majority of cases (15/18 to 83%), both of these strategies led to the expected m-value and in 13 cases (72%) to get the equation of the searched line. Following the graphical approach, the two students drew the line in a Cartesian reference and estimated by eye the values of m and q (each 1), without checking the correctness of that estimation.

For half the sample, the meaning of m was the slope of the straight line; for one the angle was formed by the straight line with the x-axis. There is a significant correlation between determining the correct value of m and the attribution of its geometric meaning ($r = 0.62$, $p < 0.001$). The attribution of a meaning to q is less frequent; in fact only seven answered: associating q to the y value of the intercept (2/7) and to the intercept point itself (5/7).

When asked to determine the equation of the vertical axis parabola passing through three points assigned [points: (0; 0) (1; 2) (2; 1)], 17 students (94%) drew the parabola graph and 15 (83%) also set the equation system in a, b, and c by inserting the known point coordinates in the equation:

$$y = a\,x^2 + b\,x + c \tag{9.2}$$

The values obtained for the coefficients a and c correspond to the expected values in nine cases (50%) ($a = -3/2 = -1.5$, $c = 0$), while only six (33%) obtained the expected value of b ($b = 7/2 = 3.5$) and the expected equation of the parabola. The other three skipped the request.

Among those who did not get the expected results, the following errors occurred: reversal of a formula (5/18), shifting a term from a side to the other of an equation (2/18 – the two students performed the same error searching the equation of the straight line), and square power lift (1/18). All these errors seem technical but are just an evidence of a lack of understanding on the geometrical/algebraic role of the quantities managed.

The students were able to assign a geometric meaning more frequently to the coefficient a (14/18–78%: "concavity" or "direction" of the parabola) than to the other coefficients (7/18–39%: c, intersection with the y-axis; 4/18–22%: b, correlated to the vertex of the parabola). The greater propensity of students to identify the slope of the straight line rather than its intercept (Hattikudur et al. 2012) and the first coefficient of the parabola seem to be linked to the greater aptitude to attribute geometric meaning to coefficient characterizing a curve. There is a positive correlation between the attribution of meaning to the coefficients and the correctness of these coefficient values reported.

The typical algorithmic-computational often followed by students uncomfortably or worse with mistakes, usually attributed to a weakness in algebraic calculation, seems to be more related to the lack of attribution of meaning to the entities with which students operate and of the procedures used. In another perspective, the technical approach usually followed by student with poor results hides a structural deficiency (Pospiech 2018). This is also linked to the fact that nobody used the graphs of the parabola drawn to address/guide or simplify the algebraic resolution of the exercise, or to verify the results obtained, or to predict the expected results. Rather, the opposite is true that there was no review of the calculations even when there were obvious differences between the obtained values and the graphical representation. According to Knuth (2000) and Redish (2005), students seem to perceive the graph and its analytical/formal representation as separate entities, as in the Recursive Plug-and-Chug game more than in the Pictorial Analysis one of Tuminaro and Redish (2007).

9.5   The Role of Real-Time Graphs to Understand Motion

9.5.1   Walking in Front of a Sensor: Analysis and Discussion of Data

This section summarizes the data on how persons of our sample analyzed the kinematic graphs for the motion of a person walking away from a sensor. Figure 9.1 synthetizes the categories of representation of graph x vs t.

The first category (GA category) includes increasing curves with steps or irregularities, as an expectation on how the walking person steps could affect the graph. The second category (GB) includes math graphs with straight increasing line (GB1), eventually connected as broken curve with an initial (GB2) and/or

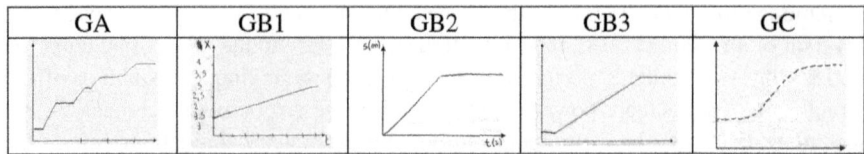

Fig. 9.1 Walking away from a sensor: Categories of graph x vs t

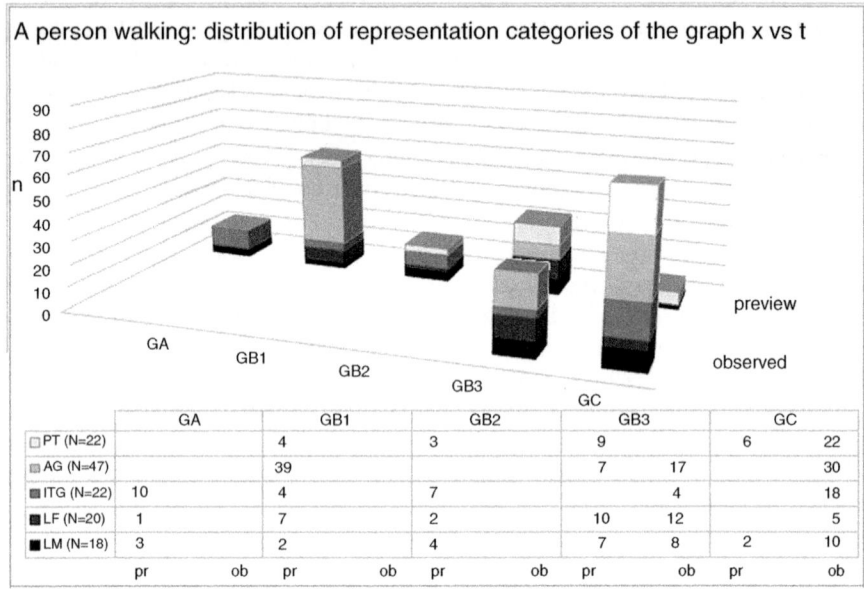

	GA		GB1		GB2		GB3		GC	
□ PT (N=22)			4		3		9		6	22
▨ AG (N=47)			39				7	17		30
▣ ITG (N=22)	10		4		7			4		18
■ LF (N=20)	1		7		2		10	12		5
■ LM (N=18)	3		2		4		7	8	2	10
	pr	ob	pr	ob	pr	ob	pr	ob	pr	ob

Fig. 9.2 Distribution of the graph x vs t categories for the person walking

a final stationary phase (GB3). All the GB graphs represent idealized uniform motions. The third category (GC) corresponds to the expected smooth graph with initial/final stationary phases and a constant/quasi-constant slope phase in between. It corresponds to a uniform/quasi-uniform motion with two accelerated phases, corresponding to smooth changes of speed.

The graphs designed as prevision were more often of the math category GB (see Fig. 9.2). The prevalence of this type of graph characterizes the lyceum secondary students, where the formal footprint given by the math teacher emerges (group LF). The frequency of the GC category is relevant, though not prevalent, for the university students (group AG) and for the prospective teachers (group PT), all of the bio-agricultural area, and characterizes them for the emphasis on the continuous variation of the variables. The GA typology characterizes students with low graphing competence both in math and physics (group IG). In the representations of the observed graphs, the GC category of representation prevailed, as expected, but there was a significant presence of GB3 math graphs (Fig. 9.2).

Some indicators, as scales and unit, are important to understand the conceptual process of students (Michelini et al. 2010). Unlike almost all of the preview graphs, the representation of the observed graphs was accompanied by units (in 62% of cases) and scales on the position/vertical axis (53%, with a peak of 70% for the LF group) and to a lesser extent on the time t axis (43%). Another important indicator of the processes activated was that the majority of persons (from 58% to 62% in each group) represented the observed and the predicted graphs with different slopes. Moreover, the different slopes were among the major differences quoted (over 55%), when comparing the preview graph with the experimental one. The comparison process activated reasoning based on the slopes of the curves, because it is recognized as the most characterizing geometrical (formal) parameters of these curves. This approach is quite similar to that of the Physical Mechanism game (Wenger 1999). This is consistent with what was previously observed about the greater students' propensity to identify the role of the angular coefficient of a straight line.

Other aspects emerged in the comparison are the following: presence of initial and final phases and/or experimental noise ("irregular," "not uniform" graph).

The reasoning based on the slopes of the curves activated by this first PEC and graph analysis sequence has led between 80% and 100% of the subjects of all the groups to make predictions on the graphs that were adequate for phases of the motions considered, for slopes and for concavities, when they were requested face the other tasks concerning the walking person, explored immediately after the first task.

A skill-building process was then activated in the use of the graphs and the correlation between the time evolution of the position of a body in motion and the type of body motion realized. The recognition of the iconic characteristics of the graph (e.g., sudden slope changes rather than continuous variations) has triggered the identification of critical instants and phases in which the observed phenomenon occurs and the recognition of processes underlying it (e.g., push at the beginning of the motion, stationarity linked to the uniformity of the walk). Almost the entire sample (between 86% and 100%) demonstrated basic level skills in reading graphs (such as reading initial and final positions and times of the motion) and intermediate level skills (such as the distance Δx and the corresponding time interval Δt). The majority of the persons of each group were able to evaluate derived quantities as the mean velocity v, which without any explicit indication was evaluated often as the ratio $\Delta x/\Delta t$ of the values of Δx and Δt determined, a resonating procedure with that used by LM students to obtain the angular coefficient of a straight line. Alongside this, the use of units was more frequent indicating the value of the speed (in 78% of cases) than other quantities, probably because it is considered the more important quantity to communicate. Finally, a significant impact was produced also in overcoming well-known difficulties (McDermott et al. 2000; Vygotsky 1962; Wainer 1992) regarding the distinction between trajectory and time graph (in 70% of cases) and the distinction between and evaluation of position, displacement, and velocity (between 70% and 80%).

Students analyzed the motion using the fit tools available in the software of acquisition. Almost the entire sample (between 92% and 100%) was able to use of the linear fit of the type $x = A\,t + B$ to determine an appropriate expression of time law for the observed motion. When asked to indicate the physical meaning of the first coefficient A among respondents (78% of total), 50% identified coefficient A with the velocity of the person and another 28% reported the geometric meaning (slope) of this coefficient, particularly in LM group with which the geometric aspect was emphasized in the test. The competence to attribute a meaning to the interpolation coefficients is a further gain generated by the conceptual process triggered by interaction with the RTL graphs: half of the sample was able to distinguish the three planes simultaneously involved (algebraic/analytic, geometric, physical). For the others, this process, while activated, remained partially open, as an unclosed Mapping Meanings to Mathematic game (Tuminaro and Redish 2007).

9.5.2 Motion of the Toy Car: x vs t Graph

This section regards the analysis of the tutorial filled by students exploring the free motion of the toy car launched on a horizontal plane and moving far away from the sensor (step EB of the motion path).

Figure 9.3 shows the categories of the preview graphs representing the time evolution of the car position (x vs t graph), from which emerges the predominance (88%) of the first three categories, including the most obvious or first-order features of the motion observed, that is, initial/final stationary phases and a gradual increase of the position in between.

This is a result of the transfer process activated by the analysis of the RTL graphs related to the walking movement (EA experiment), as specified in the following. The graphs categorized as GPA show a smooth behavior, two stationary portions connected by an oblique straight line, with similar concavity even if with opposite sign. The GPA graphs are a transposition for the car motion of the GC graphs for the walking motion (Fig. 9.1), underlying the conjecture that the characteristic of the motion of the toy car could be very similar to that of the walking person. The preview graphs of the mathematic GPB category include a linear variation of the position of the toy car, with discontinuity in the velocity at start and stop, and have been drawn as a transfer of the GB graph observed for the walk (Fig. 9.1). Graphs of the GPC category represent the expected behavior, with a sudden change

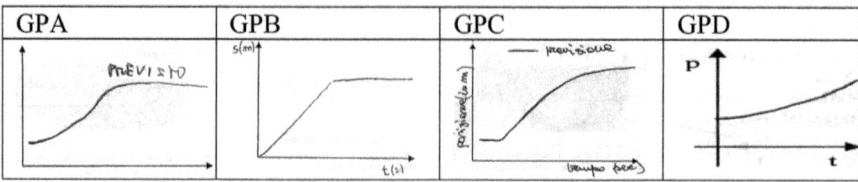

Fig. 9.3 Motion of a toy car moving on a horizontal plane: categories of x vs t graphs

and then a gradual decreasing of the slope, corresponding, respectively, to the sudden acceleration at the push stage and to the uniform accelerated free motion phase. A preview graph GPC highlights a transfer process of graph types GC for the walking case and a personal re-elaboration to adapt for the car motion. The GPD category, which accounts for 13% of the whole sample, included four different subtypes: curve growing, irregular (seven cases of IG) or concave (three cases of lyceum), a decreasing curve with regular variations, symmetrical with respect to the graph GPA (four cases of LF and PT groups), and first increasing and then decreasing curve (three cases of IG). Different causes underlying the frequent preview of a uniform motion, implemented at different formal level in the GPA and GPB graphs, are as follows: emphasis and school habits in considering only/prevalently linear relations; the students' tendency to use these relations to describe any growing monotone relationship (Pospiech 2015); and the tendency, even historically documented, to identify an accelerated motion with a motion at the same speed as the end (Trowbridge and McDermott 1980).

The presence of all categories characterizes lyceum students, predominating mathematical graphs for the lyceum students of the LF group, which still reveals the role of the formal approach of the math schoolteacher. The prevalence of the GPA category and the absence of the GPB and GPD categories characterize university students and prospective teachers of agro-bio area (AG and PT groups). The presence of GPD graphs characterizes the IG group, without previous graphing formation.

As expected, the representations of the observed graphs included only two categories, the GPA and the GPC, as summarized in Fig. 9.4. Typically, in

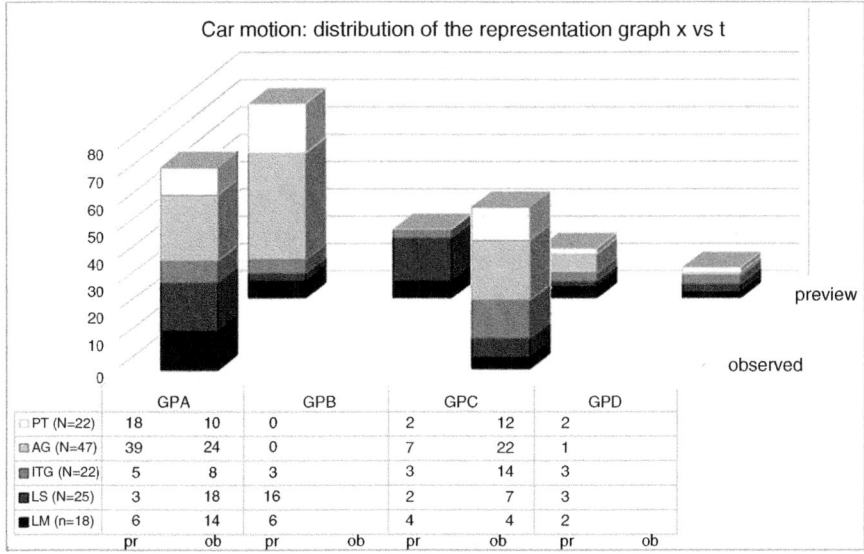

	GPA		GPB		GPC		GPD	
□ PT (N=22)	18	10	0		2	12	2	
▨ AG (N=47)	39	24	0		7	22	1	
▣ ITG (N=22)	5	8	3		3	14	3	
■ LS (N=25)	3	18	16		2	7	3	
■ LM (n=18)	6	14	6		4	4	2	
	pr	ob	pr	ob	pr	ob	pr	ob

Fig. 9.4 Distribution of the x vs t categories for the motion of the toy car on the table

the representation of the observed graphs, experimental noise was emphasized. Moreover, units and scales on axes were more frequently reported (63% in total, with peaks over 80% for the two LF and PT groups) than in the preview graphs (38%). Twenty-four percent represented the observed graph with slopes other than the preview one, without indicating units and scale.

Unexpectedly, the GPA category is prevalent in the representation of the observed graphs, as in the preview ones. In the 74 individuals (55%) representing the observed graph as a GPA type, 43 adopted an analogous representation in the preview, 17 gave a GPB-mathematical representation, 10 used a GPD graph (growing curve), and finally 4 used a GPC representation. Students' explanations can be grouped into three types: emphasis on "equality/analogy/similarity" of the preview/observed graphs, especially indicated by those who adopted in the two cases a GPA graph; failure to predict a phase of initial stasis, given by those who drew a GPB graph as prevision; and wrong preview ("dissimilarity in the graphs"), given by those who predicted a GPD graph ("the system does not go back" or "does not stop immediately").

There is an increase in the percentage of students who used the expected GPC representation (from 14% to 44%). Between the 59 subjects adopting a GPC representation: 14 (6 lyceum students, 6 of AG, and 2 PT) drew similar preview graphs, 28 (16 of AG and 10 PT) adopted a GPA type, 8 (of which 5 of LS and 3 IG) used GPB representation, and 9 (7 of IG and 2 of lyceum) adopted a GPD preview representation.

Discussing similarities and differences between the preview graph and the observed one, the reasons given to explain changes in 1/3 of the cases concerned mathematical aspects not adequately represented (concavities, different gradients). The remaining 2/3 of the cases students stressed the failure to foresee an aspect of the physical phenomenon (type of acceleration, not uniform motion). Only in two cases (PT), the mathematical and physical aspects (concavity and acceleration) were explicitly related.

Three different conceptual paths can be identified:

(A) Students adopting the GPC representation framed the salient aspects of the experimental graph in a conceptual framework, even when there was emphasis on iconic reproduction of the experimental irregularities, as in the case of most IG students' representations. In fact, it is not necessarily true that the reproduction of the observed graph includes all salient aspects of the graph, as in the case of students representing a GPC preview graph and a GPA observed graph. From the observations that the students themselves did in comparing graphs, the correlation of the features of the observed graph with the motion performed seems to be the decisive step to activate the transition from a descriptive-qualitative representation to an interpretative-quantitative one.

(B) The GPA typology was an intermediate step between student preview and the more appropriate GPC representation, for students skipping the prediction or making a prediction too far (e.g., GPD type) from what was then observed.

(C) Students adopting a GPA category both for preview and observed graph activated a confirmation model (Champagne and Gunstone 1985; Park et al. 2001; Shah 2002), based on the uniform motion, where only the first-order features found place (the initial/final stasis phase, the positive slope). The second-order differences, characterizing the specific motion (different concavities/different acceleration and slowdown times), did not find meaning in this simplified model and so did not activate revision.

Table 9.2 confirms the widespread competence in graph reading and evaluating derived quantities for over 93%, with the exception for the students of the LF group not answering or evaluating Δt as the difference between the activating/stopping times of data acquisition, as already noted in other contexts (Stefanel et al. 2002). To estimate the mean speed of the toy car, the majority of students used the expression $v_m = \Delta x/\Delta t$ and only few students (two of LS and six of IG) used the expression $v_m = x/t$, determining the lowest frequencies shown in the Table 9.2.

As anticipated, the delicate shift from graphic representation to quadratic data interpolation was addressed in different ways with the different groups. With the IG and PT groups, it was started asking if the speed of the car was always the same. All the answers to that question were negative, with motivations as the followings: "No, first it accelerates and then slows down", "Friction makes it slow down".

Discarding the possibility to find an adequate linear interpolation, four prospective teachers of the PT group suggested that the car would move at constant

Table 9.2 Reading graph and evaluation of derived quantities: initial X_i, final X_f, positions, initial t_i, final t_f times; displacement Δx and the related time interval Δt; mean velocity

	LM N = 18	LF N = 25	IG N = 22	AG N = 47	PT N = 22	TOT N = 134
Reading variables (X_i, X_f, t_i, t_f)	18 100%	19 76%	21 95%	46 98%	21 95%	125 93%
Evaluating variations (Δx, Δt)	18 100%	19 76%	19 86%	46 98%	20 91%	122 91%
Mean speed of the toy car	18 100%	14 60%	18 82%	46 98%	20 91%	116 87%
Units	14 78%	8 32%	8 36%	44 94%	20 91%	94 70%

acceleration, adding in two cases that the expected equation should be of the type:

$$x = \frac{1}{2} a\, t^2 + v_o t + x_o. \tag{9.3}$$

This hypothesis once shared has become the patrimony of the entire PT group.

Since the recognition of the uniform acceleration was the key for hypothesizing a quadratic relationship, an interpretative approach was proposed to the other students' groups. After reading the graph, it was required to explain why the toy car stops and if the force that determines the stop is constant, and then to individuate the type of motion. Almost all the students stated that the car stops "for friction," "for the friction force" over 2/3 of LM students, and almost half of AG added that this force was constant. Only few students concluded that the expected law had to be of form Eq. (9.3), but the other students gained immediately the same expectation. Again, what may be called a domino effect has been observed: when only a single student makes explicit the conceptual path to solve a problem, an interpretative challenge upon which an entire class had previously been interrogated, this result becomes a patrimony of all (Michelini 2006). This approach activated a mapping meanings game (Tuminaro and Redish 2007).

Students, after the interpolation of data with a curve of the type

$$x = A\, t^2 + B\, t + C, \tag{9.4}$$

correctly reported the value of the first coefficient of interpolation (88%), with slightly lower percentages for LS and IG groups (Table 9.3). The almost total

Table 9.3 Frequencies with which the value and the physical meaning of the first coefficient A of the interpolation (9.4) were indicated (for the different groups, not including students elapsed the section)

		LM $N = 18$	LF $N = 25$	IG $N = 22$	AG $N = 47$	PT $N = 22$	TOT $N = 134$
Coefficient A	Value	16 89%	21 84%	17 77%	46 98%	20 91%	118 88%
	Units			4 18%	1 2%		5 4%
Physical meaning	Concavity	11 61%	5 20%			13 59%	24 18%
	$A = 2a$ $A = 2a$	2 11%					2 2
	$A = a/2$	2 11%	6 24%	3 14%	33 70%	12 55%	50 27%
	$A = a$	1 6%	12 48%		10 21%	6 27%	33 24%

absence of units is a sign of a deficit in the attribution of physical meaning to the interpolation coefficients.

When asked to indicate this physics meaning (see the lower half of Table 9.3), the upper secondary students more often indicated the geometric meaning (concavity/width), as mostly LM students were probably affected by the initial test. Less frequently, the physical meaning was indicated (associated with acceleration, as in LS students). The majority of university students gave the proper physical interpretation of coefficient A. The explicit correlation between physical meaning of the coefficient A and its geometric meaning emerged only in the future teachers' PT group. The evident correlation to the formative level makes this point particularly relevant to define a learning progression.

9.5.3 Motion of the Toy Car: Representation and Analysis of the Graph v vs t

Figure 9.5 exemplifies the different typologies of the v vs t graphs grouped in terms of the following seven categories:

GV1 – graph providing the main elements of the expected one: a sudden step at increasing speed at the thrust, a successive phase with a lower gradient, during the free motion up to the arrest

GV2 – triangular graph, with similar slopes, without correlation with the graph x vs t

GV3 – bell-shaped graph with the maximum correlated to the push phase of the graph x vs t

GV4 – bell-shaped graph with the maximum corresponding to the final stage of stasis in the graph x vs t (often emphasizing different slopes)

GV5 – trapezoidal graph, with a constant velocity phase at the end of the push phase, with temporal correlation with the graph x vs t

GV6 – linear graph, for the pushing phase only

GV7 – downward graph at the free motion stage

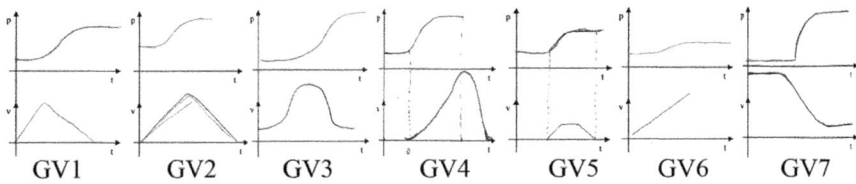

Fig. 9.5 Representation categories of the v vs t graphs for the motion of the toy machine (below in each figure), reported with the corresponding graphs x vs t (above) to appreciate how and if the temporal correlation between graphs was implemented

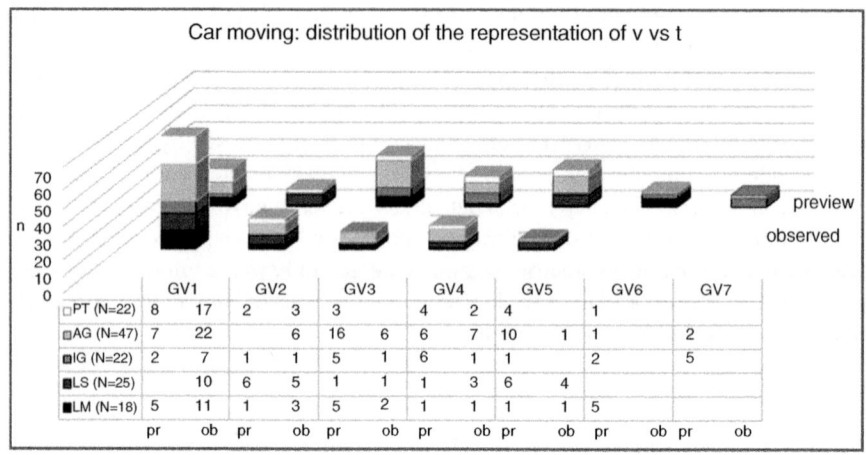

The following data appears within the figure:

	GV1		GV2		GV3		GV4		GV5		GV6		GV7	
□PT (N=22)	8	17	2	3	3		4	2	4		1			
▣AG (N=47)	7	22		6	16	6	6	7	10	1	1		2	
▪IG (N=22)	2	7	1	1	5	1	6	1	1		2		5	
▪LS (N=25)		10	6	5	1	1	1	3	6	4				
▪LM (N=18)	5	11	1	3	5	2	1	1	1	1	5			
	pr	ob	pr	ob	pr	ob	pr	ob	pr	ob	pr	ob	pr	ob

Car moving: distribution of the representation of v vs t — preview, observed

Fig. 9.6 Distribution of the representations of the v vs t graph, before (preview) and after (observation) observing the graph on the screen of the computer

Figure 9.5 GV1–GV3 graphs cover 56% of the total, as shown in Fig. 9.6. The majority of the representations (60%) were of GV1, 3, 5, and 7 categories sharing time correlation with the graph x vs t, promoted effectively by the format of the graphs one over the other as in the examples of Fig. 9.5. A similar percentage represented the different slopes for the push phase and for the free movement. These aspects were often correlated ($r = 0.33$, $p < 0.05$), evidencing a great gain in constructing graphs of the time evolution of physical quantities related with an observed phenomenon.

Three distributions characterized the groups of high school students. In the LM group, three types of graphs prevailed: GV1 and GV3, in which the time correspondence with the graph x vs t (10/18) was emphasized and GV6, where only the push phase is represented, as explained by the students itself (5/18). The LF group is characterized by the mathematical typologies GV2 and GV5 (12/25) and limited attention to time correlation with the graph x vs t. For the IG group, the three qualitative types GV3, GV4, and GV7 (16/22) prevailed. A wide dispersion and different modes (GV3 and GV1) characterize AG and PT representations.

In the representations of the observed graphs, both these aspects increased about 10%, emphasized in the GV1 category prevalent for all groups. Often, the representation of the observed graph modified or completed that of the predicted one, with better time correlation, care in differentiating the slopes, and presence of experimental noise ("the curve is broken, less homogenous"). Discussing analogy and differences between previews and observed v vs t graph, the individuals of the sample evidenced more frequently first order similarity/equality as the overall trend and less frequently second order but more specific differences of the graph (absence of a part, different slopes) or the type of motion (uniform accelerations).

Among the 16 students (11 of LS and 5 of AG) who eluded the representation of the preview graph, only one adequately reproduced the observed graph, evidencing the relevance of the preview phase to activate the skill to represent the iconic features of a graph related to the salient aspects of a phenomenon, as emerged from some student interviews. This capacity is not automatically activated observing the experimental graph in itself, but it requires an interpretative substrate that can be prepared by the preview phase.

The determination of the linear interpolation of the v vs t graph for the motion phase was proposed only to the groups LM and PT. All persons of both groups reported an appropriate value for the first coefficient, although without unit. When asked to specify the physical meaning of this coefficient, in the LM group, 3/18 (17%) identified such coefficient with the acceleration and other 10/18 (56%) indicated its geometric meaning (angular coefficient/slope of x vs t graph or concavity of the x vs t graph). In the PT group the corresponding fractions were 8/22 (36%) and 12/22 (77%), of which 9 added also the geometric-mathematical meaning. The interpolation analysis favored the PT's understanding of its physical meaning. For the majority of LM students, this recognition was triggered suggesting to analyze the characteristics of the motion, changing the initial speed. The recognition of the independency of acceleration by the initial conditions brought 12/18 (67%) to evaluate the car acceleration as average on repeated proofs, obtaining a value in the expected range and including units and an adequate uncertainty (7/18–39%).

9.5.4 Motion of the Toy Car: a vs t Graph

The representation of the graph a vs t was proposed to all groups except LF. Figure 9.7 exemplifies the six categories of representation: GA1, (expected graph) a positive peak, corresponding to the thrust phase, and a negative constant part, associated with or almost uniform acceleration for the free motion; GA2, a line at a constant negative value, representing only the free motion phase; GA3, (math

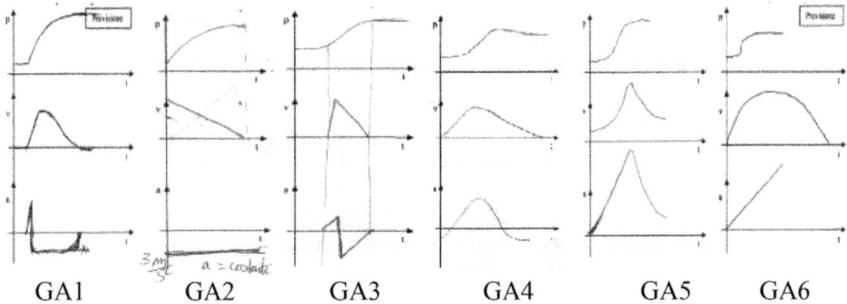

GA1   GA2   GA3   GA4   GA5   GA6

Fig. 9.7 Graph a vs t and corresponding graphs x vs t and v vs t for the toy car motion

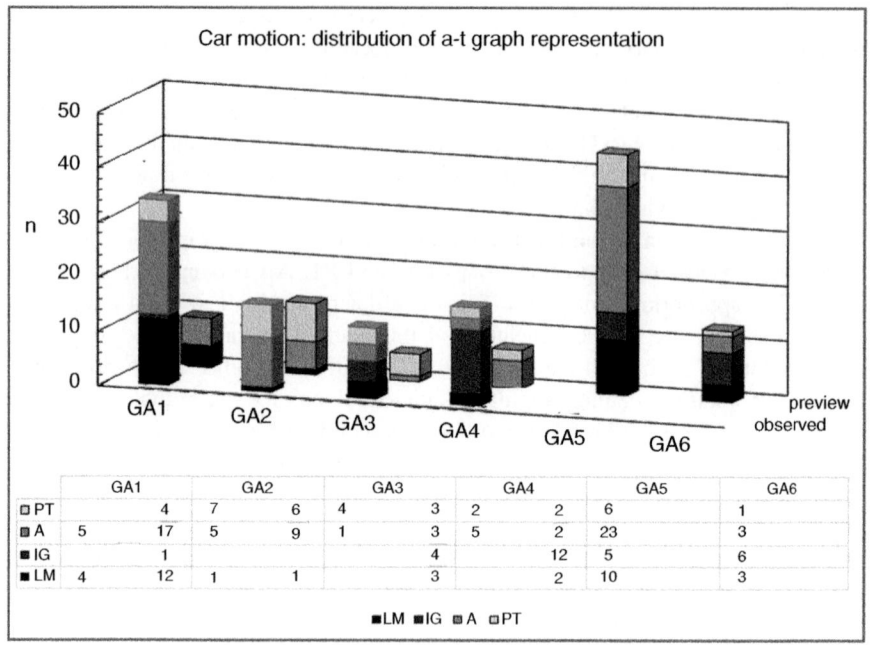

Car motion: distribution of a-t graph representation

	GA1		GA2		GA3		GA4		GA5		GA6	
□ PT			4	7	6	4	3	2	2	6	1	
▨ A	5		17	5	9	1	3	5	2	23	3	
■ IG			1				4		12	5	6	
■ LM	4		12	1	1		3		2	10	3	

■ LM ■ IG ▨ A □ PT

Fig. 9.8 Frequencies of the representation categories for the a-t graph

type) broken line, with a positive peak and a negative one corresponding to the two phases; GA4, smooth graph with a positive peak, approximatively associated to the push phase and a successive negative part, not always ending with zero value, only partially connected with free motion phase; GA5, positive asymmetrical bell-shaped graph, often with zero accelerations at the beginning and at the end, with the peak corresponding to the peak of the velocity graph; and GA6, linear growing graph.

In the previewed graphs (see Fig. 9.8), though with some differentiation, the most represented typology was the GA5, which underlies the idea that velocity and acceleration evolve alike. In most cases, the preview graphs were temporally correlated to the graphs x vs t and v vs t (i.e., the discontinuity in the trends matched on the timing axis; the peak of the a vs t graph coincided with the maximum change of the graph v vs t), in most cases (from almost all 17/18 for LM, to 19/22 for PT, to 32/47 for AG), with the exception of IG (5/22).

In the representations of the observed graphs, all typologies of representation included a phase with negative acceleration and tended to emphasize temporal correlation with the graphs x vs t and v vs t, with significant increments compared to what observed in sect. 5.4 for the groups LM and IG. A correlation between the ways to represent the observed graph v vs t and the typology of the previewed graph a vs t, very significant for LM ($r = 0.63$, $p < 0{,}01$) and in any case significant for IG ($r = 0.35$, $p < 0{,}05$), was observed. These are significant indicators of the impact of the RTL also on the comprehension of the graph a vs t, usually more problematic (Svec 1995), with differences related to the competences of the involved groups.

9.6   Student Analysis of the Light Intensity of a Diffraction Pattern

This section regards how students represented graphs of the intensity distribution of the light diffracted by a single slit and analyzed the experimental graph acquired with on-line sensors (Gervasio and Micheliani 2009; Michelini et al. 2010). At the end of a qualitative exploration of the light diffraction pattern, students were requested to represent the preview graphs they expected to detect before a quantitative measurement. Table 9.4 exemplifies the categories of representation of the light distribution previewed. The DB1 and DB2 types have the same root DB sharing the alternation of maximum and minimum and an intensity of the central peak similar to that of the secondary peaks, being quite different concerning continuity. Another discontinuity, not differentiated here, regards few representations of minima as cusps.

All the categories DA and DB graphs correspond to a point/bar drawing of the diffraction pattern and a description of graphs highlighting the decreasing of intensity from the center of the figure and in half of cases also the presence of minima/maxima. The DB2 discontinuous graphs represent the light distribution pattern represented by a set of isolated points. This means that all the DA and DB representations are quite coherent with the model of the phenomenon shown to the students.

In the case of the representations of DC, DD, and DE categories, a differentiation must be done: half of the students adopted coherently these representations, having drawn the diffraction figure as a continuous image evidencing that "intensity progressively decreases" toward the ends and the other half represented adequately the diffraction pattern as a sequence of points or segments but incoherently describing that pattern as "rarefaction of light" and representing graphically the light intensity distribution as in DC, DD, and DE pictures.

Figure 9.9 shows the distribution of these categories, evidencing a shift from the prevailing preview categories DB and DC, to the predominant observed category DA. Categories DB and DC appear also as categories of the observed graph in few cases (12% in total), as result of a single-element revision and an accommodation of incoherencies evidenced before. Typically two paths emerged: from the DC or DD typologies, where only the central peak was expected, to the DB typology representing minima and from the type DD in which the central peak was a cusp or divergent to the DC type representing a smooth curve, though not the secondary peaks. Moreover, the observed graphs were smooth lines. The category DB2 disappeared as well as the other kind of discontinuity as cusps or sharp points (Fig. 9.9).

Compared to preview graphs, there was usually (but not always) more precise reproduction of relevant details, such as the ratio of central peak width and intensity

Table 9.4 Categories of drawing of the light intensity distribution related to a diffraction pattern

Cat	Example	Description
DA	Grafico (intensità vs posizione) 	Expected graph, characterized by central peak almost three times higher than the other maxima
DB1	Grafico (intensità vs posizione) 	Graph characterized by peaks of equal/almost equal value (the intensity of the central maximum is equal or just greater than the intensity of others)
DB2		Discontinuous graph made by isolated point or vertical line corresponding to the maxima, with analogous trend of DB1
DC	Grafico (intensità vs posizione) 	Bell-shaped graph, characterized by a single peak in any case extended to cover also side peaks. Within this category also graphs where the intensity of light suddenly becomes 0 beyond a certain distance from the center of the peak
DD	Grafico (intensità VS posizione) 	Sharp/cusp/asymptote graph, where the graph is always concave (reverse power type) and the central maximum is a cusp Within this category also graphs with the only right branch
DE	Grafico (intensità VS posizione) 	Broken/triangular curve, with a sharp point corresponding to the maximum. Within this category also graphs including only the right branch

vs the first-order peaks, as well as the downward trend of secondary peaks intensity and minima/maxima equidistance. The drawing in Fig. 9.10 is an example of the better formal representation of the 2/3 of the observed graphs, compared to those of preview.

The review process triggered an approach to the physical meaning of the graph features, with greater connection to the observed physical phenomenon and better understanding of the behavior of the represented quantities, as well as better, formal, and quantitative representation.

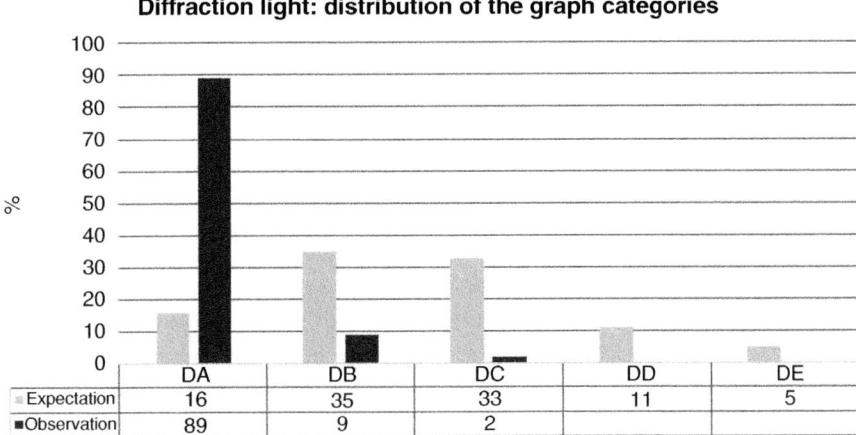

Fig. 9.9   Distribution of the categories (see Table 9.4) of the light diffraction *I* vs *x* representation

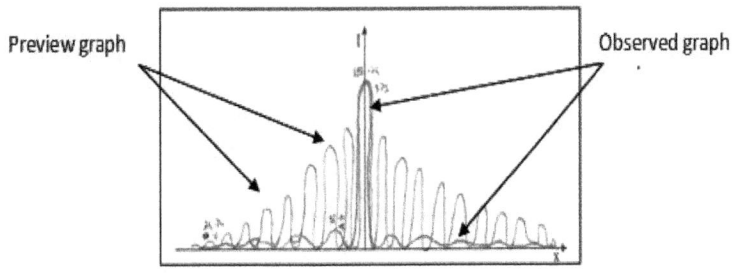

Fig. 9.10   Representation of the observed graph, performed according to cat. DA, compared with the DB graph drawn as prevision by the same student during the preview phase

Comparing the graphs, almost all the students indicated in which way their predicted graph differed from the observed one (3/5 the different relative intensity, rather than 2/5 the presence of the minima), always emphasizing also the expected aspects ("I did the graph is right, except the height of the lateral peaks intensity"; "In the initial graph there is no maxima and minima, but it was symmetrical").

At the request of designing the data analysis of distribution *I* vs *x* acquired in the experimental lab, the students proposed to read and collect the values of the following quantities: position and intensity in 2/3 of the cases of the maxima, in 1/3 of the minima (75% in total); only "maxima and minima position" (10%); only "intensity" (7%); parameters of the central peak, such as width, intensity (5%). Moreover, 1/3 of the students also stated among which magnitudes they were seeking to find some relationship (being *I* vs *x* the most frequently cited) without any preview of the formal form expected. In each class group, all meaningful relationships have been proposed, making it possible to share individual projects to realize a common analysis of the distribution I vs x, overcoming the limits shown by these less-accustomed students to experimental work.

After collecting light intensity distribution with sensors in the lab, students could analyze this distribution based on a shared project of analysis. All the graphs of the derived quantities showed the variables represented (in this case rightly without units because constructed as non-dimensional quantities). Ninety percent of the sample obtained the expected linear relationship between the angular position of the minima and the order number M; 75% identified the correct relationship between order number M and angular position of the maximum, differentiating the cases by $M > 0$ and those at $M < 0$; while 12% traced a single line, forcing the fit to pass the origin. In all cases, these analyses required expertise both in reading the graph's punctual values and in constructing a graph from the values obtained. In the face of this significant outcome, only a small fraction (15%) correlated the slope of the straight line obtained with the λ/a ratio, as, for instance, students studied in the case of Young experiment. That means that when the expected relation was linear, almost all of the students were able to determine an appropriate interpolation of the data, even in the absence of a physical model as reference.

The finding of the expected relationship between position and intensity of the maxima was obtained by 60% of the students, following either a trial procedure by first graphing I_M vs $1/X_M$ and then I vs $1/X_M^2$ or directly graphing I_M vs $1/X_M^2$, based on the hypothesis of a trend similar to that of the light intensity dependence on distances from a point source. Forty percent of the sample showed difficulties in hypothesizing other relations than $1/X$. To go beyond these minimal tools, it requires to develop alternative models that can be triggered by analogical reasoning, as the $1/X^2$ hypothesis.

9.7 Outcomes

The following discussion synthesizes the main outcomes of the two studies presented.

The way to represent a RTL graph is influenced by the different skills and characterizes the different approaches to the content of groups involved, especially in the previewing phase. Students more accustomed to a formal approach (whether by setting of their teachers or by type of school) tend to draw mathematical graphs representing ideal phenomena (see also (Corni et al. 2005)). Students and teachers of the bio-agricultural area tend to emphasize the continuity in variation of the variables, typical of the way of looking at the phenomena of this area of studies. Representations catching qualitatively some mathematical aspects associated with physical meaning are typical of students with weak formal skills.

Despite these differences, most of the representations of the graphs observed shared almost the following four common characteristics for all groups:

- Reproducing the peculiar aspects of experimental graphs. The representation of motion graphs in most cases (till 80% to 90% for most known phenomena) reproduced the peculiarities of the time evolution of the kinematic quantities. The reproductions of the light distribution of the diffraction patterns are continuous and characterizing adequately the central peak with respect to the others and the decreasing of the secondary maxima. On the contrary, some students (10–15%) observing a graph do not activate in itself the capability to reproduce the main features of a graph.

- Indication of the units and scales on the axes and time correlation. In the majority of the drawing representing the observed graph on motion, axes included units and scales, with the prevalence of the scale on the y-axis, which is more significant for understanding the observed phenomenon (Michelini 2006). The representations of the graphs x, v, a vs t, one over the other (Figs. 9.5 and 9.7), are in 60% of cases temporally correlated, aligning the discontinuities in the trends, the peaks in a graph (e.g., a vs t), and the main variation in the other (e.g., v vs t).
- Focus on slope of graph. Most of the drawings representing the observed graphs emphasized the agreement or disagreement between the slopes of predicted and observed graphs. Compared to the preview, the expected slope is one of the main differences evidenced.
- Emphasis on experimental noise. Most of the representations of the observed graphs emphasize experimental noise ("the curve is broken," "irregular," "uneven"), to distinguish it from frequently expected regular mathematics (especially in the first preview and in the simplest cases).

The combination of all these elements points out that the representation of an observed RTL graph underlies the activation of a mental model, rather than being a trivial iconic reproduction of a line observed on the screen; also when there is emphasis on the experimental noise or experimental features, that could suggest a reproductive doing. In the absence of an appropriate conceptual model of the phenomenon observed, it is very difficult to reproduce the significant features of an experimental graph (Chinn and Brewer 2001).

The models activated by RTL showed a high level of formalization based on the connection of graphical features (e.g., slope) and physical quantities (e.g., velocity). At the same time, it emerged that many students used only linear functions, and they were not able to master continuity vs discontinuity.

Students can overcome these limits by reading from the graphs the values of critical points that are significant for individuating phases of motion, constructing new variables from these values to obtain variables characterizing the phenomenon and to perform analysis based on interpolation or better fitting these models through their algebraic/analytical forms. Simple elaborations such as determining velocity from the x vs t graph activate the use of mathematical constructs, common and resonant for some students of our samples with those they used in math. For example, the expression of the angular coefficient of a straight line $m = \Delta y / \Delta x$ found transposition, use, and meaning in physics in the expression $v = \Delta x / \Delta t$, with which students evaluated the speed from a graph. The attribution of physical meaning to the coefficient of an equation fitting (based on theory) or interpolating (based on mathematics) in RTL graphs plays a very important role. A significant correlation was observed between the attribution of meaning to the coefficients of the interpolation and the determination of the values of such coefficients, both constructing relations by paper and pen and reporting equations through software instruments. The computational difficulties of many students or their tendency to perform inconclusive or worse algorithmic-computational approaches seem to be more related to the lack of attribution of meaning to the entities with which they operate than to a deficit in algebraic calculation itself. Additionally, students favor the algebraic calculus-based resolution of the exercises proposed in math courses by not activating an effective connection to the graphic representation. Rather, the opposite evidence occurred: obvious differences between results obtained and drawn graphs did not trigger always revision. The strategy adopted in dealing with RTL graphs flipped this way by favoring the construction of the meaning of the graph, to which giving analytical expression starting from the analysis of the graph itself. This process develops and activates the tendency of students of our samples to recognize more frequently the meaning of the coefficient characterizing an algebraic curve (the angular coefficient of a straight line (Hattikudur et al. 2012), the first coefficient of a quadratic relation) than the meaning of the other coefficients, as in the recalled correlations speed/slope and concavity/acceleration in the case of kinematic graphs.

The ownership of the three aspects simultaneously involved in a RTL graph (algebraic/analytic, geometric, physical), known as one of the more difficult to reach (Redish 2005), differentiates significantly the diverse groups involved in the study on motion. The activation and distinction of the three aspects emerged in some prospective teachers of the PT group also after the first PEC sequence. In the other prospective teachers and in all students (both from high school and university), the first PEC cycle in the study on motion activated an explicit connection between algebraic/analytical structures and their geometrical meaning, being the physics in the background explicitly evidenced by the increasing of the units on axes and connection between related graphs. Only few physical models were hypothesized by students in algebraic/analytical forms (see, for instance, the hypothesis of an inverse quadratic relation for light intensity in the case of diffraction).

This process was very important, because the formal structures and the algorithmic procedures are often dealt with by students only as mathematical entities, without taking into account, or neglecting altogether, their physical meaning or origin (Pospiech 2015).

As discussed in the beginning, the observation of a RTL graph enabled models. These can preexist the experimental exploration or more often are adaptation or modification of an existing model. The changes are very effective, when a person faces for the first time a new phenomenon, such as diffraction was for most of our sample, or a new way to look at a daily phenomenon, as the analysis of the walk with RTL has been for many persons. After this initial phase that often caused a strong revision process (sometimes with overload attention to experimental noise or features), an adaptation process occurs, when analyzing new situations. As seen in the case of the sequence of the walking person in front of a sensor, after the first stage, in fact, more often the reproduction of graph observed in the successive stages was a change/completion of the graph drawn in the prevision phase. The representations of the observed graphs underlined the adaptation of a preexisting model, in which a modified aspect present in the reproduction of the graph was partially framed in the model of a person. The new model became a property of the subject when he activates a process to make explicit that model (Karmiloff-Smith 1988), without which a preexisting pattern was confirmed even with obvious differences between predicted/observed graphs.

Considering important or predominantly only first-order elements (i.e., the presence of an initial and a final rest stage, a moving away phase, a positive/negative slope of the curve), the second-order differences (slope variations vs constant slope, smooth vs suddenly concavity, suddenly acceleration vs constant deceleration) became ineffective in activating an immediate revision process. This in part explains, why, in the student comparison comments, the similarities or equalities related to the global trend of the graphs predominate, rather than the differences, almost always local but characterizing the specific type of motion.

An identified dissimilarity must find place in a new model to enhance progress in learning (Chinn and Brewer 1993). The construction of a new model can be effectively developed correlating the observed graph with the motion taken, activating the reading and analysis of the graph by phases delimited by critical instants (when changing the external conditions such as at start/end of the motion). At the same time, it has been seen that also the first PEC sequence related to the experiment of a person walking in front of a sensor and the subsequent analysis of the graph impacted on the capacity to predict appropriate x vs t graphs, in the subsequent explored situations, as well as to predict significant aspects of the v vs t and a vs t. Important skills in reading and building L1-elementary, L2-intermediate, and L3-advanced graphing levels have been developed by almost all of the sample in parallel to competencies in correlating a specific motion to the corresponding graph x vs t. In the study of diffraction, analogous roles played the reflection on quantities to be detected, on relation between variables to be studied, and the sharing of hypotheses to construct a common analysis, negotiated by all the students of a class.

The outcomes of the researches here presented indicated that active learning strategies based on RTL, the review and comparison process, and reading graph and construct relation between quantities can contribute to activate or to develop models that help the graph interpretation for a conceptual construction, both for high-level students and for students with lesser basic skills. The previous competences affected the level of formalization of the models activated and the role in the interpretation of the graph. The preview phase plays an important role in activating an interpretative framework or almost a substrate on which the features of a graph can find meaning in the students' models.

Both researches showed that some students put in the field only direct or (less frequently) inverse proportionality as mathematical instruments. Both high school and university students performed the analysis of a graph in particular facing a new phenomenon and in some cases to describe generic growing or decreasing monotone relationships. The construction of these elementary mathematical instruments in the first cycle school seems to activate a strong imprinting that affects the capability of many students to develop a more complete and complex mathematical baggage. The analysis of RTL graphs can help in this perspective, activating an interpretative framework and the connection between formal constructs and their physical meaning.

9.8 Conclusion

In this paper, two studies have been discussed on how the use of graphs acquired in real time with on-line sensors can activate learning or how they develop graphing competencies that intertwine with the understanding of the concepts involved. The use of on-line sensor systems enabled the implementation of IBL-based learning approaches based on PEC strategies. The first research covered the learning processes of 60 students aged 15–16 on motion and compared with those of 47 university students of the first year of agricultural degrees and a group of 22 prospective middle school teachers. The second case involved a sample of 168 high school students aged 16–17, engaged in single-slit diffraction analysis. The analysis of tutorials has allowed to document students' learning processes activated by RTL graphs.

With regard to the first research question (RQ1), it emerges that RTL graphs produce the imaginative transduction of phenomena in graphs, revealing the physical process at the base of phenomena and opening bridges toward both its interpretation and its description in terms of algebraic/analytical expressions.

This process activates effective learning under the condition that the construction of a conceptual framework in which characterizing elements of a graph found physical meaning.

The comparison between the expected graph and the graph observed shows the development and modification of this framework, as well as providing tools for previewing graph related to new phenomena, as well as skills to preview adequate features for the temporal evolution of derivative quantities. To read the graphs to identify critical instants and analyze the phases delimited by these instants favors the link between phenomenon, its formalized description using physical quantities. The development of graphing skill and mathematical knowledge becomes active knowledge to read and interpret the world. The link between the graph and concrete contexts activated by RTL is important not only for young students, in agreement with Curcio (1987), but also to support a learning progression at every age, because these activate reasoning processes resonant with the ways people look at graph representations and thus allow to attack the conceptual learning difficulties typical of the contexts explored.

Concerning the second research question RQ2, students of our samples have been able to work with linear models on data collected with RTL. It has also been observed that there is a positive correlation between the attribution of meaning to the coefficients of an interpolation and the determination of the values of these coefficients. The development of alternative models that are supported by the geometric elements of a graph and intuitive physical model is important to go beyond linear models.

The results of the present work enrich and are coherent with that quoted in literature and confirm many of that obtained in previous researches on how students learn from graphs acquired in real time (Corni et al. 2005; Michelini 2006, 2010; Michelini et al. 2010, 2014; Stefanel et al. 2002). At the same time, future work will be needed to reinforce some conclusions that can appear too weak or too related to the samples considered.

Acknowledgment I thank prof. Marisa Michelini for the suggestions on design of the researches here presented and on the structure of the present paper and for sharing the educational proposal at the basis of this work. I also thank prof. Gesche Pospiech for the indications she has given me and for her kindness and patience.

References

Aberg-Bengtsson, L., & Ottosson, T. (2006). What lies behind graphicacy? *Journal of Research in Science Teaching, 43*(1), 43–62.

American Association for the Advancement of Science. (1993). *Benchmarks for science literacy*. New York: Oxford University Press.

Angell, C., Kind, P. M., Henriksen, E. K., & Guttersrud, Ø. (2008). An empirical-mathematical modelling approach. *Physics Education, 43*(3), 256–264.

Beichner, R. (1994). Testing student understanding of kinematics graphs. *American Journal of Physics, 62*, 750–762.

Bertin, J. (1983). *Semiology of graphics*. (2nd ed., W. J. Berg, Trans.). Madison: University of Press.

Bisdikian, G., & Psillos, D. (2002). Enhancing the linking of theoretical knowledge to physical phenomena. In D. Psillos & H. Niedderer (Eds.), *Teaching and learning* (pp. 193–204). Dordrecht: Kluwer.

Bradamante, F., Michelini, M., & Stefanel, A. (2004). The modelling in the sport for physic learning. In E. Mechlova & L. Konicek (Eds.), *Selected papers in Girep book* (pp. 206–208). Ostrava Czech Republic.

Brussolo, L. e Michelini M.. (2010). *Studiare il moto per un'educazione stradale*. At http://www.formativamente.com/files/moto_edu_strad.pdf

Carswell, C. M. (1992). Choosing specifiers: An evaluation of the basic tasks model of graphical perception. *Human Factors, 34*, 535–554.

Champagne, A., & Gunstone, R. (1985). Instructional consequences of students' knowledge about physical phenomena. In L. West (Ed.), *Cognitive structures & conceptual change* (pp. 61–90). Orlando: Academic.

Chinn, C. A., & Brewer, W. F. (1993). The role of anomalous data. *Review of Educational Research, 63*(1), 1–49.

Chinn, C. A., & Brewer, W. F. (2001). Models of data: A theory of how people evaluate data. *Cognition and Instruction, 19*(3), 323–393.

Corni, F., Michelini, M., Santi, L., & Stefanel, A. (2005). Sensori on-line per la formazione insegnanti. In M. Michelini & M. Pighin (Eds.), *Comunità Virtuale* (Vol. 2, pp. 1149–1161). Udine: Forum.

Costas, C. (2010, July). Design Based Research. In *ESERA summer school*, Udine. http://www.fisica.uniud.it/URDF/Esera2010/lecture1.pdf

Curcio, F. (1987). Comprehension of mathematical relationships expressed in graphs. *Journal for Research in Mathematics Education, 18*, 382–393.

Deacon, C. (1999). The importance of graphs in undergraduate physics. *The Physics Teacher, 37*, 270–274.

Denzin, N. K., & Lincoln, Y. S. (2011). *Handbook of qualitative research*. Los Angeles: Sage.

diSessa, A. A., & Sherin, B. (2000). Meta-representation. *The Journal of Mathematical Behavior, 19*, 385–398.

Duit, R. (2009). *Bibliography – STCSE*.http://archiv.ipn.uni-kiel.de/stcse/

Duit, R., Gropengießer, H., & Kattmann, U. (2005). Toward science education research: The MER. In H. E. Fisher (Ed.), *Developing standard in RSE* (pp. 1–9). London: Taylor.

Educational studies in Mathematics. (2004). Bodily activity and imagination in mathematics learning. *PME Special Issue, 57*(3).

Erickson, F. (1998). Qualitative research methods for Sci. Educ. In B. J. Fraser & K. G. Tobin (Eds.), *International handbook* (pp. 1155–1174). Dordrecht: Kluwer.

European Commission. (1995). *White paper on education and training* (COM (95) 590). Brussels: Author.

Fan, J. E. (2015). Drawing to learn: How producing graphical representations enhances scientific thinking. *Translational Issues in Psychological Science, 1*(2), 170–181.

Fisher, B., Green, T. M., & Arias-Hernández, R. (2011). Visual analytics as a translational cognitive science. *Topics in Cognitive Science, 3*, 609–625.

Friel, S., Bright, G., & Curcio, F. (1997). Understanding students' understanding of graphs. *Mathematics Teaching in the Middle School, 3*, 224–227.

Friel, S. N., Curcio, F. R., & Bright, G. W. (2001). Making sense of graphs. *Journal for Research in Mathematics Education, 32*(2), 124–158.

Gal, I. (2002). Adult statistical literacy: Meanings, components, responsibilities. *International Statistical Review, 70*(1), 1–25.

Gervasio, M., & Micheliani, M. (2009). *Lucegrafo. a simple USB Data Acquisition System for Diffraction* (M. Michelini, Ed.). Prooc. MPTL 14, at http://www.fisica.uniud.it/URDF/mptl14/contents.htm

Gilbert, J. K. (2007). Visualization: A metacognitive skill. In J. J. Gilbert (Ed.), *Visualization in science education* (pp. 9–27). Dordrecht: Springer.

Glazer, N. (2011). Challenges with graph interpretation: A review of the literature. *Studies in Science Education, 47*(2), 183–210.

Griffiths, T., & Guile, D. (2003). A connective model of learning: The implications for work process knowledge. *The European Educational Research Journal, 2*(1), 56–73.

Guttersrud, Ø., & Angell, C. (2014). Mathematics in physics: U.S, physics students' competency. In W. Kaminski & M. Michelini (Eds.), *Teaching and learning physics today* (pp. 84–89). Udine: Lithostampa.

Hale, P. (2000). Kinematics and graphs. *Mathematics Teacher, 93*(5), 414–417.

Hattikudur, S., Prather, R., Asquith, P., Knuth, E., et al. (2012). Constructing graphical representations. *School Science & Mathematics, 112*(4), 230–240.

Hofstein, A., & Lunetta, V. N. (2004). The laboratory in science education: Foundations for the twenty-first century. *Science Education, 88*(1), 28–54.

Jones, P. L. (1993). Realizing the educational potential of the graphics calculator. In L. Lum (Ed.), *Proceedings of the IC-TCM* (pp. 212–217). Reading: Addison.

Karmiloff-Smith, A. (1988). The child is a theoretician, not an inductivist. *Mind & Language, 3*, 183–195.

Klein, U. (2001). *Tools and modes of representation in the laboratory sciences*. Dordrecht: Springer.

Knuth, E. (2000). Understanding connections between equations and graphs. *The Mathematics Teacher, 93*(1), 48–53.

Krajcik, J. S., & Layman, J. W. (1993). MBLs in the science classroom. In *NARST research matters, no. 31*. Retrieved January 2, 2017, from http://www.narst.org/publications/research/microcomputer.cfm.

Leinhardt, G., Zaslavsky, O., & Stein, M. K. (1990). Functions, graphs, and graphing. *Review of Educational Research, 60*, 1–64.

Lemke, J. L. (2003). *Teaching all the languages of science*. At. http://www-personal.umich.edu/~jaylemke/papers/barcelon.htm

Linn, M. C., Layman, J. W., & Nachmias, R. (1987). Cognitive consequences of MBL: Graphing skills. *Contemporary Educational Psychology, 12*, 244–253.

Lumbelli, L. (1996). Focusing on text comprehension. In C. Cornoldi (Ed.), *Reading comprehension difficulties* (pp. 301–330). Mahwah: Erlbaum.

McDermott, L. C. (1991). What we teach and what is learned. *American Journal of the Physics, 59*, 301–315.

McDermott, L. C. (1993). How we teach and how students learn-A mismatch? *American Journal of Physics, 61*(4), 295–298.

McDermott, L. C., Rosenquist, M. L., & van Zee, E. H. (1987). Student difficulties in connecting graphs and physics. *American Journal of Physics, 55*, 503–513.

McDermott, L. C., Shaffer, P. S., & Costantiniou, C. P. (2000). Preparing teachers to teach physics by inquiry. *Physics Education, 35*(6), 411–416.

McDonald-Ross, M. (1977). How numbers are shown: A review of research on the presentation of data. *Audiovisual Communication Review, 25*(4), 359–409.

Meltzer, D. E. (2002). The relationship between mathematics preparation and conceptual learning in physics. *American Journal of Physics, 70*(2), 1259–1268.

Michelini, M. (1988). L'elaboratore nel laboratorio di fisica: alcune considerazioni di carattere generale. *La Fisica nella Scuola*, XXI, 2, IR, p. 159.

Michelini, M. (2006). The learning challenge. In G. Planinsic & A. Mohoric (Eds.), *Informal learning* (pp. 18–39). Ljubljana: Girep Book.

Michelini, M. (2010). Building bridges between common sense ideas and a physics description of phenomena. In L. Menabue & G. Santoro (Eds.), *STE* (Vol. 1, pp. 257–274). Bologna: CLUEB.

Michelini, M., Santi, L., & Sperandeo, R. M. (Eds.). (2002). *Proposte didattiche su forze e movimento*. Udine: Forum.

Michelini, M., Santi, L., & Stefanel, A. (2010). Thermal sensors interfaced with computer as extension of senses. *Il Nuovo Cimento, 33C*(3), 171–179.

Michelini M., Santi L.., & Stefanel, A. (2014). Upper secondary students face optical diffraction. In E. Kajfasz & R. Triay (Eds.), *Proceedings of the FFP14*.http://pos.sissa.it/archive/conferences/224/240/FFP14_240.pdf

Mokros, J. R., & Tinker, R. F. (1987). The impact of MBL on children's ability to interpret graphs. *Journal of Research in Science Teaching, 24*, 369–383.

Mossenta, A., Michelini, M., & Stefanel, A. (2014). Context- based physics. In F. Tasar (Ed.), *Proceedings of the WCPE 2012* (pp. 941–950). Ankara: Pegem.

National Council of Teachers of Mathematics. (2000). *Principles and standards for school mathematics*. Reston: NCTM.

National Research Council. (1996). *National science education standards*. Washington, DC: National Acc.

Nersessian, N. J. (2007). Mental Modeling. In S. Vosniadou (Ed.), *International handbook of conceptual change* (pp. 391–416). London: Routledge.

Niedderer, H. (1989). *Qualitative and quantitative methods of investigating alternative frameworks of students*. Presented to the AAPT-AAAS meeting.

Padilla, J. M., McKenzie, L. D., & Shaw, L. E., Jr. (1986). An examination of line graphing ability of students. *Scholl Science & Mathematics, 86*, 20–26.

Park, J., Kim, I., Kim, M., & Lee, M. (2001). Analysis of students' processes of confirmation and falsification. *International Journal of Science Education, 23*(12), 1219–1236.

Perkins, D. N., & Grotzer, T. A. (2000). *Models and moves: Focusing on dimensions of causal complexity to achieve deeper scientific understanding*. AERA conference, New Orleans, LA.

Planinic, M., Ivanjek, L., Susac, A., & Milin-Sipus, Z. (2013). Comparison of university students' understanding of graphs in different contexts. *Physical Review ST Physics Education Research, 9*, 020103.

Pospiech, G. (2015). Interplay of mathematics and physics in physics education. In *Proceedings of the International Symposium MACAS – 2015*, pp. 36–43.

Pospiech, G. (2018). Framework of mathematization in physics from a teaching, in this book.

Redish, E. F. (2005). *Problem solving and the use of math in phys courses*. ICPE, Delhi, Invited Talk.

Roth, W. M., & Bowen, G. M. (2003). When are graphs worth ten thousand words? An expert-expert study. *Cognition and Instruction, 21*(4), 429–473.

Roth, W., & Lee, Y. (2004). Interpreting unfamiliar graphs: A generative, activity theoretic model. *Educational Studies in Mathematics, 57*(2), 265–290.

Sassi, E. (1996). Addressing learning/teaching difficulties in basic physics. In S. Oblack (Ed.), *Proceedings of the Girep* (pp. 162–179). Ljubljana: Girep.

Sassi, E., Monroy, G., & Testa, I. (2005). Teacher training about real-time approaches. *Science Education, 89*(1), 28–37.

Scott, P., Asoko, H., & Leach, J. (2007). Student conceptions and conceptual learning in science. In S. Abell & N. G. Lederman (Eds.), *Handbook of research on science education* (pp. 31–54). Hillsdale: Erlbaum.

Shah, P. (2002). Graph comprehension: The role of format, content, individual difference. In M. Anderson, B. Mayer, & P. Olivier (Eds.), *Diagrammatic representation* (pp. 207–222). London: Springer.

Shah, P., & Carpenter, P. A. (1995). Conceptual limitations in comprehending line graphs. *Journal of Experimental Psychology, 124*, 43–61.

Shah, P., & Hoeffner, J. (2002). Review of graph comprehension research: Implications for instruction. *Educational Psychology Review, 14*, 47–69.

Shah, P., Freedman, E. G., & Vekiri, I. (2005). The comprehension of information graph displays. In P. Shah & A. Miyake (Eds.), *The Cambridge handbook of visuospatial thinking* (pp. 426–476). Cambridge: Cambridge University Press.

Sokoloff, D. R., Lawson, P. W., & Thornton, R. K. (2004). *Real time physics*. New York: Wiley.

Stefanel, A., Moschetta, C., & Michelini, M. (2002). Cognitive labs in an informal context. In M. Michelini & M. Cobal (Eds.), *Developing formal thinking in physics* (pp. 276–283). Udine: Forum.

Suri, H., & Clarke, D. (2009). Advancements in research synthesis methods. *Review of Educational Research, 79*(1), 395–430.

Svec, M. T. (1995). *Effect of MBL on graphing interpretation skills*. Paper presentation at NARST, San Francisco. http://files.eric.ed.gov/fulltext/ED383551.pdf

Testa, I., Monroy, G., & Sassi, E. (2002). Students' reading images in kinematics: The case of real-time graphs. *International Journal of Science Education, 24*(3), 235–256.

Theodorakakos, A., & Psillos, D. (2010). PEC task explore. In C. Constantinou (Ed.), *CBLIS 2010* (pp. 75–83). Warsaw: Oelizk.

Thornton, R. K. (2004). Uncommon knowledge. In E. F. Redish & M. Vicentini (Eds.), *Research on PER* (pp. 591–601). Amsterdam: IOS.

Thornton, R. K., & Sokoloff, D. R. (1990). Learning motion concepts using real-time MBL based laboratory tools. *American Journal of Physics, 58*, 858–867.

Thornton, R. K., & Sokoloff, D. R. (1998). Assessing student learning of Newton's laws. *American Journal of Physics, 66*, 338–352.

Trowbridge, D. E., & McDermott, L. C. (1980). Investigation of student understanding of the concept of velocity. *American Journal of Physics, 48*, 1020–1028.

Trowbridge, D. E., & McDermott, L. C. (1981). Investigation of student understanding of the concept of acceleration. *American Journal of Physics, 48*, 242–253.

Tufte, E. R. (2001). *The visual display of quantitative information*. Cheshire: Graphics Press.

Tuminaro, J., & Redish, E. F. (2007). Elements of a cognitive model of physics problem solving: Epistemic games. *Physical Review Special Topics – Physics Education Research, 3*, 020101.

Von Korff, J., & Rebello, N. S. (2013). Student epistemology about mathematical integration in a physics context. In *PER conference proceedings* 17–18, pp. 353–356.

Vygotsky, L. S. (1962). *Thought and language*. Cambridge: MIT Press.

Wainer, H. (1992). Understanding graphs and tables. *Educational Re searcher, 21*(1), 14–23.

Wellington, J., & Osborne, J. (2001). *Language and literacy in science education*. Buckingham: Open University Press.

Wenger, E. (1999). *Communities of practice: Learning, meaning & identity*. Cambridge: University Press.

Windschitl, M., Thompson, J., & Braaten, M. (2008). Beyond the scientific method: Model-based inquiry. *Science Education, 92*(5), 941–967.

Woolnough, J. (2000). How can students learn to apply math knowledge to interpret graphs in physics? *Research in Science Education, 30*(3), 259–268.

Wu, H.-K., & Krajcik, J. S. (2006). Inscriptional practices in two inquiry-based classrooms. *Journal of Research in Science Teaching, 43*(1), 63–95.

Chapter 10
Comparing Student Understanding of Graphs in Physics and Mathematics

Maja Planinic, Ana Susac, Lana Ivanjek, and Željka Milin Šipuš

10.1 Introduction and Background

Research suggests that many students at high school or introductory university level lack the ability to understand and interpret graphs. This has been documented in several physics education studies (e.g., McDermott et al. 1987; Brasell and Rowe 1993; Beichner 1994; Forster 2004; Araujo et al. 2008; Nguyen and Rebello 2011; Christensen and Thompson 2012), as well as mathematics education studies (Dreyfus and Eisenberg 1990; Leinhardt et al. 1990; Swatton and Taylor 1994; Graham and Sharp 1999; Kerslake 1981; Hadjidemetriou and Williams 2002; Habre and Abboud 2006). Student difficulties with calculating and interpreting slope of a graph and area under a graph were common.

The concept of slope (gradient) is very important for physics since many physical quantities are defined as gradients (e.g., velocity, acceleration) and represented with line graphs. The concept of slope is also important for mathematics as a necessary

M. Planinic (✉)
Department of Physics, Faculty of Science, University of Zagreb, Zagreb, Croatia
e-mail: maja@phy.hr

A. Susac
Department of Physics, Faculty of Science, University of Zagreb, Zagreb, Croatia

Department of Applied Physics, Faculty of Electrical Engineering and Computing, University of Zagreb
e-mail: ana.susac@fer.hr

L. Ivanjek
Austrian Competence Center Physics, University of Vienna, Vienna, Austria
e-mail: lana.ivanjek@univie.ac.at

Ž. Milin Šipuš
Department of Mathematics, Faculty of Science, University of Zagreb, Zagreb, Croatia
e-mail: milin@math.hr

© Springer Nature Switzerland AG 2019
G. Pospiech et al. (eds.), *Mathematics in Physics Education*,
https://doi.org/10.1007/978-3-030-04627-9_10

prerequisite for the development of the concept of derivation. Students study line graph slope in both mathematics and physics, but because of the differences in contexts, they may not necessarily realize that they are studying the same concept.

Student difficulties concerning the interpretation of the area under a graph may be even stronger than those concerning graph slope, since interpretation of slope is usually more emphasized by school mathematics and physics teaching than the interpretation of area under a graph. Yet, the area interpretation, with the idea of accumulation of infinitesimal quantities, underlies the concept of definite integral, important for both mathematics and physics teaching.

Student difficulties with graphs were identified through both physics and mathematics education research. Most of the research on student understanding of graphs in physics was done in the context of kinematics, because of the very broad use of graphical representations in kinematics. Students were found to have difficulties with linking the graph and the verbal descriptions of a given event and with understanding graphs as symbolic representations of relationships among variables (Brasell and Rowe 1993; Beichner 1994). They often have trouble discriminating the slope and height of a graph and interpret changes in height as changes in slope (McDermott et al. 1987; Beichner 1994). Many students are unable to choose which feature of the graph represents the information that is needed to answer the question (e.g., they calculate slope when they should have been calculating the area) (McDermott et al. 1987; Beichner 1994). Very few students seem to be able to interpret the area under an a vs. t graph as a change in velocity, whereas they have far less problems interpreting the area under a v vs. t graph as a distance travelled (Beichner 1994). Some of the research in mathematics education was also based on kinematics motion graphs and had similar general findings (Graham and Sharp 1999; Kerslake 1981), while studies in purely mathematical context have in addition shown that student understanding of mathematical concepts (such as functions) tends to be typically algebraic and not visual. Visual information, including graphs, seems to be more difficult for students to learn and is considered by them to be less mathematical (Dreyfus and Eisenberg 1990; Habre and Abboud 2006).

Student difficulties with graphs are sometimes classified as interval-point confusions, slope-height confusions, and iconic confusions (Leinhardt et al. 1990). The iconic confusion is usually characteristic of younger students, although traces of it can also be found in older populations, sometimes even university students (McDermott et al. 1987; Beichner 1994). It consists in students' incorrect interpretation of the graph as an actual picture of the motion. Students who show this difficulty will tend to interpret, for example, a curved v vs. t graph as representing the motion along a curved trajectory. Such students do not yet see the graph as a symbolic representation of an abstract relationship between the variables on its axis but as a concrete picture of body's motion. It is therefore difficult for them to see why the graph should change if the variables on the axes change, and they will generally expect the graph to remain the same.

The slope-height confusion happens when students mistake the height of the graph for its slope (McDermott et al. 1987; Beichner 1994; Leinhardt et al. 1990). For example, when asked to reason about the slope of a graph, students sometimes just read off the y coordinate (the height of the graph at the point of interest). If they observe, for example, the constant diminishing of the y coordinate of the graph, they usually conclude that the slope of the graph shows the same behavior (e.g., the slope of the straight line constantly diminishes, because the y coordinate constantly diminishes).

The interval-point confusion refers to the cases where students focus on a single point of the graph when they should be using an interval. This difficulty will be displayed, for example, when students attempt to determine the slope of a graph from one point only, instead of choosing two points and calculating $\Delta y/\Delta x$. Slope-height and interval-point confusions are quite common among students at high school and university level (McDermott et al. 1987; Beichner 1994; Forster 2004; Leinhardt et al. 1990; Hadjidemetriou and Williams 2002; Wemyss and van Kampen 2013). Overall, the findings of both physics and mathematics education research are rather similar and point to the presence of similar student difficulties in both domains.

The important issue of transfer of knowledge between mathematics and physics (usually expected to occur from mathematics to physics) was also tackled in several studies on graphs (Christensen and Thompson 2012; Wemyss and van Kampen 2013; Woolnough 2000), with mostly negative results. It was suggested in one of the studies that most secondary students, even those who do well in mathematics and physics, do not make substantial links between the two domains and that some students may even think that it is not appropriate to transfer concepts from mathematics to physics (Woolnough 2000). For transfer to occur, it is necessary that students possess the required mathematical knowledge, but this is not always the case, especially when advanced concepts such as derivative or integral are concerned (Nguyen and Rebello 2011; Christensen and Thompson 2012). The problem of transfer of knowledge between mathematics and physics was addressed in cognitive psychology, unrelated to graphs. One study that investigated interdomain transfer between isomorphic topics in algebra and physics (kinematics) found very high transfer from algebra to physics, but almost no transfer from physics to algebra, and suggested that "transfer from physics to other domains is blocked by the embedding of physics equations within a specific content domain" (Bassok and Holyoak 1989). The problem of domain specificity of knowledge is not limited to physics; it is also present in mathematics. Michelsen (2005) suggests that it is not just the mathematical formalism that presents a barrier in learning physics but that the problem lies in the missing link between mathematics and physics. He suggests that the mathematical domain should be expanded by using examples from physics and from everyday life contexts in mathematics teaching, in order to solve the problem of domain specificity. In such an expanded domain, modeling of real-life situations could be a way of bridging the gap between mathematics and physics. We will take a closer look at the key issue of transfer of learning from theoretical viewpoint in the following chapter.

10.2 Theoretical Framework

Transfer of learning is usually defined as the ability to extend what has been learned in one context to new contexts (Bransford et al. 1999) and is sometimes regarded as one of the ultimate goals of education. Hammer et al. (2005) suggest that it would generally be more appropriate to speak of activation of cognitive resources than of transfer, since knowledge and reasoning abilities are comprised of many resources that may, or may not, be activated in a particular context. They oppose the view of knowledge and abilities as objects which are acquired, manipulated, and transferred as intact units, with the exception of locally coherent sets of resources which activate together and possess internal structural stability. Such cognitive units, whose mechanism of stability is structural rather than contextual, can be viewed as transferable (Hammer et al. 2005). In our opinion, students' concepts of the graph slope and of the area under a graph can be examples of such transferable units in cases when they are well formed and stable.

Whether or not transfer will happen depends not only on the presence or absence of relevant resources but also on students' framing of the situation (Hammer et al. 2005). Framing means that students have to interpret what is going on in a certain situation or in a certain problem and decide accordingly which resources to use or which epistemic game to play (Tuminaro and Redish 2007). In physics education we usually expect students to transfer their mathematical knowledge from mathematics to physics. There are several reasons why the expected transfer could fail: either the required resource does not exist, or the resource exists, but is not activated due to the wrong framing of the problem, or the resource is activated, but its mapping to the problem is not appropriate (Tuminaro 2004). Research suggests that transfer is more likely to happen when students have seen the given idea in at least two separate contexts or when they receive metacognitive scaffolding (Bransford et al. 1999).

Many studies that have looked for transfer of knowledge have usually come up with mostly negative results, which may be due, among other things, also to the design of those studies (Bransford and Schwartz 1999). Bransford and Schwartz (1999) have suggested to shift the view on transfer from the direct application perspective (successful application of knowledge acquired in one context to similar problems in different contexts) to a more dynamical view of preparation for future learning (PFL). The PFL perspective can be demonstrated through questions about and approaches to the new problem, which were shaped and influenced by the previous learning, even if students are not able to completely solve the new problem. The PFL perspective is very important for learning, because it reveals more about students' useful learning trajectories than the direct application perspective. The focus is not only on what students can or cannot directly transfer and solve but whether students are able to learn while they transfer. In this way transfer can be considered a dynamical way of reconstructing knowledge (Cui 2006) rather than just an application of previously acquired knowledge in a different situation. This dynamical view of transfer is in agreement with knowledge-as-elements perspective, because it assumes activation of different knowledge elements in a new context and dynamical creation of the response on the spot.

Theories of transfer of knowledge are based upon the idea that knowledge can be transferred from one situation to another and linked with a new situation (Potgieter et al. 2008). Some researchers disagree and argue that learners' mental processes are structured by the context and the implemented activities and tools (Lave 1988). Teachers often expect students to rise above the context, but that is not easy for students. Recognizing mathematics in a different context requires good understanding of the context (which is often missing), along with mathematical knowledge (Potgieter et al. 2008). To investigate transfer of knowledge in more detail, some comparative studies in mathematics and physics were conducted and produced interesting results.

10.3 Results of Comparative Studies on Graphs in Mathematics and Physics

Few studies attempted to compare student reasoning difficulties about graphs in different contexts and domains (Wemyss and van Kampen 2013; Woolnough 2000). Such comparison, on the other hand, can provide interesting and important insights in student knowledge and learning and the issue of possible transfer between domains. An example is the study of Wemyss and van Kampen (2013), in which first-year university students solved three different context problems including line graphs, found that the number of students' correct answers to a problem involving water level vs. time graph, which students had not encountered in the formal educational setting before, was much higher than the number of correct answers to the supposedly more familiar problem of determining the speed of object from a distance-time graph. The reason for students' poorer performance on physics problems was attributed to students' reliance on learned procedures in physics (e.g., use of formulas). This study also found evidence that students' mathematical knowledge of slope does not guarantee their success on problems involving slope in kinematics.

In our first study on graphs (Planinic et al. 2012), we compared second-year high school students' (N = 114) understanding of the line graph slope in the domains of physics and mathematics. Student answers to two pairs of parallel (isomorphic) questions regarding line graph slope from mathematics and physics (kinematics) were analyzed and compared. Also, a sample (N = 90) of Croatian physics teachers were asked to rank the isomorphic questions according to their expected difficulty for students. Physics teachers largely thought that the physics questions would be easier for students because they were regarded as less abstract than the mathematics questions. Many also expressed the belief that the lack of mathematical knowledge would present the main problem for students when solving physics questions. It was found however that, contrary to the prevalent belief of physics teachers, students did better on mathematics than on physics questions. The main source of student difficulties with the concept of line graph slope in physics seemed not to be their lack of mathematical knowledge but rather their lack of ability to interpret the meaning of the line graph slope in physics context. Many students successfully solved the mathematical questions but were unable to solve parallel physics questions or used different strategies for solving analogous mathematics and physics problems. It was observed that the transfer of knowledge from mathematics to physics did not always occur, even though many students possessed the needed mathematical knowledge. (Interestingly, beside the expected transfer from mathematics to physics, which was relatively weak, some occasional cases of transfer from physics to mathematics were also observed.) Also, the same student difficulty known as slope-height confusion was detected in both domains, but it occurred far more frequently in physics than in mathematics (about twice as often).

After this study it was natural to pose the question about the reason for the observed higher difficulty of physics questions relative to parallel mathematics questions: Is the higher difficulty of physics questions the consequence of students' lack of relevant physics conceptual knowledge, or would the same effect be observed to the same extent also in parallel questions situated in different contexts, which did not require additional content knowledge? We attempted to investigate this issue by using sets of three parallel (isomorphic) questions and to analyze and compare item difficulties as well as student strategies in different domains. The three domains were mathematics without context, physics (kinematics), and mathematics in contexts other than physics, which did not require additional conceptual knowledge. Eight such sets of parallel (isomorphic) mathematics, physics, and other context questions about graphs were developed by the authors and administered to 385 first-year students at Faculty of Science, University of Zagreb in Zagreb, Croatia, and later also to 417 first-year students at University of Vienna. Students were either prospective physics or mathematics teachers or prospective physicists or mathematicians. Students were tested at the beginning of the first semester, before any formal instruction on graphs, so their knowledge on graphs came only from high school mathematics and physics instruction. Five sets of questions referred to the concept of graph slope and three to the concept of area under a graph. Four sets were in a multiple choice format, and four sets were open-ended (the whole test can be accessed through the link in reference (http://journals.aps.org/prstper/supplemental/10.1103/PhysRevSTPER.9.020103/Pl aninic_TEST_PRST_PER.pdf)). In addition to choosing the correct answer in multiple choice questions, or providing the answer in open-ended questions, students were asked to provide explanations for their answers and/or necessary calculations where appropriate, so that insight into the underlying student reasoning could be obtained. Rasch analysis (Linacre 2006, n.d.; Bond and Fox 2001) was performed to evaluate the functioning of the test and obtain linear measures of item difficulties. Both sets of data (Croatian and Austrian students) seemed to fit the Rasch model. The functioning of the test as a whole for Croatian students was found to be satisfactory with very high item reliability (0.99) and somewhat lower, but satisfactory, person reliability (0.85) and Cronbach alpha (0.88) (Planinic et al. 2013). For Austrian students the test functioned similarly: item reliability was found to be 0.99, person reliability 0.86 and Cronbach alpha 0.90 (Ivanjek et al. 2015). The analysis of item fit showed that no test items in either data set were degrading for measurement (all had infit and outfit MNSQ values within the range of 0.5–1.5). The point-biserial correlations of items were all positive and greater than 0.3 (Planinic et al. 2013; Ivanjek et al. 2015). It can be concluded that all items worked together in defining the underlying variable (student understanding of graphs) and that a reliable scale of item difficulties was obtained for the items in the test, which allowed further analysis of difficulties of different groups of items. Interestingly, parallel questions of the same set usually differed quite significantly in difficulty.

In order to compare the difficulties of items in each investigated context, the average values of item difficulties over three different domains (mathematics without context, physics, mathematics in context) and two investigated concepts (slope, area) were calculated. The comparison of average difficulties of slope and area items for the two samples is presented in Fig. 10.1. Since in Rasch analysis the average difficulty of items in the test is usually assigned the value of zero logits, positive values in the graph indicate higher than average difficulty (harder items) and negative values lower than average difficulty (easier items).

The comparison of the results of the two samples indicated the stability of the construct of the test. Although some differences in the performance of students in the two groups were noticed, the general trends were the same. For both groups of students, it was noticed that mathematics without context was the easiest domain. Adding context to questions generally had the effect of increasing the difficulty. Kinematics was found to be a difficult context for the students in both samples, in

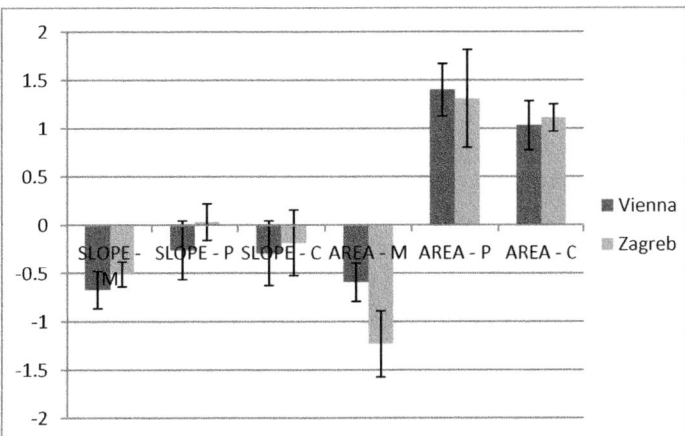

Fig. 10.1 Average difficulties of slope and area items in three different domains for the two groups of students – students from University of Zagreb and from University of Vienna (*M* stands for mathematics without context, *P* for Physics, and *C* for other contexts) (Ivanjek et al. 2015). Error bars indicate the combined uncertainties of each average value

spite of the presumed students' familiarity with the type of questions (kinematics questions used in the test were of the type that is often used in physics teaching, whereas other context questions were typically new to students). When comparing student understanding of slope and area, it was found that, on the average, slope seemed to be better understood. It was also found to be more homogenous – the differences between domains were less pronounced than in the case of area under a graph. Interpretation of area in kinematics and other context was the most difficult aspect of graph interpretation for the students in the both samples (Fig. 10.1). The difficulty of the concept of area under a graph differs dramatically between mathematics on one side and physics and other contexts on the other. This is consistent with the findings of Nguyen and Rebello (2011) that very few students are able to apply this concept in physics problems.

Another aspect of the study conducted on Croatian sample was the analysis of students' strategies and expressed difficulties, obtained through the analysis of their explanations and procedures provided with the answers to questions (Ivanjek et al. 2016). The main findings can be summarized as follows:

1. *Strategies used on parallel questions are often context-dependent and domain-specific. The preferred strategy on physics questions seems to be the use of physics formulas.*

Only a small fraction of students typically used the same strategy on all three questions of the same set of questions, although some have used the same strategy on two of the three questions. It seems that in many cases, students perceived the questions from the same set as different and approached them in different ways. The strategy that was used usually depended on the domain and the context of the problem.

It seems that if students acquire domain-specific procedures for solving a certain class of problems (such as determining the slope of the straight line in mathematics with the use of mathematical formulas or calculating acceleration in physics with the use of physics formulas), they will tend to stick to those procedures and will generally not seem to recognize the mathematical similarity of the problems in different domains. This may be an indication of the absence of transfer of knowledge between the domains, but it could also be a consequence of students' different learning experiences in different school subjects, where they had implicitly learned that each discipline has its own language and conventions and that they have to answer questions in the way that the particular discipline requires. How students framed the problem (Hammer et al. 2005) may have determined their choice of strategy for its solving.

Even though students demonstrated that they were capable of using different strategies for reasoning about graphs, the preferred strategy in physics domain tended to be the use of kinematics formulas. On all area problems and some slope problems, students chose the use of formulas as the main strategy for solving physics problems. The application of the incorrect or inappropriate formulas led them to many incorrect conclusions on physics questions, even on the questions where calculations were not necessary. At the same time, it was not uncommon for students to give correct answers to parallel questions in mathematics and other contexts domains, demonstrating that they were able to reason correctly about the same problem in a different context. The very extensive use of the formula $a = v/t$ on the test indicates, for example, that many students may not have understood the very meaning of the concept of acceleration (as the *rate* of change of velocity) and therefore cannot be expected to understand its representation as the slope of the v vs. t graph. All these findings suggest that students not only have problems with graph interpretation but also with the understanding of the meaning and applicability of physics formulas.

2. *Students use a wider spectrum of strategies on other context problems than on physics problems. Other context problems could be potentially useful in physics and mathematics teaching.*

Other context problems seemed to activate more of students' cognitive resources, and students displayed a wider variety of strategies on those problems than on physics problems. Some students came to the idea that multiplication is needed and others to calculate the area under a graph on other context questions by using some form of dimensional analysis. Dimensional analysis is an approach primarily developed in physics, but surprisingly students did not use the same approach on physics questions.

Many of the student approaches to other context problems could have helped them to solve physics problems as well, but the reliance on formulas as the primary strategy in physics prevented students from using other approaches of which they were capable. Some instances of transfer of knowledge in the sense of preparation for future learning were evidenced in students' use of knowledge and techniques (e.g., dimensional analysis, modeling), acquired in one domain (usually physics), in some other domain (usually other context questions). Some students seemed to think more creatively and used more of the available resources on other context questions than on physics questions, where they seemed to be bound too much by how they perceived the conventions of the discipline. Other context problems could therefore be a potentially useful tool in teaching of both mathematics and physics.

3. *Students show similar difficulties with graph interpretation in all domains, but there are differences between their understanding of graph slope and area under a graph.*

The same patterns of naïve reasoning (slope-height confusion and interval-point confusion) were present in all three domains, but not equally often in each one of them (more frequently in physics than mathematics domain). This is something that we had already noticed in a previous study on high school students' understanding of line graph slope (Planinic et al. 2012).

However, differences were found in students' understanding of the concepts of slope and area and their interpretation. Student explanations on mathematics slope items revealed that for many students, slope may not be more than the vague notion of how steep a straight line is, sometimes identified with the angle that the straight line forms with one of the coordinate axis. In problems which demand only qualitative comparison of slopes, this may often be enough to produce the correct answer. However, when it comes to calculating slope, this vague idea no longer helps. Even though students did not do too well on determining slope in mathematics domain, they did even worse in other domains. The percentage of Croatian students in the study who knew how to determine slope mathematically (54%) was roughly the same as was found in two other studies on first-year university students (Beichner 1994; Wemyss and van Kampen 2013), whereas the respective percentage for Austrian students was somewhat higher (66%). Calculation of slope, as some other studies also suggest (Hadjidemetriou and Williams 2002), may be the most difficult aspect of the concept of slope.

An important aspect of the understanding of the concept of slope is the understanding of the meaning of negative slope. Negative slope is obviously more difficult to understand than positive slope. It seems that students who used vague explanations of negative slope on the basis of graph appearance (e.g., "*straight line is going down*") do not fully understand the concept but have some visual rule for recognizing it.

When it comes to area under a graph, most students know how to determine it, but the interpretation of the meaning of that area seems to be a much bigger problem. Few students seem to be able to interpret areas under graphs in new situations. Unlike slope, whose meaning is more often discussed during teaching, and which is encountered in a greater variety of situations than the area under a graph, interpretation of area seems to be limited to a few isolated examples in physics and learned without sufficient understanding and without necessary reasoning required to transfer that knowledge to other situations. It is interesting that students are more likely to come to the correct interpretation of area in other context questions than in physics, because in physics they often seem to be blocked in their thinking by their overreliance on physics formulas.

10.4 Conclusions and Implications for Teaching

We have attempted through several studies to compare student performance on mathematically similar problems in different domains. The results suggest that students interpret graphs best in mathematics without context. Even though mathematics questions appear more abstract, they are more direct and require less processing of information and less conceptual understanding than parallel physics (kinematics) questions. Kinematics was found to be a difficult context for students, even though it was rather extensively covered in high school. It can be concluded that context generally seems to increase the difficulty of items. Context added to the mathematical slope or area problem will usually increase the cognitive demand on the students, acting as an additional barrier in the problem, and will therefore also increase the difficulty of the item. The only exception may be very familiar contexts for students. Teachers should realize that it is very important to work on students' conceptual understanding and interpretation of physical and mathematical quantities as well as on building stronger links between the two subjects.

Many physics teachers attribute student difficulties with graphs in physics to their presumed lack of mathematical knowledge. But even if students have the needed mathematical knowledge, which was generally the case in our studies (although some problems were noticed in that area too), the transfer to a different domain is not guaranteed. The interpretation of the mathematical quantities in physics or in other contexts is a crucial step which most students in our sample were not able to perform. Some cases of transfer of the problem-solving strategies from physics to other contexts were found on the area items (e.g., dimensional analysis). During teaching of kinematics, the interpretation of slope is usually much more emphasized than the interpretation of area under a graph. An important implication for physics teaching is that we should work more on building student reasoning which leads toward the interpretation of area (which is essentially the idea of integral) and not only provides ready-made interpretation for specific cases in physics. That could also help later to strengthen student understanding of the concept of a definite and indefinite integral in mathematics.

Student reasoning about problems is often very much bound by the contexts and conventions of the disciplines in which their knowledge was acquired. The observed dependence of student strategies on the domain and context of the questions seems to support the knowledge-in-pieces framework, which explains this dependence through context-dependent activation of cognitive resources and the importance of framing. Students seemed to think more freely and creatively, and to transfer more of their knowledge, in problems which in their perception probably did not fall in the category of either physics or mathematics (other context problems). Other context problems may have a potential to expose and develop student reasoning more than the standard domain-specific mathematics and physics questions. They should be used more, in both mathematics and physics teaching. Both disciplines should work more on establishing links between common concepts and procedures in mathematics and physics and promote their integration in students' minds to a much larger extent than is the case now. Students' almost exclusive reliance on formulas in physics presents, in our opinion, an important obstacle for the development of students' deeper reasoning in physics and sometimes even an obstacle for the application of their already existing knowledge and reasoning developed in other domains.

The comparison of the results of Croatian and Austrian students has confirmed the stability of the test and its relevance beyond just Croatian educational system. Currently we continue the research on graphs on other groups of students, besides physics and mathematics students, using also other techniques, such as eye tracking. Some preliminary results suggest better success of nonspecialist groups of university students (e.g., psychology students) on qualitative than quantitative slope and area questions and higher transfer of strategies from physics to finance problems for physics students.

On the basis of our findings, we can summarize some teaching recommendations that might help in the effort of building stronger and more unified student knowledge about graphs:

- Use of other context problems in both mathematics and physics teaching
- Use of multiple strategies on graph problems, which can help remove emphasis from the use of physics formulas as the primary strategy
- Promoting conceptual understanding of graph slope and area in both mathematics and physics teaching
- Building better understanding of the meaning and applicability of physics formulas (and their graphical interpretations where possible)
- Encouraging transfer between mathematics and physics by using and linking different contexts when teaching graphs (e.g., using kinematics examples in mathematics teaching and relating kinematics graphs and formulas to their mathematical origin and meaning in physics teaching)
- Strengthening and operationalizing student understanding of the concept of slope and its calculation (the practice of drawing the rise and run triangle on a line graph – *Steigungsdreieck* in German – seems to help for calculation of slope)
- Promoting interpretation of area under a graph in physics teaching wherever possible by leading students to the idea of accumulation

The presented findings confirm once again that human knowledge is very complex and multifaceted. Students' answers to questions and problems are influenced by the context and formulation of the question, students' framing of the question, the procedures and conventions of the domain in which a certain piece of knowledge was first acquired, the existing or missing links between the domains, as well as many other factors. Using many contexts during teaching and constantly building and strengthening links between different domains could be a good way to building stronger student knowledge. This could help education efforts in both mathematics and physics.

References

Araujo, I. S., Veit, E. A., & Moreira, M. A. (2008). Physics students' performance using computational modeling activities to improve kinematics graphs interpretation. *Computers in Education, 50*, 1128.

Bassok, M., & Holyoak, K. J. (1989). Interdomain transfer between isomorphic topics in algebra and physics. *Journal of Experimental Psychology: Learning, Memory, and Cognition, 15*(1), 153.

Beichner, R. J. (1994). Testing student interpretation of kinematics graphs. *American Journal of Physics, 62*, 750.

Bond, T. G., & Fox, C. M. (2001). *Applying the Rasch model: Fundamental measurement in the human sciences*. Mahwah: Lawrence Erlbaum.

Bransford, J. D., & Schwartz, D. L. (1999). Rethinking transfer: A simple proposal with multiple implications. *Review of Research in Education, 24*, 61–100.

Bransford, J. D., Brown, A. L., & Cocking, R. R. (1999). *How people learn: Brain, mind, experience, and school*. Washington, DC: National Academy Press.

Brasell, H. M., & Rowe, B. M. (1993). Graphing skills among high school physics students. *School Science and Mathematics, 93*, 63.

Christensen, W. M., & Thompson, J. R. (2012). Investigating graphical representations of slope and derivative without a physics context. *Physical Review Physics Education Research, 8*, 023101.

Cui, L. (2006). *Assessing college students' retention and transfer from calculus to physics* (PhD Thesis). Kansas State University.

Dreyfus, T., & Eisenberg, T. (1990). On difficulties with diagrams: Theoretical issues. In G. Booker, P. Cobb, & T. N. De Mendicuti (Eds.), *Proceedings of the Fourteenth Annual Conference of the International Group for the Psychology of Mathematics Education* (Vol. 1, pp. 27–36). Oaxtepex: PME.

Forster, P. A. (2004). Graphing in physics: Processes and sources of error in tertiary entrance examinations in Western Australia. *Research in Science Education, 34*, 239.

Graham, T., & Sharp, J. (1999). An investigation into able students' understanding of motion graphs. *Teaching Mathematics and its Applications, 18*, 128.

Habre, S., & Abboud, M. (2006). Students' conceptual understanding of a function and its derivative in an experimental calculus course. *Journal of Mathematical Behavior, 25*, 57–72.

Hadjidemetriou, C., & Williams, J. S. (2002). Children's' graphical conceptions. *Research in Mathematics Education, 4*, 69.

Hammer, D., Elby, A., Scherr, R. E., & Redish, E. F. (2005). Resources, framing, and transfer. In J. Mestre (Ed.), *Transfer of learning from a modern multidisciplinary perspective* (pp. 89–120). Greenwich: Information Age Publishing.

Ivanjek, L., Planinic, M., Hopf, M., & Susac, A. (2015). Student difficulties with graphs in different contexts. In K. Hahl, K. Juuti, J. Lampiselkä, A. Uitto, & J. Lavonen (Eds.), *Cognitive and affective aspects in science education research – Selected papers from the ESERA 2015 conference* (pp. 167–178). Cham: Springer International Publishing AG.

Ivanjek, L., Susac, A., Planinic, M., Milin-Sipus, Z., & Andrasevic, A. (2016). Student reasoning about graphs in different contexts. *Physical Review Physics Education Research, 12*, 010106.

Kerslake, D. (1981). Graphs. In K. M. Hart (Ed.), *Children's understanding of mathematics: 11–16* (pp. 120–136). London: John Murray.

Lave, J. (1988). *Cognition in practice: Mind, mathematics and culture in everyday life*. Cambridge: Cambridge University Press.

Leinhardt, G., Zaslavsky, O., & Stein, M. K. (1990). Functions, graphs, and graphing: Tasks, learning, and teaching. *Review of Educational Research, 60*(1), 1–64.

Linacre, J. M. (2006). *WINSTEPS Rasch measurement computer program*. Chicago: Winsteps.com.

Linacre, J. M. *A user's guide to WINSTEPS*. www.winsteps.com

McDermott, L. C., Rosenquist, M. L., & van Zee, E. H. (1987). Student difficulties in connecting graphs and physics: Examples from kinematics. *American Journal of Physics, 55*, 503.

Michelsen, C. (2005). Expanding the domain – Variables and functions in an interdisciplinary context between mathematics and physics. In A. Beckmann, C. Michelsen, & B. Sriraman (Eds.), *Proceedings of the 1st International Symposium of Mathematics and its Connections to the Arts and Sciences*. The University of Education, Schwäbisch Gmünd, Germany, pp. 201–214.

Nguyen, D. H., & Rebello, N. S. (2011). Students' understanding and application of the area under the curve concept in physics problems. *Physical Review Physics Education Research, 7*, 010112.

Planinic, M., Milin-Sipus, Z., Katic, H., Susac, A., & Ivanjek, L. (2012). Comparison of student understanding of line graph slope in physics and mathematics. *The International Journal of Science and Mathematics Education, 10*(6), 1393.

Planinic, M., Ivanjek, L., Susac, A., & Milin-Sipus, Z. (2013). Comparison of university students' understanding of graphs in different contexts. *Physical Review Physics Education Research, 9*, 020103.

Potgieter, M., Harding, A., & Engelbrecht, J. (2008). Transfer of algebraic and graphical thinking between mathematics and chemistry. *Journal of Research in Science Teaching, 45*(2), 197–218.

Swatton, P., & Taylor, R. M. (1994). Pupil performance in graphical tasks and its relationship to the ability to handle variables. *British Educational Research Journal, 20,* 227.

Tuminaro, J. (2004). *A cognitive framework for analyzing and describing introductory students' use and understanding of mathematics in physics.* PhD thesis, University of Maryland, College Park.

Tuminaro, J., & Redish, E. F. (2007). Elements of a cognitive model of physics problem solving: Epistemic games. *Physical Review Physics Education Research, 3,* 020101.

Wemyss, T., & van Kampen, P. (2013). Categorization of first-year university students' interpretations of numerical linear distance-time graphs. *Physical Review Physics Education Research, 9,* 010107.

Woolnough, J. (2000). How do students learn to apply their mathematical knowledge to interpret graphs in physics? *Research in Science Education, 30,* 259.

Chapter 11
Combining Physics and Mathematics Learning: Discovering the Latitude in Pre-service Subject Teacher Education

Terhi Mäntylä and Jaska Poranen

11.1 Introduction

Most physics teachers in Finland teach mathematics and vice versa. This situation is rather natural, because physics and mathematics have substantial overlapping knowledge, although knowledge has a different role in the disciplines and the disciplines often approach and apply the knowledge differently. Therefore, it is essential for a teacher that he/she is able to operate fluently on both subjects, knows the common ground of the subjects, and at the same time knows the differences between the subjects—for example, the epistemological differences and different goals and targets of the subjects. In physics teaching, the mathematics is an integral part of the subject, particularly the deeper one goes into the subject of physics. In mathematics teaching, the area of physics offers fruitful examples and implications for mathematical concepts. However, occasionally, instead of discussing mathematics in physics or physics in mathematics, it is fruitful to encounter situations where both subjects are needed without the other being a subordinate to the other—particularly nowadays, when interdisciplinary or multidisciplinary teaching approaches (e.g. Barton and Smith 2000) are becoming more general in curricula. For example, in the latest reform of Finnish national core curriculum of basic education, multidisciplinary learning modules are one issue to be taken into account in teaching (Opetushallitus 2015a). Similar reforms have also taken place in the national core curriculum of general upper secondary schools (Opetushallitus 2015b).

An obligatory course entitled "multidisciplinary thinking" has been added, where the boundaries of two subjects overlap in order to develop upper secondary school

T. Mäntylä (✉) · J. Poranen
Faculty of Education and Culture, Tampere University, Tampere, Finland
e-mail: Terhi.Mantyla@tuni.fi; Jaska.Poranen@tuni.fi

© Springer Nature Switzerland AG 2019
G. Pospiech et al. (eds.), *Mathematics in Physics Education*,
https://doi.org/10.1007/978-3-030-04627-9_11

247

students' creative and critical thinking. In this course, a real-life phenomenon is the starting point of teaching and is examined from the perspective of different school subjects. These reforms appear to share many features with the multidisciplinary approach or interdisciplinary teaching, with a concentration on real-life phenomena or problems and where knowledge of different disciplines is needed and applied. The challenge with interdisciplinary teaching is that the learning often remains shallow and fragmented (Applebee et al. 2007; Feng 2012). The interdisciplinary approach could lend some new perspectives to traditional mathematics and physics teaching, and it could complement teaching and learning of the subjects in a valuable manner. Here, we concentrate on an example from pre-service physics and mathematics teacher education, where both physical and mathematical knowledge are needed and where they are not subordinate to each other; instead both are applied in an interrelated manner. Simultaneously, the example demonstrates what is needed in order to use the interdisciplinary approach meaningfully.

11.2 Background

11.2.1 Physics and Mathematics

Physics and mathematics are, in many cases, profoundly interrelated, as is exemplified in the historical development of these disciplines (Pietrocola 2008). In physics, the distinction between physical and mathematical knowledge is often impossible; therefore, it is appropriate to discuss, for example, physical-mathematical models (Uhden et al. 2012). The deep understanding of physical concepts is not possible without the mathematical meanings and formalism involved and defining them. However, it is possible to recognize the different levels of mathematization (Uhden et al. 2012; Karam and Mäntylä 2015) or different phases of mathematization (Mäntylä and Hämäläinen 2015) in different cases of physical concepts.

Ataíde and Greca (2013) have classified three stages of the relationship between physics and mathematics: (1) Mathematics is used to bridge the physical objects and phenomena of the real world to the idealized and abstract world of physical structures. This is particularly done using geometry. (2) Mathematics is the language that physics uses in describing the world. (3) Mathematics and mathematical structures have a profound role in forming physics knowledge in guiding this knowledge formation process. The stages show also the different epistemological roles that mathematics can play in physics. Ataide's and Greca's (Ataíde and Greca 2019) research of problem-solving in physics highlights that the epistemic views students have affect their problem-solving strategies in physics and that the epistemological role of mathematics in physics knowledge construction should be explicitly discussed in teacher education.

The teaching of physics without mathematics is impossible; the further physics is studied the more important and profound the role mathematics plays in it. However, mathematics in physics is rather different from 'pure' mathematics (Redish 2006). Therefore, good performance in mathematics class does not automatically lead to good performance in physics class.

In mathematics teaching, physics offers real-world situations, where mathematical modelling can be applied (Uhden et al. 2012). There are also didactical solutions, where mathematics is heavily rooted in physical grounds and the emphasis is on mathematical modelling (Radtka 2015). Senn-Fennell (2000) discusses the mathematization of the world—most different disciplines use mathematics or have a robust mathematical base (e.g. physics, chemistry, medicine, ecology, sociology); therefore, a fluent mathematical communication is a necessity. He states that at school, students should become competent in using mathematical (technical) language (Senn-Fennell 2000). Thus, mathematics teachers should be good at 'speaking mathematics', that is, using correct concepts and formulations.

The specific pedagogical content knowledge (PCK) focusing on the relationship between physics and mathematics could offer some valuable insights. Pospiech, Eylon, Bagno, and Lehavi (Pospiech et al. 2019) have suggested a PCK model for addressing this issue based on an interview study of experienced teachers. They found that teachers differed in terms of their teaching principles and teaching strategies. For example, certain teachers preferred to begin with teaching the physical concepts before introducing the mathematical treatment of the topic (concept-related), while some preferred to approach the learning of the topic from its mathematical formulation (math-related). For some, the teaching began from concrete and mathematical aspects that appeared as a means to apply to ideas of physics. The examination-specific teaching strategies revealed that certain teachers emphasized technical aspects of the mathematics interplay in physics (technical-oriented) and others, for example, emphasized the use and understanding of concepts (concept-related). The results of Pospiech et al. (2019) show that even experienced teachers have, for certain aspects, limited views of the role of mathematics in physics.

11.2.2 Designing Teaching for Combining Physics and Mathematics Knowledge

The starting point for interdisciplinary teaching is that real-life phenomena or problems seldom fall into the knowledge base of one discipline; instead understanding them often requires knowledge of several disciplines. The important requirement for interdisciplinary teaching is that it requires applying knowledge and methods of different disciplines (Spelt et al. 2009). Often, there is a unifying idea or phenomenon, which is examined from the perspective of different disciplines (multidisciplinary teaching). The danger in this is that the perspectives do not coincide and what is learned includes bits and pieces of knowledge of different disciplines that do not form a coherent whole (Barton and Smith 2000). Therefore, it is important to design teaching, where the interdisciplinary aspects complement each other and thus make it possible for a learner to form a coherent whole of the idea or phenomenon under inspection.

In interdisciplinary approach in particular, it can be expected that the learning does not follow a chronological path, nor can the teaching sequence be designed as a chronological, continuous path. Therefore, the idea of the exemplary in teaching suggested by Wagenschein (2000) provides a fruitful framework for designing teaching. The exemplary in teaching implies that the teacher chooses an interesting example, which also enables one to learn essential knowledge and skills and that is thoroughly handled in teaching, even if it leaves gaps in the so-called continuous path of learning. It is believed that profound learning of the chosen example provides more satisfaction to the learner and provides such skills to the learner that she is able to fulfil any important gaps later if necessary. The framework, where the example is examined profoundly, fits into interdisciplinary teaching in order to leave time to combine different perspectives from different disciplines. In addition, the interdisciplinary thinking should develop during few carefully chosen examples instead of implementing different projects without time to build coherent wholes and reflect on them.

11.2.3 Teaching Design: Discovering the Latitude During the Autumn Equinox

The latitude of a place X on the Earth (see Fig. 11.1) is defined as the angle between a plumb from the point X and the plane of the equator (see, e.g. Karttunen et al. 1984, 25). If we know that angle, we also know the parallel of the latitude.

During the autumn equinox, the Sun is in the zenith position above some point (E) at the equator (or the spring equinox). In particular, let an observer be at a point (place) X on the northern side of the Earth when the Sun is in the south. Then, point E is the intersection of the meridian, which goes through point X and the equator. If place X is, for example, the Finnish town Tampere, that intersection point E appears to be somewhere on the borderline between Congo and Ruanda, close to the Congo River. In the year 2015, the date of the autumn equinox was 23 September, and the Sun was then in the south 13:18 h in the horizon of Tampere.

On grounds of the astronomical and physical facts above, we can think up of a measurement system to find the latitude of the point X in the following manner, for example, the next figure provides an understanding of the measurement system (Fig. 11.1). By examining Fig. 11.1 geometrically, we see that *as corresponding angles, the angle EOX is equal to the angle NKX, because the lines a and b are parallel.* Then, by measuring the length of stave (KX) and its shadow (XN), we can determine the angle NKX. This angle equals the latitude of point X.

There appear to be mainly geometric elements in Fig. 11.1: points, a circle, lines, particularly parallel lines, angles, etc. However, the real meaning of Fig. 11.1

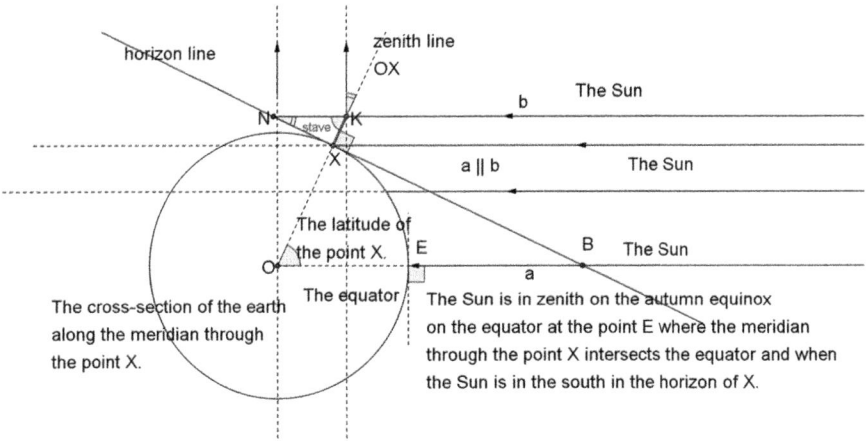

Fig. 11.1 The main conceptual elements in discovering the latitude (the angle EOX) of place X through the measurement on the Earth of the angle NKX in the right-angled triangle NKX. (Of course, for example, the length of the stave and the length of the radius of the Earth cannot be in any reasonable scale here)

comes from the possibility of physical interpretation of these elements. With the geometrical points in our circle, we can model the location of cities or some other places on the Earth's surface—for example, standing straight in place X on the Earth's surface implies that we are standing straight on a (geometric) horizontal plane, that is, on the plane which is perpendicular to the zenith line OX and which has just one common point X with the earth (sphere). Figure 11.1 comprises (geometrically) a cross-section of the Earth along the meridian through the point X, and, of course, that is the reason why in Fig. 11.1, it depicts a horizon line, etc.

As is evident from Fig. 11.1, there is a natural and reasonable interplay between geometry and physics. The geometry there reflects or models the real world. For example, putting a stave physically in balance on the ground (at the place X) means that geometrically we place it at a right angle to the horizon plane/horizon line. In Fig. 11.1, this is described with the segment KX. The Sun is somewhere far away in the space. But we know experimentally that it reaches its highest daily position above the horizon in the same direction always, which is called the south; on the other hand, there are certain physical reasons for this phenomenon. This is why we have made the cross-section along the so-called meridian in Fig. 11.1. Further, we know by experience that if we put, for example, two staves vertically on the floor, their shadows on the floor caused by the sunlight will be parallel. Observations like these give us grounds to model the sunlight geometrically using parallel lines. The reason for shadows is that light proceeds directly and therefore is blocked by an opaque object.

There are several rather simple astronomical measurements that one can apply and where the disciplines of physics (astrophysics) and mathematics (geometry) are combined in an interrelated manner. For example, Camino and Gangui (2012) have also developed a similar instruction sequence for discovering the latitude. Eratosthenes was the first who obtained a rather accurate determination of the Earth's circumference using similar knowledge of the solstice and the Sun's rays in the third century BC (Bekeris et al. 2011; de Hosson and Décamp 2014). Similar measurements like that of Eratosthenes are used at schools (Božić and Ducloy 2008; Bekeris et al. 2011) and in teacher education (de Hosson and Décamp 2014).

11.3 Methodology

The interest here is to see how pre-service physics and mathematics teachers combine the knowledge of physics and mathematics and still preserve the distinct features of these disciplines. The research questions are related to discovering the latitude:

1. How do pre-service teachers explain and justify the discovery of the latitude from a:

   (a) Mathematical perspective?
   (b) Physical perspective?

2. How are physics and mathematics interrelated in pre-service physics teachers' reports?
3. How do they evaluate the applicability of the task at the school level?

11.3.1 Context and Participants

The context of the study was a course on subject didactics (5 cr), which is part of pre-service teachers' pedagogical studies (60 cr) in School of Education. One credit (cr) approximately equals 27 h of work. This course is the first course on subject didactics, and the current study was conducted before the pre-service teachers' first teaching practice. The course lasted for 1 semester (14 weeks) and included 50 h of seminars and approximately 80 h of independent work, for example, seminar diary and 6 assignments on various aspects of teaching science and mathematics. The case of this study is one of those six assignments.

The participants of the study are 21 pre-service teachers from the University of Tampere and Tampere University of Technology. There were three pre-service teachers, whose major subject (field of engineering or information science) was not a subject taught at school. However, they are classified according to the subjects taught at school. Thus 15 of the pre-service teachers had mathematics, 4 had physics, and 2 had chemistry as their major subject to be taught at school. Those who had chemistry as a major subject also had physics and mathematics as their minor subjects to be taught at school, and those who had physics as a major had mathematics as a minor subject to be taught at school. Ten of the mathematics major students had physics as a minor subject to be taught at school, and one had chemistry. There were four mathematics major students, who did not have physics (or chemistry) as a subject to be taught at school. Overall, 16 pre-service teachers out of 21 (76%) had both physics and mathematics as their subjects to be taught at school[1].

11.3.2 Task

We assigned a measuring task of Tampere's latitude to our students:

> Put vertically a stave on the horizontal level going through the point X on the Earth, that is, on a floor, on a field, etc., during the autumn equinox. At the moment the Sun is in the south, you have to measure the length of the shadow of your stave. Now, you have a right-angled triangle whose legs are the lengths of the shadow and the stave. Draw it on paper on some suitable scale and measure the angle opposite to the length of the shadow (or calculate it using trigonometry). This angle is the latitude of the point X.

We also asked them to explain and justify why the measurement and its result gives us the latitude of Tampere. This required explaining why the angle EOX = the angle NKX (see Fig. 11.1). Further, students also had to contemplate how such a task could be used at school and what could be a suitable context (school subject, school level) for it. The reason for this was to make the shift from the learner's perspective to teacher's perspective easier. Figure 11.1 presented earlier was also included in the task. However, the knowledge that the lines a and b are parallel was added afterwards for this article.

We conducted a seminar on 22 September 2015. We went to the park next to the university to do the measurement at 13:18 h. In their measurements, students mainly either used themselves and measured the length of their own shadows (Fig. 11.2 left) or a pencil on paper and its shadow (Fig. 11.2 right).

11.3.3 Data and Analysis

This is a case study concerning how pre-service science and mathematics teachers are able to explain and justify an interdisciplinary phenomenon of discovering the latitude. The data comprises 21 students' reports of the assignment of discovering the latitude of Tampere. The students consented to the use of their reports. The length of the reports varied from one page (only one student) to three pages, and the average length of reports was two pages.

---

[1]In Finland, the major subject to be taught at school is studied in minimum 120 cr, and the minor subject(s) is studied minimum 60 cr (1 cr = 1 ects = 26.7 h of studying).

Fig. 11.2 Pre-service teachers doing the measurement. On left, the shadow of a student is measured. On the right, the shadow of a pen is measured

As an analysis method of reports, content analysis was applied because it allows the examination of relative frequencies of topics of interest across written data (Cohen et al. 2007). First, students' reports were read through several times. Related descriptions and notions were written down and were further reduced, thereby resulting in a coding sheet. Next, students' reports were coded and compared with each other, and categories of mathematical and physical explanations were formed on the basis of the most evident similarities and differences. The features that were paid most careful attention to and that were coded included (1) the use of mathematical justifications, (2) the description and explanation of the physical situation and measurement, and (3) how these necessary parts are interrelated in students' reports. As it was discussed earlier, the distinction between mathematical and physical knowledge is not often clear or even necessary. However, here the distinction is made: the knowledge is physical if it involves a real-life situation or typical empirical methodology or tradition of physics; the knowledge is mathematical if it does not directly include or require any real-life situation (i.e. instead of a shadow's length, a student is using line XN). The mathematical knowledge here includes the general semantic structure and principles of geometry and trigonometry, and the mathematical justifications do not require any specific physical or other system. After this iterative process, categories of mathematical and physical explanations and justifications were created. The other author crosschecked the entire data using the coding sheet. The inter-rater agreement of 84% was considered adequately good.

In order to evaluate the overall quality of the report, the different categories were scored and added together. The score table is presented in Table 11.5. The scores varied between 4 and 17. Because of the wide spread of the overall scores and the differences in the quality of reports, we chose to divide the overall quality into four different categories. Three categories would have been too rough, and there were no grounds for five categories.

In addition, the students had to reflect on how this kind of task of measuring latitude could work in actual school settings. The school level and the subjects that students believed to be appropriate were categorized. Further, if a student reflected on the task to the goals of national curriculum or a method to implement it in actual instruction, it was recognised from the reports.

11.4 Results

Next, the results of analysis of 21 reports are presented and discussed.

11.4.1 Mathematical Description

Students had to explain why the measurement yields the latitude of Tampere. From the mathematical perspective, students had to justify why the angle EOX = the angle NKX using Fig. 11.1 for help. Four different subcategories of mathematical justification were created and are presented in Table 11.1.

Absent Student has not justified why the angle obtained from the measurement equals to the latitude of Tampere.

Trigonometry Is Evident from the Figure Student just states either that with the help of trigonometry, the obtained angle equals to the latitude or that it is evident from the figure (Fig. 11.1):

> Generally, the latitude gives the location of the given point (Tampere in this case) on the Earth's surface from the equator. The forming acute angle corresponding to the length of the shadow is exactly of the same degree as the angle of the triangle which is drawn into the centre of the earth (see Fig. 11.1). This is easy to justify by trigonometry using the properties of the vertical angle and the right-angled triangle. (S13)

Table 11.1 Categories of mathematical explanations

Category	Science majors (N = 6)	Mathematics majors (N = 15)	All (N = 21)
Mathematical justification			
Absent	33% (2)	–	10% (2)
Trigonometry/evident from figure	–	27% (4)	19% (4)
Similar triangles	50% (3)	20% (3)	29% (6)
Parallel lines, corresponding angles	17% (1)	60% (9)	48% (10)
Equation of latitude/protractor	67% (4)	67% (10)	67% (14)

We see from the Figure that the latitude, that is, the angle O, and the angle K formed by the pencil [= "stave"] and its shadow are of the same size. (S10)

Similar Triangles Student states that the triangles EOX and NKX are similar without actually justifying the statement:

By analysing the figure, we find by means of trigonometry that the right-angled triangles OKB' and KNX, which is formed by the pencil placed vertically on the ground, and of its shadow, are similar. (S3)

Parallel Lines and Corresponding Angles Student justifies why the obtained angle equals to the latitude using parallel lines and corresponding angles:

Let us denote by the letter B the intersection of the line, which is continued from the segment XN, and the ray of sunlight a. The triangles XNK and XBO are congruent (angle-angle):

1. The angles OXB and NXK are right angles and are thus equal.
2. The segments NK and OB are parallel (two rays of sunlight a and b); therefore the segment KO meets them in the same angle. Thus, the angles NKX and XOB are equal. Therefore, the triangles are congruent.[2] (S8)

It was also checked, if the student had explicitly stated, how the latitude was obtained from the measurement, that is, did the student present the trigonometric equation for calculating the latitude or stated it using a protractor (last row of Table 11.1).

The latitude of the point X corresponds to the angle XOB, that is, to the angle NKX. Let us measure the lengths of the pencil (XK) and its shadow (XN). The desired angle

$$\angle(XOB) = \arctan\frac{(XN)}{(XK)} = \arctan\frac{The\,length\,of\,the\,shadow}{The\,length\,of\,the\,pencil}.\,(S8)$$

As evident from Table 11.1, it was not easy for our students to find this geometric explanation even though they had the illustration of Fig. 11.1 in use. Only 48% of the students provided an adequate mathematical justification. It was also seen that science major students struggled more in providing an appropriate justification (17%). One-third of the students did not explain how they calculated or solved the angle corresponding to the latitude from the measurement results.

11.4.2 Physical Description

The examination of the task required several physical aspects from students (Table 11.2). First, the physical situation required explanation. The following were the aspects of the physical situation that required explanation:

Defining the Latitude It is the object of the task, and, naturally, it is expected that a careful explanation be provided:

---

[2]Two comments: (1) Although the mathematical justification is correct, there is also unnecessary information. (2) The student uses the term *congruent*, although the term *similar* is the appropriate one.

Table 11.2 Categories of physical explanations

Category	Science majors (N = 6)	Mathematics majors (N = 15)	All (N = 21)
Physical situation			
Defining latitude	50% (3)	40% (6)	43% (9)
Justification of measurement time			
Absent	17% (1)	27% (4)	24% (5)
Directly from the task	33% (2)	27% (4)	29% (6)
Explicitly discussed	50% (3)	47% (7)	48% (10)
Parallel sunrays	17% (1)	40% (6)	33% (7)
Earth's flat surface	33% (2)	33% (5)	33% (7)
Shadow	17% (1)	7% (1)	10% (2)
Measurement			
Can be replicated	67% (4)	60% (9)	62% (13)
Results			
No results	–	7% (1)	5% (1)
One result	50% (3)	53% (8)	52% (11)
Two results	33% (2)	40% (6)	38% (8)
Several results, average value	17% (1)	–	5% (1)
Examination of results			
Comparison to real value	100% (6)	73% (11)	81% (17)
Discussion of factors influencing the result	33% (2)	33% (5)	33% (7)

> The latitude of the point X means the angle between the zenith line OX, which goes through it and the centre of the Earth, and the equatorial line, which goes through the centre of the Earth. (S9)

Justification of Measurement Time The reason for conducting the measurement at midday during the autumn equinox is important to ensure that the measurement produces sensible results. This also gives meaning to the physical/astronomical context. Three subcategories – Absent, Directly from the Task and Explicitly Discussed – were formed:

Absent The measurement time is not justified.

Directly from the Task Student has copied the text from the assignment but has not pondered it more deeply.

Explicitly Discussed Student provides an explicit justification for the reason of conducting the measurement during autumn equinox:

> Because the axis of the Earth is inclined relative to the Sun at approximately 23.4 degrees, we cannot see directly from the Sun which is our latitude. [. . .] If we would measure the angle between the rays of sunlight and the surface of the Earth during the summer solstice, when the Sun is in the south, we would get as a result the measure of 90 degrees—the latitude +23.4 degrees; during the winter solstice, the result would be 90 degrees—the latitude –23.4 degrees.

During the autumn equinox (and the spring equinox), the Earth is in a such position that measuring in the midday the angle between the rays of sunlight and the surface of the Earth gives us exactly our latitude. [...]. The angle of altitude of the Sun is then the latitude, because then the angle between the axis of the rotation of the earth and the line defined by the earth and the Sun is normal. Practically, this means that in the pole area the Sun shines exactly from the horizon (90 degrees latitude).

If we then begin to move from the pole along the surface of the Earth towards the Sun, we move directly along the meridian. Simultaneously, we remain in the position where the Sun shines from its highest position. Therefore, the Sun shines from the south if we are coming from the North Pole. The angle decreases exactly in relation to latitude parallel to where we are (in reality, the Earth's shape is ellipsoid, not a sphere). On the 0 parallel of latitude, we are on the equator and during the autumn equinox (and spring equinox), the Sun shines at midday exactly from the zenith.[3] (S1)

Parallel Sunrays The astronomical scales are not easy to understand, and the Sun is treated as a point source, particularly by the students at school. Therefore, an explanation of parallel Sun's rays is expected from future teachers, that is, the explanation for why the lines a and b (Fig. 11.1) are parallel:

The Sun is very far away from the Earth, so we can approximate that its rays are parallel. (S15)

Earth's Flat Surface The surface is assumed to be flat, or the student is stating that unevenness or inclination of the surface might have caused some error in the measurement:

The road where I put the paper was not necessarily quite horizontal. (S14)

Shadow The measurement is based on measuring the length of a shadow. This offers an opportunity to discuss the behaviour of light:

Some objects, say a pencil or a lamp post, generate a shadow.[4] The object and its shadow are measured and interpreted as two sides of a triangle. (S16)

Measurement Next, the aspects of measurement were categorized to Can Be Replicated and Results:

Can Be Replicated The student has described the measurement process in such a detailed manner that someone else with similar base knowledge can replicate the measurement. This is the general 'thumb rule' of describing measurements. Most often, the reason for not fulfilling the criteria for this category was that the student did not explicitly state that the pen, etc. should be orthogonal to the Earth's surface. One student dangled a pen attached to a string in order for it to be orthogonal towards the surface.

Results There were three distinct subcategories of measurement results:

---

[3]Although the measurement time is explained, there is a misconception: the latitude is confused with the altitude of the Sun.
[4]The forming of the shadow is not explicitly connected to the straightforward path of the sunlight.

No Results The student describes the procedure of measurement and discovering the latitude but lacks the actual measurement.

One Result The student makes one measurement and calculates the latitude based on the measurement results.

Two Results The student has two values for the latitude from the measurement of two different objects.

Several Results, Average Value The student has several different measurements for different objects. The value of the latitude is the average value based on the measurement results. This is the physical way of doing measurements in school settings and minimizes errors compared to a single measurement result.

Examination of Results The final category of physical explanations is the examination of results with two subcategories Comparison to Real Value and Discussion of Factors Influencing the Result:

Comparison to Real Value The student has compared the obtained latitude to the actual value of Tampere's latitude. In some cases, the source of the latitude's value is not revealed, but in most cases, students 'Googled' the actual latitude of Tampere. Some students also calculated the deviation between the obtained results and Tampere's latitude.

Discussion of Factors Influencing the Result The student has contemplated the different factors affecting the obtained result, such as the accuracy of the measurement.

The results presented in Table 11.2 show that, overall, the physical explanations were slightly more challenging to students than mathematical ones. Only 43% discussed what is implied by the identified latitude. One student with a physics background put effort in discussing the different types of latitudes—geocentric and geodetic ones; he also calculated from the obtained value the geodetic latitude. Less than half (48%) discussed why the measurement is conducted at midday during the autumn equinox. Parallel rays are essential for the measurement design and only one-third of the students discussed it. Here, the mathematics major students were more likely to explain this (40% compared to 17%).

Only one student had conducted the measurement and its treatise in a proper manner; most students were satisfied in one (52%) or two (38%) results. In case of two results, they were treated separately, and the average value was not calculated. One student was so interested in the task that he repeated the measurement in St. Petersburg during his vacation. One-third of the students contemplated the factors influencing the result, such as (in the order of commonness) the accuracy of measurement, the orthogonality of the pen towards the surface (using the plumb line is suggested), uneven or inclined surface, and that the atmosphere bends the light a little.

Students managed best in describing the measurement in such a manner that it can be replicated (62%) and almost all (81%) compared the obtained latitude to the actual latitude of Tampere. The four mathematics majors who did not compare the obtained value were those who also relied only on one measurement result.

11.4.3 Interplay of Physics and Mathematics in Students' Reports

Next, the reports were examined as a whole. The interest was to see how students describe the process of discovering the latitude of Tampere and if in their descriptions either the physical or the mathematical language is emphasized. The main emphasis was placed on the description of the situation, because in describing the measurement, the physical aspect is naturally rather strong. Four different categories were formed (Table 11.3):

Only Physical Description The report concentrates on describing the physical situation and the measurement, thus leaving the mathematical treatment at a vague level.

Only Mathematical Description The physical situation is described in a mathematical manner, and the description of measurement fulfils the minimum requirement. The mathematical aspect is clearly emphasized.

Both, but Separately The mathematical and physical aspects are discussed in separate paragraphs, and the overlaps are not obvious.

Both in a Combined Manner In the description of the situation, the mathematical and physical aspects are interrelated in an inseparable manner. The report proceeds in a seamless manner.

Table 11.3 shows that approximately half of the students' reports (52%) indicate that physical and mathematical language and aspects are combined seamlessly. However, half of the science major students concentrated only on the physical aspects of the task. One-third of the mathematics major students had both aspects, but they were not able to combine them. One mathematics major student, who did not have any science minor subjects, described mainly the mathematics of the situation.

Table 11.3 Categories of intertwinement of mathematical and physical descriptions

Category	Science majors (N = 6)	Mathematics majors (N = 15)	All (N = 21)
Only physical description	50% (3)	7% (1)	19% (4)
Only mathematical description	–	7% (1)	5% (1)
Both, but separately	–	33% (5)	24% (5)
Both in combined manner	50% (3)	53% (8)	52% (11)

11.4.4   Overall Quality of Reports

Finally, the overall quality of describing the process of discovering the latitude of Tampere was evaluated. The different categories presented in Tables 11.1, 11.2, and 11.3 were scored. The scoring is presented in Table 11.5. The maximum score was 18 points. The reports were classified into four different categories according to their overall scores (see Table 11.4):

Poor The report does not meet the requirements. The latitude is obtained, but the fruitful mathematical and physical aspects are omitted. Thus, students in this category have not yet begun to think and argue like teachers.

Moderate There are already necessary pieces of information, but overall the report is still fragmented or deficient.

Good The students already have a rather good sense of necessary knowledge, and they argue it in a reasonable manner. The versatility of the report shows that the teachers' perspective has begun to form.

Excellent The reports treat the task in a versatile manner but simultaneously in a coherent and balanced manner. The level of explanation is at the level necessary for a teacher.

As is evident from Table 11.4, the students are scattered rather evenly into different categories. One has to bear in mind that most of the students are still quite new to the teaching and, therefore, are not experienced in explaining what a necessary skill is for a teacher. In their subject studies, the explanations and justifications of the solution processes are rarely required; instead, often only the correct solutions are emphasized. In the reports (classified into the Good and Excellent categories, in all except one report, the physical and mathematical aspects were treated in a balanced and interrelated manner; in reports classified as Poor, the interrelated treatment of physical and mathematical aspects was lacking.

11.4.5   Applicability of Task in School

All students evaluated that this task could be applied to school. Twelve students stated that depending on the manner in which it is implemented, it could be done

Table 11.4 Categories of the overall quality of reports

Category of quality	score	Science majors (N = 6)	Mathematics majors (N = 15)	All (N = 21)
Poor	4–6	33% (2)	13% (2)	19% (4)
Moderate	7–9	33% (2)	27% (4)	29% (6)
Good	10–12	–	47% (7)	33% (7)
Excellent	13–18	33% (2)	13% (2)	19% (4)

Table 11.5 The scoring of
students' reports. Maximum
score is 18 points

Category	Score
Mathematical justification	
Absent	0
Trigonometry/evident from figure	1
Similar triangles	2
Parallel lines, corresponding angles	3
Equation of latitude/protractor	1
Physical situation	
Defining latitude	1
Justification of measurement time	
Absent	0
Directly from the task	1
Explicitly discussed	2
Parallel sunrays	1
Earth's flat surface	1
Shadow	1
Measurement	
Can be replicated	1
Results	
No results	0
One result	1
Two results	2
Several results, average value	3
Examination of results	
Comparison to real value	1
Discussion on factors influencing the result	1
Only physical description	0
Only mathematical description	0
Both, but separately	1
Both, combined	2

either in lower or upper secondary school. Some of them could also see this task
done in elementary school. Six students said that the task could be implemented
only in lower secondary school and two said that it could be implemented in upper
secondary school.

Students had also to think about which school subject lesson the task could
be implemented as part of. All except one believed that it could be done within
two to four different subject lessons. The most common school subjects were
mathematics (20), physics (15), and geography (14); ten students suggested that the
task combines these three subjects. One student suggested combining history, and
one stated physical education (orienteering) to augment the task. There was also a
suggestion to expand it further:

> It can be done in small groups or as whole group demonstration. It can be augmented as
> a part of process drama to bigger school project. You can add, for example, a trip to the
> observatory. (S21)

Further, students also recognized the interdisciplinary potential of the task:

> The phenomenon can be examined from different disciplinary viewpoints. The pupil sees
> how one can determine important knowledge through simple methods and calculations and
> how geometry can be applied in real life. (S18)

> Every subject has its own viewpoint on the topic and it would be interesting to combine
> them as a whole. (S13)

11.5   Discussion

From the mathematical perspective, approximately half of the pre-service teachers' explanations in the reports were far from what is expected from a teacher, although the required subject content of mathematics was not particularly difficult. This was regardless of the amount of support and scaffolding given to pre-service teachers in the tasks, for example, the geometric construction in Fig. 11.1. This is also evident from the lack of precise mathematical language or in the ability to understand what counts as the mathematical explanation of why the angles NKX and EOX are equal. This suggests that more efforts in speaking and writing mathematics should be put in mathematics teacher education in a manner discussed by Senn-Fennell (2000). Here, the results also suggest a difference between the science and mathematics major students: only 1 out of 6 science (17%) students provided adequate mathematical justification, and in the case of mathematics students, 9 out of 15 (60%) managed this.

Overall, the physical perspective in explaining and justifying the discovery of the latitude was weaker in students' reports. The essential concept of latitude was explained by less than half of the students, and other important features of the situation (shadow, etc.), which are the premises of the measurement or the factors influencing the measurement results, were either not recognized or not seen necessary in most of the reports. Only one student presented an appropriate method to treat measurement results, and almost all students were satisfied with either one or two separate measurement results. It is possible that this happened because the pre-service teachers have just entered the world of mathematics and science education and they have barely begun their process of shifting from learners to teachers. This implies that they have to begin playing a more active role in constructing scientific explanations and knowing what is required in these explanations.

Approximately half of the students managed to combine the mathematical and physical description in an interrelated manner. One-fourth provided only a physical or mathematical description, and one-fourth of the students provided both but in a separate manner. Although the pre-service teachers' PCK on this issue is not directly examined, instead, the representation of the subject matter content knowledge is emphasized more; in this context, there is a resemblance with the results of Pospiech et al. (2019)—for example, in presenting only the physical description resembles the case of concept-related views of teaching strategies. The pre-service teachers' views of the different roles of physics and mathematics can affect the manner in which they represented their reports; based on the reports, the views were not yet very developed. Therefore, in teacher education, the PCK model for mathematics in physics teaching proposed by Pospiech et al. (2019) could provide structure for discussing the interplay. However, there were also very good reports, where the interplay of physics and mathematics was in balance and a coherent story of discovering the latitude of Tampere was presented. This is a sign that the selected example fulfils the exemplary and interdisciplinary requirements set for it. It also inspired some of the pre-service teachers:

> I have to admit that these exemplary tasks truly inspires one to use creativity in designing tasks. The tasks are fun and very instructive compared to the mechanical and "boring" textbook tasks. (S11)

> Fun and functional method (S19)

> The measurement was in my opinion a great way to combine the topics of different subjects. They support each other extremely well in the measurement. I have never done such measurement; therefore, it was very interesting measurement for a student of my age. (S20)

There was no drastic difference between science and mathematics major students, which implies that, in this case, the major subject was not overemphasized in students' reports.

The interplay of physics and mathematics in the example is identified in the first stage in Ataíde's and Greca's (2013) classification: the real-world situation is abstracted to geometric construction. This stage was the first in history of physics and mathematics where mathematics began growing as an inherent part of physics. Most pre-service teachers recognize this and believe that the example could be applied at lower secondary level or even at elementary level. The task given to students did not require them to reflect on the relationship between physics and mathematics, but some of them did it:

> The task trains both the skills of physics and mathematics. The task also interrelates mathematics and physics usefully in everyday life. It also helps to bring the space closer to concrete. (S21)

Perhaps, as Ataide and Greca (Ataíde and Greca 2019) have suggested, an explicit instruction on the epistemological roles of physics and mathematics, in this case, could improve students' skills of meaningfully applying physics and mathematics knowledge in a combined manner.

11.6   Conclusion

The selected activity of discovering the latitude of Tampere enables the meaningful, interdisciplinary teaching: knowledge of physics and mathematics has to be applied and combined; consequently, experience and understanding about the real-world phenomena (autumn equinox), concepts (latitude), and objects (Earth and Sun) increase. Pre-service teachers also see that this example could be applied at school, combining the subjects of mathematics, physics, geography, and even history. However, it appears that the approach is rather new to students, and many of them could have benefited from additional support from either the instructors or their peers. Most of these pre-service teachers will be teaching both mathematics and physics, and it is essential for them to be able to bridge the gap between these subjects, because in real life, these subjects are often deeply interrelated.

References

Applebee, A., Adler, M., & Flihan, S. (2007). Interdisciplinary curricula in middle and high school classrooms: Case studies of approaches to curriculum and instruction. *American Educational Research Journal, 44*(4), 1002–1039.

Ataíde, R., & Greca, I. (2013). Epistemic views of the relationship between physics and mathematics: Its influence on the approach of undergraduate students to problem solving. *Science & Education, 22*, 1405–1421.

Ataíde, R., & Greca, I. (2019). Pre-service teachers' theorems-in-action about problem solving and its relation with epistemic views on the relationship between physics and mathematics in understanding physics (This book).

Barton, K., & Smith, L. (2000). Themes or motifs? Aiming for coherence through interdisciplinary outlines. *The Reading Teacher, 54*(1), 54–63.

Bekeris, V., Bonomo, F., Bonzi, E., García, B., Mattei, G., Mazzitelli, D., Dawson, S., de la Vega, C., & Tamarit, F. (2011). Eratosthenes 2009/2010: An old experiment in modern times. *Astronomy Education Review, 10*, 1–9.

Božić, M., & Ducloy, M. (2008). Eratosthenes' teachings with a globe in a school yard. *Physics Education, 43*(2), 165–172.

Camino, N., & Gangui, A. (2012). Diurnal astronomy: Using sticks and threads to find our latitude on Earth. *The Physics Teacher, 50*(1), 40–41.

Cohen, L., Manion, L., & Morrison, K. (2007). *Research methods in education.* London: Routledge.

de Hosson, C., & Décamp, N. (2014). Using ancient Chinese and Greek astronomical data: A training sequence in elementary astronomy for pre-service primary school teachers. *Science & Education, 23*, 809–827.

Feng, L. (2012). Teacher and student responses to interdisciplinary aspects of sustainability education: What do we really know? *Environmental Education Research, 18*(1), 31–43.

Karam, R., & Mäntylä, T. (2015). The influence of mathematical representations on students' conceptualizations of the electrostatic field. In F. Claudio & R. M. Sperandeo Mineo (Eds.), *Teaching/learning physics: Integrating research into practice. Proceedings of the GIREP-MPTL 2014 International Conference* (pp. 819–826). Palermo: Dipartimento di Fisica e Chimica, Università degli Studi di Palermo.

Karttunen, H., Kröger, P., Oja, H., & Poutanen, M. (1984). *Tähtitieteen perusteet.* (Introductory astronomy). Tähtitieteellinen yhdistys Ursa. Helsinki.

Mäntylä, T., & Hämäläinen, A. (2015). Obtaining laws through quantifying experiments: Justifications of pre-service physics teachers in the case of electric current, voltage and resistance. *Science & Education, 24*(5–6), 699–723.

Opetushallitus. (2015a). Perusopetuksen opetussuunnitelman perusteet 2014. (Finnish National Agency Education. (2015a). National core curriculum of basic education 2014). Juvenes Print – Suomen Yliopistopaino, Tampere.

Opetushallitus. (2015b). Lukion opetussuunnitelman perusteet 2015. (Finnish National Agency Education. (2015b). National core curriculum of general upper secondary school 2015). Next Print Oy, Helsinki.

Pietrocola, M. (2008). Mathematics as structural language of physical thought. In M. Vicentini, & E. Sassi (Eds.), *Connecting research in physics education with teacher education*. International Commission on Physics Education. https://web.phys.ksu.edu/icpe/Publications/teach2/index.html

Pospiech, G., Eylon, B., Bagno, E., & Lehavi, Y. (2019). Role of teachers as facilitators of the interplay Physics and Mathematics. This book.

Radtka, C. (2015). Negotiating the boundaries between mathematics and physics: The case of late 1950s French textbooks for middle schools. *Science & Education, 24,* 725–748.

Redish, E. (2006). Problem solving and the use of math in physics courses. In *Proceedings of the Conference, World View on Physics Education in 2005: Focusing on Change*. Delhi, August 21–26, 2005, arXiv:physics/0608268

Senn-Fennell, C. (2000). Oral and written communication for promoting mathematical understanding: Teaching examples from Grade 3. In I. Westbury, S. Hopmann, & K. Riquarts (Eds.), *Teaching as a reflective practice—The German Didaktik tradition* (pp. 223–250). Mahwah: Lawrence Erlbaum Associates, Inc.

Spelt, E., Biemans, H., Tobi, H., Luning, P., & Mulder, M. (2009). Teaching and learning in interdisciplinary higher education: A systematic review. *Educational Psychology Review, 21,* 365–378.

Uhden, O., Karam, R., Pietrocola, M., & Pospiech, G. (2012). Modelling mathematical reasoning in physics education. *Science & Education, 21,* 485–506.

Wagenschein, M. (2000). Teaching to understand: On the concept of the exemplary in teaching. In I. Westbury, S. Hopmann, & K. Riquarts (Eds.), *Teaching as a reflective practice—The German Didaktik tradition* (pp. 161–176). Mahwah: Lawrence Erlbaum Associates, Inc.

Part III
Teaching Mathematization

Chapter 12
Role of Teachers as Facilitators of the Interplay Physics and Mathematics

Gesche Pospiech, Bat-Sheva Eylon, Esther Bagno, and Yaron Lehavi

12.1 Introduction

Doing physics is unthinkable without mathematical elements and structures to describe and predict physical processes. Therefore, not only in higher education or high school but also at secondary school the fundamental structural role of mathematics in physics should be taught and learned. However, many lecturers at college or university complain about rote calculation with plug'n chug techniques that are applied by students and do not further physics understanding. These views were confirmed in many studies on students' problem-solving strategies with physical-mathematical problems. So the question arises what happens in the corresponding teaching-learning processes as early as in physics lessons of secondary school. In this context the role of teachers comes into play. Teachers, their stances and their teaching methods play the decisive role for learning as is confirmed by many studies. The specific knowledge teachers need for identifying their goals, for shaping their lessons and the instructional strategies and assessing the learning progress of their students is summarized in the pedagogical content knowledge (PCK).

Concerning mathematics education, it was found that high content knowledge and high pedagogical content knowledge (PCK) of teachers were positively related to better learning success among their students (Baumert and Kunter 2013). Studies with prospective physics teachers indicated that pedagogical content knowledge is indeed a special body of knowledge (Riese 2010). It may be correlated with content

G. Pospiech (✉)
Technische Universität Dresden, Dresden, Germany
e-mail: gesche.pospiech@tu-dresden.de

B.-S. Eylon · E. Bagno · Y. Lehavi
The Weizmann Institute of Science, Rehovot, Israel
e-mail: bat-sheva.eylon@weizmann.ac.il; esther.bagno@weizmann.ac.il

© Springer Nature Switzerland AG 2019
G. Pospiech et al. (eds.), *Mathematics in Physics Education*,
https://doi.org/10.1007/978-3-030-04627-9_12

knowledge but is nevertheless different and specific for physics teachers implying that it is also not only pedagogical knowledge as teachers of other subjects do not show elements of PCK for physics teachers (Riese 2010).

As the interplay of mathematics and physics covers a special aspect of physics teaching relating two school subjects, it seemed appropriate to establish a specific PCK model. This should rely on generally accepted models of PCK in physics but allow for defining the necessary specific knowledge of teachers in the interrelation of mathematics and physics. This normative determination of a PCK-phys-math-interplay model then provides the hull containing different aspects or categories. The next step would be to explore what could reasonably be expected from teachers concerning insight into the interplay and its realization in teaching. The benchmark for this would be the views, experiences and knowledge of teachers with many years of teaching experience and with special qualifications. This would provide an empirical basis for filling the normative hull with content.

12.2 Theoretical Background

The goal of this section is to specify a known and acknowledged model of PCK with respect to teaching the interplay of mathematics and physics.

12.2.1 General Aspects of the Role of Mathematics in Physics

An important source for shaping the sought PCK-model is a theoretical view on the role of mathematics. Pietrocola (2008) distinguished a technical and a structural role. In a further refinement, Ataide and Greca (2013) also added the role of mathematics as language of physics. This is insofar an important category as it allows for a more fine-grained analysis of procedures in teaching: the role of mathematics as language refers mainly to representations. Their appropriate use is clearly beyond the purely technical role but on the other hand does not yet arrive at a deeply structural insight, e.g. for derivations, deductions or exploitation of analogies. So the role of mathematics as a language of physics plays an intermediate role between the technical and the structural role. For a more detailed description, see the contribution "Framework of Mathematization in Physics from a Teaching Perspective" in this book.

These distinctions allow to frame not only the knowledge of teachers but also their deeper convictions and viewpoints that guide their overall shaping of instruction. We want to explore these convictions and possible relations to the teachers' description of instructions. The questions are: What views and attitudes do teachers have? How do teachers realize their views in the classroom?

12.2.2 Views, Knowledge and Teaching

Generally not only cognitive aspects play an important role for learning on all levels but also epistemic or affective viewpoints (see, e.g. Bodin and Winberg 2012). As we have seen, we can distinguish several roles of mathematics in physics: technical, as a language and structural. The relative weight of these aspects depends on the individual teacher and might influence his or her students. Therefore epistemic views underlying the teachers' views on the interplay and their choice of teaching patterns are important for thinking about physics teaching. This relation was confirmed in a study finding that traditionally oriented teachers tend to view the role of mathematics as instrumental, whereas more conceptually oriented teachers viewed mathematics as the language of physics and as suitable to derive models or new insights into physics, corresponding to the structural role (Mulhall and Gunstone 2007).

Concerning the importance of the structural role, it could be shown that the conscious "self-made" connection between the physical phenomenon and the mathematical model in kinematics offers great learning opportunities for teacher students with respect to their view on the interplay (Carrejo and Marshall 2007). According to a study with 34 teacher students, "a strong relationship between students' problem solving strategy and their epistemological perception to the role mathematics plays in physics, learning and understanding physics, and solving problems in physics" exists (Al-Omari and Miqdadi 2014). Similar studies were conducted by Başkan et al. (2010) and Ataide and Greca (2013). They identified three types of teacher students with a certain alignment of their attitudes towards the role of mathematics in physics and their problem-solving strategies (see contribution in this book). In a quantitative study, it was found that prospective teachers prefer a constructivist stance on mathematical-physical modeling (Fazio and Spagnolo 2008). However, it is not clear how teacher students reflect their views on the interplay of mathematics and physics in a concrete teaching sequence. That the merging is difficult was seen in a case study where it was observed that the conceptions and actual teaching practices of teachers might diverge (Freitas et al. 2004).

In order to shape the teaching-learning process, teachers have also to be aware of the students' ideas. Only if they know the typical difficulties, they can think of appropriate teaching strategies and implement them. In a study by Khalili (2016), the teachers stated that they achieve the best results by abandoning the "number crunching" and instead emphasizing careful explaining and reasoning as well of the mathematical operations as of the physical processes. These teachers were focussing on structural elements of the interplay.

The findings of these studies mostly concerned small groups of teacher students and hence still need confirmation with respect to completeness and reliability. So additional studies will be necessary in order to reach the ultimate goal to find ways of promoting the understanding and practice of future teachers. As experience is an important component of the teachers' professionalization, it is an interesting question how experience is shaping the teachers' views, their knowledge of students' ideas and how they are framing it and how this corresponds to their daily practice. In order to get a feeling in which direction the teacher students should develop their competences during their teaching career, we explored the views of experienced teachers on the interplay physics with mathematics. As a first step, we defined and detailed a PCK-model and then validated it with the help of interviews with teachers.

12.2.3   PCK-Model for Teaching Mathematics in Physics

The definition of teachers' PCK is the first step towards a systematic description of teachers' views, knowledge and experience. The construction of a frame of PCK has proven to be fruitful for domain-specific characterization of teachers' views and their teaching strategies. Nevertheless, there are many models of PCK in physics with different focus and elements depending on the goals of research (Gramzow et al. 2013). A model in agreement with most authors is the model of Magnusson (Magnusson et al. (1999), modified by Etkina (2010)). As PCK cannot be separated from the content area (Loughran et al. 2012), we adapt this model slightly to the topic of mathematics and physics. Even if mathematics is no separate explicit topic in physics education, the teaching of the interplay mathematics and physics nevertheless requires considerable care. If we look at this special area, the need arises to capture the necessary specific competences, e.g. the distinction of technical and structural role of mathematics. A suitably modified model will serve as the starting point for the study of teachers' PCK (see the introductory chapter). We highlight its most important elements in Fig. 12.1 adding the feature of experience. Besides profound explicit knowledge, teachers command rich experience. This consists of implicit knowledge which is not described but condensed in stories about what is working in class and how to adapt the teaching goals and strategies flexibly with differing students.

In addition we also take into account a PCK model of mathematics education. Inspection of the PCK model proposed by COACTIV (Baumert and Kunter 2013) shows that there are some parallels. The COACTIV-model contains – similar to the Magnusson model – students' cognitions and curriculum but distinguishes between knowledge of short-term and of long-term teaching strategies as well as insight into overarching aspects of teaching mathematics such as the role of representations. This implies quite a strong focus on teaching strategies with an extension on more general aspects of teaching. Merging both PCK models by inserting the additional category of "Teaching Principles" into the Magnusson model, we arrived at a more detailed model of pedagogical content knowledge specific for the interplay of mathematics and physics in physics lessons (see Fig. 12.2). This model should mirror the competence and reflection of teachers by distinguishing and interrelating several facets of PCK: the overarching views and attitudes, the general teaching principles informed by the conception of teachers about the learning process and the detailed teaching strategies partly clarified by concrete examples from class

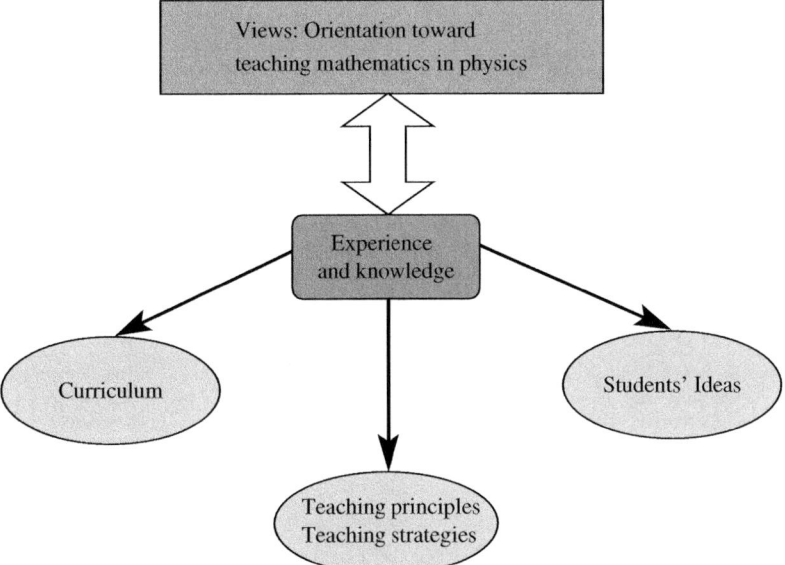

Fig. 12.1 General model of teachers' PCK based on the model of Magnusson. It is slightly modified by stressing the interrelation of teachers' views and their experience and knowledge. The foundation of the teaching and the epistemic views of teachers are included in "Orientation towards Teaching the interplay". This is in interrelation with their experience and knowledge that is shown in the choice of instructional strategies which focus on the technical or structural role, respectively. Furthermore the requirements by the curriculum and the awareness of students' ideas play an additional role

(Loughran et al. 2012). Furthermore the model includes the teachers' perception of students' views and learning. The most important feature is the interrelation of linking the abstract definition and understanding of PCK to the actual practices of experienced teachers.

12.3 Research Questions

The goal of the study presented here was to find evidence of the described PCK-model and to characterize experienced teachers' views on the role of mathematics and their teaching physics as basis for further studies and developments in teacher education and professional development. So the research questions are:

1. Can the PCK-model be confirmed in the interviews?
2. How can the teachers be characterized according to

   (a) their teaching principles?
   (b) their teaching strategies and realized teaching patterns?
   (c) their awareness of students' knowledge and attitudes?

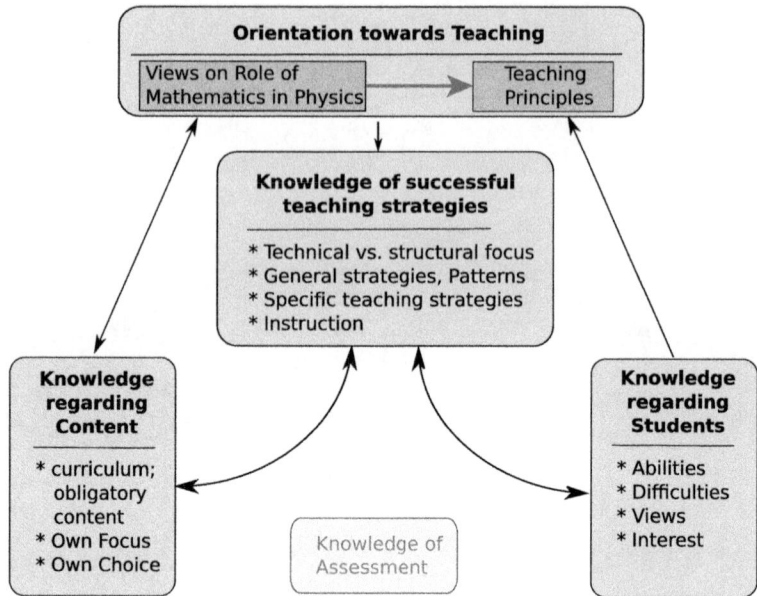

Fig. 12.2 The specification of the PCK-model with more details especially concerning teaching strategies. The aspect "Knowledge of Assessment" only is indicated as it is contained in the Magnusson model but not treated in this study. In elaborating on this model, we see an interrelation between the overall "views" of teachers and their knowledge about what is required by the curriculum and their individual preferences. The overall "views" influence on the basis of the general "Orientation towards Teaching" the individual "Teaching Principles". These on their part contribute to the choice of "successful teaching strategies", which on the other hand are influenced by the requirements of the "Curriculum" and the teachers' awareness of "Students' Views". Even if the "Teaching Principles" are informed by the "Orientation towards Teaching" they are also shaped by experience, especially the perception of students' views, competence and knowledge. The indicated details serve as a starting point for deriving the categories used for analysis of the interviews (see Sect. 12.4.2.1)

12.4   Design and Method of Study

In order to learn about the teachers' views and attitudes towards the interplay of mathematics and physics, an interview study has been conducted in Israel (8 teachers) and Germany (15 teachers). Mostly very experienced and distinguished teachers have participated in the interviews. The semi-structured interviews followed a guideline with seven questions. The first question addressed very generally the teacher's personal view on the role of mathematics in physics teaching. Other questions focused on the support or hindrance of mathematics for understanding physics or vice versa. In addition the teachers' views on students' difficulties and their corresponding instructional strategies were asked for. In order to contextualize the theoretical answers, the teachers should also describe concrete examples from their lessons.

The teachers received the questions one week in advance so that they had the opportunity of thinking about them and preparing answers. They took this chance with only one exception. The interviews took 30 to 60 minutes. From the interviews it can sometimes be seen how intense the preparation has been. They were audiotaped and completely transcribed.

The results of the interviews of the Israeli teachers have been published elsewhere (Lehavi et al. 2015). From their interviews and additional classroom observations, guiding strategies could be identified, named as "patterns". Four patterns that teachers realized in the classroom could be identified: construction pattern, exploration pattern, application pattern and broadening pattern (Lehavi et al. 2017). These are explicitly included into the model described above as the realization of these patterns is in close relation to the teaching principles and more general strategies (see Fig. 12.2).

12.4.1 Description of Participants

During the study 15 teachers (male as well as female) have been interviewed. In one interview the teacher did not agree with the audiotaping, while in another interview, the audiotaping did not work. Therefore the interviews with 13 teachers remained for analysis.

Two of the teachers could be called master teachers because of their role and official functions, e. g. in curriculum development. The 11 other teachers were all very expert teachers with at least 15 years of teaching experience, some even more than 30 years. All teachers have the license to teach physics in secondary school including high school. However, some teachers specialized either in lower secondary school or in high school (see Table 12.2). They also had as a second subject mathematics, meaning that they also have the license for teaching mathematics in secondary school including high school.

All the teachers had a very similar teacher education in which the focus was on good explaining, on clear structure of instruction and close monitoring of students' progress with high expectations of their success in assessments. The fulfillment of curriculum was very important for them all.

12.4.2 Procedure of Analysis

Interviews were analysed with qualitative content analysis. The free R-package RQDA together with simple basic functions of R (version 3.2.3–4) was used. This software allows the use of main categories, subcategories and the codings themselves. In a first step, main categories from the description of the chosen PCK model were defined deductively and refined by subcategories and codings. During the coding process, some subcategories and codings were added inductively from the data.

The category system was discussed several times with members of the group, refined and made more precise. After a first coding of the material, the descriptions of some codes were adapted and clarified which led to a recoding of some of the codings resulting in a bigger coherence. In the end there results a system of 10 main categories, 22 subcategories belonging to 8 of the main categories and comprising 98 codings. These codings are described in a coding manual and illustrated by an anchor example. Overall there are about 1100 coded text fragments, partially double coded.

12.4.2.1   Description of Main Categories

For analysis the main categories as given in the PCK model above are described (see Fig. 12.2).

The main categories mirroring the general views of teachers on the interplay are:

- Role of Mathematics in Physics
- Teaching Principles

These categories are related to the "Orientation towards Teaching" of the PCK-model of Magnusson but have been distinguished here (see Sect. 12.2.3). The Teaching Principles rely on the deeper convictions and thus serve as an intermediate step between those and the more specific teaching strategies. Hence they guide the teaching as a whole. The preference of certain aspects of the interplay is coded into the following main categories:

- Technical role
- Structural role
- Mathematics as language

These categories describe the teachers' view on these roles with a strong relation to their teaching. They are not as general as both categories above but do not yet form a strategy for lessons even if they may be regarded as a basis for the choice of "General Strategies", related to teaching patterns, identified by Lehavi et al. (2017). In a further step, they influence on a more concrete level the specific teaching strategies actually used. These teaching strategies may correspond to the technical or to the structural role as well as the role of mathematics as a language. They are attributed to

- Teaching strategies

These strategies are used by teachers to introduce and facilitate the use of mathematics. Here strategies can be distinguished that are being described as more general underlying strategies and very specific strategies applying to concrete examples which are closely related to topics from the curriculum. Concrete examples are coded in the category:

• Curriculum

In describing their strategies, teachers also discuss the views and competences of students ("Students' Views"), resulting in the following main categories:

• Students' Views on Mathematics as problem
• Students' Views on Mathematics as supporting
• Students' Attitudes

Herewith it has to be stressed that these categories refer to the teachers' views on the students' thinking.

12.5 Results of Data Analysis

First the results are described before they are interpreted in the following section. The most interesting results concerning

• Views on the interplay mathematics and physics
• Preferred teaching strategies
• Awareness of students' knowledge and attitudes

are presented.

12.5.1 General Observations

The interviews had a clear focus as can be seen in the distribution of codings among the main categories. Most coded statements belong to: "teaching strategies", "curriculum" and "perceived problems of students in applying mathematics" (see Fig. 12.3). This may partly be caused by the incitation in the interviews to support possibly general statements with concrete examples from own lessons as one aim was to shed light on the relation between views and their realization in the classroom. Only relatively few statements are related to possible support or increasing the understanding by application of mathematical elements in physics, although this topic has been explicitly addressed in the interview questions. One of the master teachers has a number of codings clearly above average. Four other teachers also have many codings, the other less codings (mainly connected with shorter interviews).

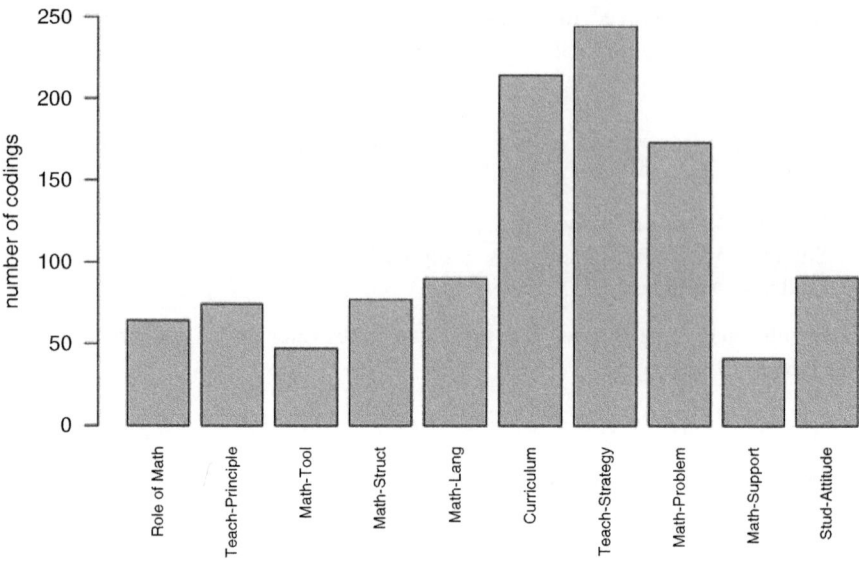

Fig. 12.3 Number of codings per main category

12.5.2 Analysis of Special Viewpoints

The analysis of the data is led by the different aspects of the role of mathematics in physics as described in Sect. (12.2.1). These serve as pivotal elements.

As the duration of the interviews varies quite significantly, the teachers have a very different number of statements and hence codings (Fig. 12.4). Therefore we base the analysis not on the absolute numbers of codings but on the percentage of specific codings in relation to the total number of codings for each teacher. This reflects more the importance of a certain point for the individual teacher and prevents that a teacher with overall many statements dominates the interpretation of results.

12.5.2.1 Views on the Interplay of Mathematics and Physics

The basis of teaching is the "Orientation towards Teaching" (see Fig. 12.1) comprising teachers' beliefs about content and goals of teaching. Therefore the teachers were first asked about their "Views on the role of mathematics in physics" and corresponding "Teaching principles".

Generally all the interviewed teachers regard the interplay as important for physics and also for teaching physics at least at high school. In analysing their views in more detail, we see that both teachers 13 and 8 give the most extensive and multifaceted view on the role of mathematics in physics and emphasize its structural

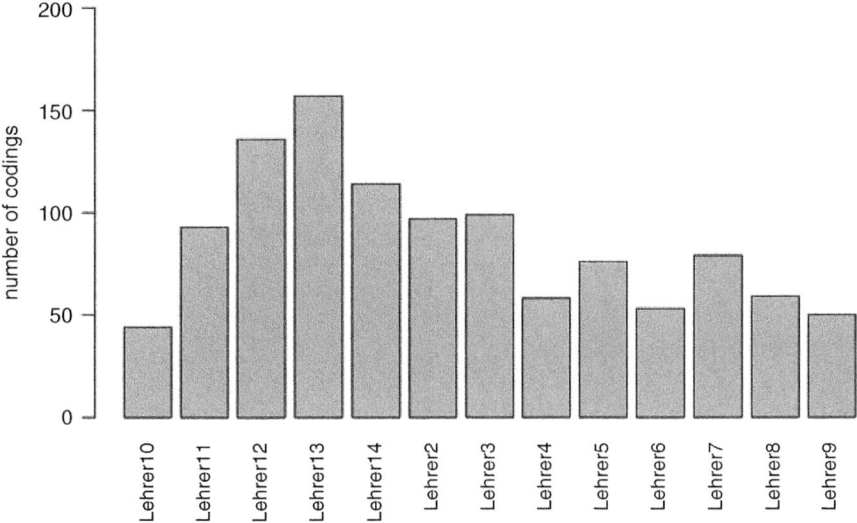

Fig. 12.4 Number of codings per teacher

role. Also teachers 14, 3 and 7 mention a variety of different aspects. Some teachers (teachers 4, 5, 9) stress the technical role of mathematics as an ancillary science

> .. For me mathematics is, I say it always so nicely ...... that it is an ancillary science. [teacher4]

On the whole a gap is seen between the very general view that mathematics is important (stated by all teachers) and more precisely what role mathematics plays in the physical method (elaborated on by 7 out of 13 teachers). Interestingly, only three teachers (teachers 3, 13, 14) mention "prediction" as one of the traits of the mathematical description of physics.

> One aspect is that one can arrive from qualitative statements, e.g. "the more – the more" at exact quantitative predictions by calculating something. [teacher 14]

The structural role implies that mathematics might be a support of physics and, e.g. can help to derive new insights. But only about half of the teachers mention these aspects explicitly. What seems more at hand for them is the necessity and possibility of the physical interpretation of mathematical elements, the function of mathematics as a language:

> ...diagrams as such. Out of them you can gain a lot. E. g. when I recognize – in the s(t) diagram – and know that the slope is a measure for velocity, ... [teacher2]

Nearly all teachers stress this for the interpretation of diagrams but only about half of them also for formula and units. Only two teachers (2, 13) appear to be conscious of all these facets.

Only one of the teachers (13) addresses all aspects of the role of mathematics and is the only one to explicitly mention the use of analogies in terms of mathematics:

Table 12.1 Percentage of statements concerning the role of mathematics in physics and method of physics in relation to all codings per teacher

Teacher	2	3	4	5	6	7	8	9	10	11	12	13	14
Percentage	9	16	10	7	11	15	20	14	9	1	4	25	16

> In the end they [formula and diagram] are also a help in recognizing analogical structures because the mathematical structures reveal analogies because of the stringent formalized language. [teacher13]

12.5.2.2 Teaching Principles

In order to find possible characteristics of teachers and their teaching, we refer to their statements concerning the guiding teaching principles which also imply general teaching goals.

In the following it is assumed that teachers with a deeper reflection refer more often to the different roles of mathematics in physics at several points of the interview, e.g. in describing teaching strategies. As the interviews differ in length, hence also in the overall number of codings per teacher, this effect should be removed in comparing teachers with respect of the weight they gave to the role of mathematics in physics. Therefore as a simple measure, the number of codings of the general role of mathematics and physics is related to all codings in the interview of a teacher (Table 12.1). This number varies significantly among the teachers. One of the master teachers (teacher13) has a bigger percentage (25%) of "reflective" statements, i.e., statements in the main categories "Role of mathematics" and "Teaching principles" than the other teachers. But also the teacher mainly teaching at high school level (teacher8) shows above-average percentage (20%). However, these numbers can only be hints that, e. g. the "master teachers", as perhaps might be expected, tend to show more awareness of the complexity of the interplay.

Overall it can be observed that most teachers (9 out of 13) explicitly mentioned building the relation between mathematics and physics as a goal. In order to refine this general teaching goal, additional codings are introduced: "concept-related", "math-related" or "application-related". Then the teachers are grouped according to how many (measured in percentage) of each teacher's statements refer to the respective focus. If more than half of their statements concerning the teaching principles belong to one coding, these are grouped accordingly. Four groups can be identified tentatively:

concept-related: This teaching principle is characterized by statements such as *"I like it more first to induce an understanding before I treat it with math."* Two teachers mention such a viewpoint more often and regard it especially important to first treat the concepts before they go to the mathematical description. The focus lies on the physics side with some structural elements of the interplay.

math-related: This teaching principle is characterized by statements such as *"I always try to explain it again and again starting from math. So that they* *understand it there also."* Also two teachers make the use of mathematics very strong. These teachers tend to stress the technical role, but this does not imply that they neglect the structural or language aspect.

application-related: This is the biggest group of teachers. Six out of 13 pronounce especially the importance of relating physics to applications or visualization, e. g. by statements like *"It is important that the practical aspect of physics does not fall short."* Often the motivation of students and shaping the learning process from the concrete are the reason for this aspect.

multifaceted: Three teachers show no specific focus in their goals but seem to cover several aspects equally.

Besides describing these teaching principles, some teachers (4/13) also allude to philosophical viewpoints, in the sense that a physics world view should be reached at the end of high school. One of the master teachers makes this point very strong.

Relating this analysis to the theoretical background from Sect. 12.2.1, it can be noticed that the structural role in the sense of balancing mathematics and physics occurs from time to time in the interviews but does not seem to be dominant among the teachers.

Patterns

Concerning physics teaching Lehavi et al. (2015) identified teaching patterns mentioned by teachers in describing their practices in physics lessons. Therefore the question arose if those teaching patterns could also be retrieved in the interviews analysed here. Indeed some teachers of the interviewed sample mentioned similar patterns. Of course the small sample does not allow to deduce how often these are used in general but only if they might occur. The most often pattern mentioned by teachers is the "Construction" pattern as often a formula is being derived by means of an experiment. Five teachers also use the "Exploration" pattern. Especially one teacher very explicitly reflects on the problems of teaching when using mathematics with respect to limiting cases going beyond "Anschauung":

> …That both [limits] will not be reached, even if it approaches more and more and that these are rather limiting cases and mathematical models which do not at all agree with the practice. [teacher7]

The "Broadening" pattern is described by three teachers. Four teachers mention the "Application" pattern (where it can be assumed that more teachers really use it in class). Only teacher 13 has described all four patterns. From the definition of the patterns, one could say that the "Broadening" and the "Exploration" patterns are especially well suited to transport the structural role of mathematics for physics.

Broadening The teachers mention that similar arguments or techniques, most often the use of "area under graph" or "proportionality", can be transferred to other physics contexts.

Table 12.2 Characterization of teachers. Here the teaching principles are given in the first four columns. It is also indicated if the teachers are master teachers and who is teaching only in lower secondary school or only in high school (last three columns). We also indicated whether they mentioned examples of using a "Broadening" or "Exploration" pattern (see section "Patterns")

Teacher	Concept	Math	Appl.	Mult.fac.	B/E-Pattern	Master	Only lower	Only high
Teacher2			x		x			
Teacher3	x							
Teacher4			x					
Teacher5				x				
Teacher6			x					
Teacher7			x		x		x	
Teacher8		x			x			x
Teacher9				x				
Teacher10			x				x	
Teacher11				x			x	
Teacher12		x			x		x	
Teacher13	x				x	x		
Teacher14			x		x	x		

Exploration   Some teachers use mathematics for exploring limits: "$v \to c$", "$V \to 0$" in order to arrive at results they cannot get by an experiment. Another strategy concerns the appropriate use of the electronic calculator where the students should recognize the nature of regression in evaluating measuring data.

Construction   This pattern is mostly used for deriving formula from experiments via use of different representations, as a rule resulting in a proportionality.

On the whole the data show how complex the interaction between the teachers' teaching principles and goals and their views towards the general role of mathematics and physics is (Table 12.2). From the statements it becomes obvious that the teachers as practitioners very strongly relate their general opinions to their teaching experiences and vice versa.

12.5.2.3   Teaching Strategies

An important element of the PCK is the knowledge of successful teaching strategies. From the interview material, it evolved that it makes sense to distinguish more general strategies and their concretized forms, in the following called "specific teaching strategies".

The teachers describe as well their general strategies which can partly be attributed to the patterns identified by Lehavi et al. (2015) as some specific strategies, sometimes even the precise instruction sequence.

General Strategies

Under "General Strategies" we understand that the teachers describe their general procedures resulting from their teaching principles. They give hints to their focus in selecting or shaping the specific instruction from a more elevated perspective. They address, e. g. the proportionality and how they are introducing and using it, or they describe how they are balancing mathematical techniques and physical concepts.

In this context some teachers describe their strategies with a focus on technical aspects (teachers 10, 2, 3, 5, 7), e.g.

> It is quite important that the basics are well founded, and then I think it [use of math] won't hinder anything. [teacher 2]

or with a focus on concepts first (teachers 13, 6, 9).

> not only calculating, and not only describing [teacher 6]

Some teachers describe as well more technical and also conceptual-oriented strategies (teachers 11, 12, 14, 4, 8)

> Drawing diagrams and then deriving from these that e. g. distance corresponds to the area under the curve in the velocity-time diagram. [teacher12]

On the whole it can be inferred that most teachers value highly the development of routine with their students in order that the students master the mathematical techniques and are sure in selected types of reasoning, mostly concerning proportionality. The extensive treatment of proportionality is due to the requirements of the curriculum, but it also serves as an example for reasoning mathematically and relating mathematical results to physics.

Specific Strategies

Here the teachers describe how they shape their lessons regarding the use of mathematics. Again we perform the analysis in the light of Sect. 12.2.1. Overall the teachers show a broad range of different strategies where it became nevertheless obvious that each teacher has his or her own preferences. A strategy mentioned by all teachers is, e.g. the use of diagrams:

> Then it makes sense if I use diagrams. [teacher4]

Also many teachers use the electronic calculator, as it opens new ways for teaching and reduces the requirements of technical-mathematical abilities on the side of students.

A very common feature is the procedure "deriving formula from experiments" (corresponding to the "Construction" pattern) as this is considered the most appropriate way to enable students to connect the abstract formula with concrete experience. This approach has the potential of demonstrating the structural aspect of the interplay. Teachers with high reflection and regard of the structural role also tend to state more often that they use strategies such as "making meaning of formula" or "qualitative reasoning".

That mathematics also has a communicative function was addressed by three teachers (teachers 2, 9, 6):

.. graphical forms anyway, mathematical elements, and also to describe things by words, that plays a role. [teacher 6]

So some teachers make explicit use of language for imparting the relation between mathematics and physics.

Besides calculating during problem-solving – which would correspond to the technical aspect – some teachers also adopt the strategy of letting the students calculate numerical examples in order to enhance understanding:

When you are doing calculations here from time to time where the student realises: 'o. k. at the term scheme there is this quantum jump. Then this number results, this frequency and I can assign this to the colours.' [teacher8]

This strategy might focus on the technical aspect but nevertheless could enhance insight into the predictive power of mathematics for physics.

Starting from these descriptions, preferences of teachers could be identified that make up their "style of teaching". It is determined by their individual emphasis, their environment (mainly the students, resp., the grades they are teaching) and the requirements of the curriculum. Teachers try to support students' learning also by sticking to few repeated strategies in order to concentrate on certain competences. We characterize these repeated strategies according to our framework.

Technically oriented: Especially many answers relating to technically oriented strategies were given by five teachers (10, 11, 13, 2, 8). An example would refer to the use of units considered important by some teachers:

.. because of that I make quite a point of discussing the units, that they really see there must emerge e. g. a force. [teacher2]

Representation-oriented: Strategies, taking into account the function of mathematics as a language (use of language, of graphs or different representations), were described above average by three teachers (11, 6, 9):

...let the students run, the students measure the times by themselves and we represent it graphically on the calculator; then they see by means of the curve that it is indeed so. [teacher 6]

Conceptual-oriented: Six teachers (13, 14, 3, 4, 7, 5) focus above all on strategies stressing the use and understanding of concepts. These are often set into relation to real-life applications, or the role of math is made clear by use of analogies:

Mathematically you need a proportionality factor and then I do not ask: " How could we name it?" but: "What could it be in reality?". [teacher 3]

or

Many analogical examples in physics: $v \sim t$ for the uniform motion, $m \sim V$ for density, change of length $\sim$ change of temperature or to original length. [teacher 7]

These somehow characteristic strategies may be interwoven, e. g. teacher 12 does not show a specific pattern in the answers. Furthermore, it can be noticed that statements on general and specific strategies do not completely agree in each case. This can be due to differences in the interpretation of statements or some discrepancy between the general opinions of teachers and the realization in class. Sometimes a "shift" is observed from conceptual general strategies to technical specific strategies or to a focus on representations (mathematics as a language) with teachers (6, 8, 9, 11). However, a consistency along the levels of strategies can be seen with four teachers (2, 4, 10, 14). In the case of teachers (3, 5, 7), the specific strategies tend more to the conceptual side, whereas the general strategies seem to be more on the technical side. This can be an artefact of the coding of the handling of "proportionality".

Herewith we have to state also a caveat as the sample was very small and teachers might adapt to the classes, e.g. they might distinguish their teaching style between beginners' lessons and lessons on high school level.

12.5.2.4   Awareness of Students' Views and Knowledge

The description of the students' views and prior knowledge focusses mainly on the problems students have with the application of mathematics in physics, mainly technical difficulties but also difficulties in the transfer and structural role of mathematics. The positive aspects are mentioned also but by far less (see Fig. 12.3). Also emotional aspects are considered by the teachers.

Problems

The teachers observe their students very carefully. Those teachers with a long experience and the master teachers tend more often to state more general problems such as a cognitive overload by the complexity of bringing together the mathematical as well as physical aspects, especially when also the calculator comes into play. Here some teachers stress that the interplay and combination are indeed very complex for students and hence its implementation needs time:

> There you notice that it is unfamiliar to the children and that it needs a certain exercise and a certain time until some insight is reached. [teacher 14]

On the whole there are no big differences between the teachers. However, most aspects were stated by teacher 12 who is following the most explicitly math-related teaching principle.

Most teachers complain – as might be expected – generally about the lack of knowledge or technical problems, mainly the rearranging of formula or inappropriate use of routines in solving physics problems. They also state that there is not a sufficient coordination between the physics and the math curriculum. But at some occasions, they even see it as an advantage or at least as a chance that in physics they can introduce some mathematical elements – regression, functions or similar – in a concrete way thus enabling the students to make experiences before the abstract treatment in mathematics lessons later on. Here the electronic calculator opens up new possibilities.

Some teachers observe the lack of physical understanding in handling math:

> They often describe a relation, but without giving a reason how they arrive at the relation. [teacher10]

or

> But to do this by themselves, to come up on their own with the idea of doing what, that is very, very difficult for the students. [teacher14]

Only one teacher sees the complication in an unsufficient understanding of formula and its relation to the physical world:

> If they cannot imagine anything with this quotient, then they fail at those places where no calculation is done. [teacher5]

Also the important validation of results is often not done by students:

> There, at many places, the students are very uncritical towards their own results. [teacher12]

So only some teachers clearly mention difficulties of students on the structural side.

Positive Aspects

Also if teachers often complain of the students' problems, sometimes there are other observations:

> But then the students have auch a good knowledge that you can simply use such things [meant: proportionality] [teacher12]

From this quotation we get to the positive aspects of mathematics. Teachers mention that at a certain stage, the students have developed a routine that gives them kind of assurance that they can cope with the requirements and have some basic competence in solving physics problems. As formula are unambiguous, they can be a help for memorizing. Especially graphs are quite easy for the students to remember.

Mostly in high school, the teachers stress that besides physics instruction, students should also have an advanced level in mathematics as this mostly helps in developing the thinking skills necessary for physics:

> .. I find it an advantage because – according to experience – the students from the advanced course in mathematics can better draw these logical connections as those from the arts or creative subjects, . . . [teacher2]

So teachers often stress the necessity of certain logical or mathematical abilities of students they should bring as prerequisite into the lessons.

Emotional Aspects

The teachers also describe emotions students can have towards the application of mathematics in physics. One teacher (11) sees the emotions quite differentiated, on the positive as well as on the negative side. Three teachers (teachers 14, 10, 8) mention more positive than negative aspects. On the negative side, they most often notice rejection, such that mathematics is boring or that students just cannot cope with it and do not want to do it. On the other hand, as soon as some routine is developed, students might see the possibility of calculation as advantage, or they get insight into the usefulness of mathematics in physics.

Students' Interest

All the teachers are eager to find ways how they could motivate students, which topics they could choose or how they could show the relevance of physics as a whole and the role of mathematics in detail.

> That one should teach physics in schools always related to applications. you first arrive at qualitative statements. If I want to examine exactly how this is related, the force of the magnetic field as a function of the number of turns, ... in some places you can arrive only with the help of equations at insights .. only by this theoretical pursuit, by dealing with these theoretical equations we came to these findings, which subsequently can be checked in practice, ... Now you can see on the basis of this equation [Thomson's formula] what needs to be changed in this resonant circuit. [teacher14]

Herewith they adapt the strategies to the abilities of the given class.

12.5.3   Characterization of Single Teachers

The detailed analysis above indicates the interrelation of the personal views, the requirements of curriculum and the difficulties to achieve certain levels with the given classes influencing the realized teaching. As the sample is small, the picture is not yet very clear but shows a variability and hints to necessary compromises teachers have to make. Furthermore it proved to be difficult of getting a complete and consistent account in the interviews. Therefore in this section, some teachers are characterized showing that they nevertheless have developed for themselves a coherent principle for teaching.

12.5.3.1   Master Teacher: Structural-Oriented Teaching Principle and Strategies

The master teacher has a high level of theoretical reflection and stresses that lessons, especially the use of mathematics, should contribute to a physical world view. In his strategies it is important to him that the students relate knowledge of different topics. He weighs his goals and chooses the strategies accordingly. Concerning the students he is aware of possible cognitive overload and sees that they tend to focus on calculating and neglecting the transfer.

12.5.3.2   Teacher with Focus on Technical Aspects

Mathematics is seen as important in teaching physics because mathematical procedures are needed for calculating or for representations. In that the teacher regards graphs as important, also as a possibility of visualization. On the whole he sees a calculation as a prerequisite of understanding physical relations. The teacher perceives that students prefer calculating with routines and avoid explaining or reasoning and mentions the lack of knowledge in the basics of mathematics. Therefore he tends to provide the students with instrumental aids for mastering the procedures. Aside from this he regards applications as important for motivation.

On the whole the teacher shows a coherent reasoning in his views, teaching principles and strategies.

12.5.3.3   Teacher with Focus on Representational Aspects

The teacher views mathematics as important but stresses that the physics has to be in the centre of teaching. In derivations or evaluating experiments, the mathematics is a tool with a permanent relation to the physics content. Mathematically formulated laws are being derived from the physical process with careful reasoning. Therefore verbal explanations play an increasingly important role. In contrast to calculating with a formula, only reasoning and explaining show if the students understood the physics. On the other hand, the students should master the mathematical procedures, know basic function diagrams and be able to apply them, e.g. for regressions. This implies that the students should not only calculate something but also reflect on the result and validate it.

So on the whole the teacher emphasizes the complementing roles of mathematics: as tool for calculating, the importance of relation to physics including explaining and representations.

12.6   Interpretation of Results and Conclusion

What can we learn from these interviews with experienced teachers? Before we answer this question, we want to give two caveats.

First of all because of the low number of participants, we want to stress that only some hints can be derived which should be explored in subsequent studies. Secondly, one has to be careful to infer from a good PCK that the teachers have corresponding success in their teaching as, e.g. measured in students' learning (see, e.g. Cauet et al. 2015; Kirschner et al. 2016). This can only be verified by separate tests and analyses.

However, some insights can be gained. In the views of the teachers, mathematics as a tool or an instrument in physics is prevalent. But the teachers also share the view that mathematics could contribute to the understanding of physics. Aside from this general remark, we observe a broad continuous spectrum of views and strategies. We were interested if we could find indications that teachers elaborating on the role of mathematics in physics, especially the structural role, also show a specific pattern in choosing teaching strategies and enacting them. This could not be confirmed.

With many teachers working as practitioners, the views remain quite focussed on practical teaching, shaped by the everyday practice and what is required by curriculum and what can be reached with the given class. The complex interrelation of their experiences, the explicit knowledge and the individual views form an amalgam determining the actual teaching practice. So often we see a reduced awareness of structural aspects in teaching. Only few teachers stay aware of the broadness and the whole complexity of the role of mathematics in physics beyond the daily routine. These are mostly master teachers or teachers often teaching the last classes before the final school leaving exam ("Abitur").

In lower secondary school and even in high school, the teachers mainly struggle with the students' mathematical-technical abilities. They often see it as their first task to ensure the necessary technical competences and procedural knowledge as a presupposition of more conceptual aspects. But even in this stage, some teachers emphasize that the mathematical techniques have always to be connected to the physics. This is very important to them and relates to the so-called perceptual knowledge which combines the mastery of procedures with conceptual insight (Brahmia 2014). Some teachers even have the impression that intuitive physical understanding or mathematical knowledge gets blocked during the school career. This could be due to a higher degree of formalization or to the fact that the interplay of mathematics and physics gets more and more involved as the curriculum develops.

The higher the mathematical and cognitive abilities of the students are, the more the teachers focus on structural connection between mathematics and physics or treat deeper aspects of the interplay. However, it also happens that teachers state that teaching structural aspects is sometimes difficult because of students abilities. In Sect. 12.5.3.2 we have an indication that some teachers just apply technical or instrumental teaching as it seems to be the most adequate for their students. So in an actual classroom, it cannot be expected to conclude from the observations to the complete background of the teacher as he or she will proceed according to the possibilities of her students. The other way around, we also see from the analyses that sometimes the theoretical awareness or the intentions are more structural oriented than visible in everyday teaching.

On the whole we see the views of the teachers are shaped strongly by their experiences. Therefore it will be necessary to develop materials that enable students as well as teachers to fully exploit their competences in order to broaden their experience, e. g. by providing materials allowing to go beyond the technical level and inviting to include more structural aspects in the teaching. From some teachers we can learn how they try to adopt those strategies and to increase flexibility in thinking with their students.

So the developed model of PCK proved to be suitable for catching the different strategies of teachers with their knowledge, shaped by their experience. We found some characterization of the teachers and their teaching, but because of the low number of participants, these cannot be generalized.

What is open? We see that the teachers are aware of the students' difficulties and their attitudes. But what do students really learn and what would be possible in a general classroom? Do students gain additional insights, motivation or knowledge and capabilities if teachers are using patterns with more structural-oriented elements in a more conscious way? Those are important questions which have to be answered in future research.

References

Al-Omari, W., & Miqdadi, R. (2014). The epistemological perceptions of the relationship between physics and mathematics and its effect on problem-solving among pre-service teachers at Yarmouk university in Jordan. *International Education Studies, 7*(5), 39–48.

Ataide, A. R. P. D., & Greca, I. M. (2013). Epistemic views of the relationship between physics and mathematics: Its influence on the approach of undergraduate students to problem solving. *Science & Education, 22*, 1405–1421.

Baumert, J., & Kunter, M. (2013). The COACTIV model of teachers' professional competence. In M. Kunter, J. Baumert, W. Blum, U. Klusmann, S. Krauss, & M. Neubrand (Eds.), *Cognitive activation in the mathematics classroom and professional competence of teachers* (pp. 25–48). Boston: Springer.

Başkan, Z., Alev, N., & Karal, I. S. (2010). Physics and mathematics teachers' ideas about topics that could be related or integrated. *Procedia – Social and Behavioral Sciences, 2*(2), 1558–1562.

Bodin, M., & Winberg, M. (2012). Role of beliefs and emotions in numerical problem solving in university physics education. *Physical Review Special Topics – Physics Education Research, 8*(1), 010108.

Brahmia, S. M. (2014). *Mathematization in introductory physics.* Ph.D. Thesis, Rutgers University-Graduate School, New Brunswick.

Carrejo, D. J., & Marshall, J. (2007). What is mathematical modelling? Exploring prospective teachers' use of experiments to connect mathematics to the study of motion. *Mathematics Education Research Journal, 19*(1), 45–76.

Cauet, E., Liepertz, S., Borowski, A., & Fischer, H. E. (2015). Does it matter what we measure? Domain-specific professional knowledge of physics teachers. *Schweizerische Zeitschrift für Bildungswissenschaften, 37*(3), 462–479.

Etkina, E. (2010). Pedagogical content knowledge and preparation of high school physics teachers. *Physical Review Special Topics – Physics Education Research, 6*(2), 020110.

Fazio, C., & Spagnolo, F. (2008). Conceptions on modelling processes in Italian high-school prospective mathematics and physics teachers. *South African Journal of Education, 28*(4), 469–487.

Freitas, I. M., Jiménez, R., & Mellado, V. (2004). Solving physics problems: The conceptions and practice of an experienced teacher and an inexperienced teacher. *Research in Science Education, 34*(1), 113–133.

Gramzow, Y., Riese, J., & Reinhold, P. (2013). Modellierung fachdidaktischen Wissens angehender Physiklehrkräfte- Modelling Prospective Teachers' knowledge of Physics Education. *ZfDN (Zeitschrift für Didaktik der Naturwissenschaften), 19*, 7–30.

Khalili, P. (2016). *Mathematical needs in the physics classroom*. Ph.D thesis, Education: Faculty of Education.

Kirschner, S., Borowski, A., Fischer, H. E., Gess-Newsome, J., & von Aufschnaiter, C. (2016). Developing and evaluating a paper-and-pencil test to assess components of physics teachers' pedagogical content knowledge. *International Journal of Science Education, 38*(8), 1343–1372.

Lehavi, Y., Bagno, E., Eylon, B., Mualem, R., Pospiech, G., Böhm, U., & others (2015). Towards a PCK of physics and mathematics interplay. In C. Fazio, S. Mineo, & R. Maria (Eds.), *The GIREP MPTL 2014 Conference Proceedings* (pp. 843–853). Palermo: Università degli Studi di Palermo.

Lehavi, Y., Bagno, E., Eylon, B.-S., Mualem, R., Pospiech, G., Böhm, U., Krey, O., & Karam, R. (2017). Classroom evidence of teachers' PCK of the interplay of physics and mathematics. In T. Greczylo et al. (Eds.), *Key competences in physics teaching and learning* (pp. 95–104). Cham: Springer.

Loughran, J., Berry, A., & Mulhall, P. (2012). Portraying PCK. In J. Loughran et al. (Eds.), *Understanding and developing science teachers' pedagogical content knowledge* (pp. 15–23). Dordrecht: Springer.

Magnusson, S., Krajcik, J., & Borko, H. (1999). Nature, sources, and development of pedagogical content knowledge for science teaching. In J. Gess-Newsome & N. G. Lederman (Eds.), *Examining pedagogical content knowledge* (pp. 95–132). Heidelberg: Springer.

Mulhall, P., & Gunstone, R. (2007). Views about physics held by physics teachers with differing approaches to teaching physics. *Research in Science Education, 38*(4), 435–462.

Pietrocola, M. (2008). Mathematics as structural language of physical thought. In M. Vicentini & E. Sassi (Eds.), *Connecting research in physics education with teacher education*. ICPE – Book (Vol. 2). New Delhi: International Commission on Physics Education.

Riese, J. (2010). Empirische Erkenntnisse zur Wirksamkeit der universitären Lehrerbildung – Indizien für notwendige Veränderungen der fachlichen Ausbildung von Physiklehrkräften. *PhyDid A-Physik und Didakt. Schule und Hochschule, 9*(1), 25–33.

Chapter 13
A Case Study of the Role of Mathematics in Physics Textbooks and in Associated Lessons

Lena Hansson, Örjan Hansson, Kristina Juter, and Andreas Redfors

13.1 Introduction and Background

In relation to the teaching of physics, mathematics skills among students are widely discussed (cf. Uhden et al. 2012). For example, the TIMSS advance study discusses the decrease in students' mathematics knowledge as an explanation for the decline in Swedish students' results in physics (Angell et al. 2011). Despite this, research focusing directly on the role of mathematics in ordinary physics classrooms is scarce. In Hansson et al. (2015), we presented an analysis of different organisational forms of physics teaching: lectures, problem-solving in groups and labwork. For this purpose, we developed an analytical model where the focus is on communicated relations between *Theoretical models*, *Reality* and *Mathematics* (Hansson et al. 2015; Redfors et al. 2016). The reason for this was to make it possible to analyse the role of mathematics in connection to how theoretical models are communicated in relation to other entities in the teaching of physics.

Theoretical models and the complex relation between them and reality are central for the scientific research process. Observations and experiments are by necessity embedded in theory and therefore "theory laden" (Hanson 1958). Empirical and theoretical work is thus interwoven leading to construction, confirmation or refinement of theories and theoretical models. This is an interactive process of discussions, experiments and observations made within the science community (Adúriz-Bravo 2012; Giere 1988; Koponen 2007). Communicating this in physics class is part of making the nature of science (Erduran and Dagher 2014; Lederman 2007) explicit, which has been found central for the teaching of science.

L. Hansson · Ö. Hansson · K. Juter · A. Redfors (✉)
School of Education and Environment, Kristianstad University, Kristianstad, Sweden
e-mail: lena.hansson@hkr.se; orjan.hansson@hkr.se; kristina.juter@hkr.se;
andreas.redfors@hkr.se

© Springer Nature Switzerland AG 2019
G. Pospiech et al. (eds.), *Mathematics in Physics Education*,
https://doi.org/10.1007/978-3-030-04627-9_13

The theoretical models of physics are in most cases framed in mathematics, and mathematics is sometimes described as the language of physics (Pask 2003). In this sense the central role of mathematics could be viewed as part of the nature of physics (see also Krey 2019, chapter in this book). This, in combination with the problems students are reported to have handling the mathematics in the physics classroom in productive ways (cf. Planinić et al. 2019 chapter in this book), makes this a central topic for further studies. This has also previously been pointed out by, for example, Karam (2014) who emphasises the importance of more research on "the role of mathematics in physics from the teaching perspective – a facet rather overlooked in current physics education research" (p. 1). It is also emphasised that mathematical concepts are used in a different way in physics and in mathematics (Karam et al. 2019, chapter in this book). At the same time, we want to keep in mind the research showing the importance of focusing on discussions of relations between theoretical models and observations of the real world for students' meaning making (Hansson et al. 2015).

The aim of this chapter is to further develop the framework from Redfors et al. (2016) and test the framework in a case study focused on synchronous analysis of textbook sections and associated physics lessons. The results are discussed both in terms of relations made between *Theoretical models*, *Reality* and *Mathematics* in the physics textbook[1] and by the teacher and in regard to opportunities and constrains of the developed framework.

13.2 Design of the Study

13.2.1 Data and Context for the Study

The study is executed in the context of Swedish upper-secondary physics. Upper-secondary school in Sweden is a voluntary school of 3 years following the 9-year compulsory school, i.e. school years 10–12. Physics is in Swedish upper-secondary school studied by students in the Natural Science programme and the Technology programme. These programmes are studied by just under 20% (Skolverket 2016) of the upper-secondary students. There is a national physics curriculum (Skolverket 2012), where aims and content are established for the physics courses. However, there are no textbooks sanctioned by the authorities, and the teacher has a far-reaching responsibility to plan the teaching following the aims set by the authorities. Most teachers use textbooks in their teaching (e.g. Nelson 2006), for example, a majority of the teachers often use the students' textbooks when planning their teaching (Frejd 2012; Bachmann 2005; Sánchez and Valcárcel 1999). In this chapter, a developed model for analysis of physics teaching and textbooks is tested through a

---

[1]Gottfridsson, D, Jonasson, U, & Lindfors, T. Nexus – Fysik A & B, Malmö: Gleerups.

synchronised analysis of two sections of a physics textbook and associated physics lessons.

The first section is about the introduction of a chapter called "Energy" introducing students to the energy concept, the work concept, different forms of energy (potential and kinetic energy) and the energy principle. This is covered in five pages in the textbook, covering text, figures, images, as well as solved and unsolved problems. The associated lesson is from the first physics course in year one in an upper-secondary class with students from both the Science and the Technology programmes. During this lesson, the teacher covers (more or less) the content presented in the textbook.

The second section is about introduction and teaching of a unit on electromagnetism starting with electric fields, which was partly covered in the preceding physics course. Hence, the teacher sometimes refers back to what they did in a previous year. The unit is covered in 12 pages in the textbook, with text, figures, images, as well as solved and unsolved problems. The associated lessons are from a third-year class of students in the Science programme.

13.3 The Theoretical Framework

In Redfors et al. (2016), we presented and discussed the uses of an analytical model based on relations made between *Reality, Theoretical models and Mathematics*, during classroom communication. An analysis of classroom conversation was presented in Hansson et al. (2015) where results were presented on how students and teachers communicated relations between the three entities *Reality, Theoretical models* and *Mathematics* (see Fig. 13.1). Hence, the focus of the analysis was on the links made between the three entities, and the framework was used to analyse different organisational forms: lectures, problem-solving in groups and labwork.

In this previously published model, we made distinctions concerning whether mathematics was used in a technical or structural way (Karam 2014; Pietrocola 2008; Uhden et al. 2012). In this chapter, we present a further developed model, where we also make distinctions within the *Reality* and the *Theoretical model* entities. These additional discriminations evolved from the analysis presented in this chapter but are also related to earlier work as described below.

Fig. 13.1
Reality-Theoretical models-Mathematics in physics teaching. (Adapted from Hansson et al. 2015)

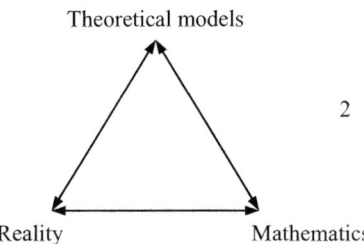

Reality refers to objects or phenomena (or observations of them) in the real world. Here we differentiate between *Reality (R)* and *Reality School (RS)*, *the latter with two subcategories.*

13.3.1 Reality (R)

Reality is used in a broad sense comprising well-known objects, phenomena and events of which students have experiences in their everyday life. This category includes situations such as determining changes in breaking distance if the speed limit is changed or costs for energy in a home depending on different habits and decisions.

13.3.2 Reality School (RSr/a)

There are two different types of *Reality School* in the model: *Reality School-reduced* and *Reality School-altered.*

Reality School-reduced (RSr) is a form of "reduced reality" often encountered in physics teaching, where factors influencing the real-world phenomenon are held constant, minimised or disregarded, e.g. a frictionless air track to study motion. It also encompasses phenomena only observed (or observed for the first time) during demonstrations and labwork, sometimes through the use of complex measuring equipment, i.e. objects and events usually not part of the students' everyday lives. When categorised as *RSr*, there is no starting from or reference to real-world situations, but the focus is entirely on the idealized situation.

Reality School-altered (RSa) includes situations from everyday life (outside the physic classroom). However, the given and/or sought information is something normally not known or asked for. An example of this is when questions are asked on how much work is done when a certain amount of force (given in Newton) is applied when pushing a sledge. This category thus has similarities to the category *Reality* in respect of the situations as such but differs in the information at hand or in the questions asked. Thus, through the given and/or sought information, the situation is formed in a specific way – away from *Reality* and instead altered into a school (physics) reality.

There are similarities between *reduced* and *altered reality* – both are constructed in the context of physics teaching for specific purposes different from those of the everyday life or the life as a citizen. This is why we present them as subcategories to the category *Reality School.*

Furthermore, R, RSr and RSa have been used with an internal structure distinguishing between recalled and systematically described realities following the description in Triantafillou et al. (2016), based on Bliss et al. (1983).

13.3.3 Recall Experience, R1, RSr1, and RSa1

The text/discussion makes reference to students' everyday life (school) experiences. An object or an event from everyday or school life is mentioned (often as a motivation or background), but not explained or elaborated on.

13.3.4 Systematic Description, R2, RSr2, and RSa2

The text/discussion makes reference to a systematic description, e.g. demonstration of an instrument's operation, necessary conditions for an experiment or measurements in everyday life (or school experience).

Theoretical models refer to a semantic view of theoretical models in physics and concepts related to them (Hansson et al. 2015 and references therein). The theoretical models could be formulated with or without use of mathematics. In this chapter, we introduce a distinction between an *instrumental approach* and a *relational approach* to theoretical models in line with what Skemp (1976) introduced for mathematics. Johansson et al. (2016) have previously found how an instrumental physics culture is dominating a higher physics quantum mechanics course. In that case, this was coupled to a classroom discourse where "calculating physics" dominated.

13.3.5 Instrumental Approach to Theoretical Models: TM1

Constituents of the theoretical model (concepts, representations) are used without focusing on describing or motivating the theoretical model. For example, concepts or formulas are used without making the theoretical context explicit. Instead, the focus is on using the model for a specific purpose.

13.3.6 Relational Approach to Theoretical Models: TM2

The theoretical model with connected concepts and representations is present in the discussion. Meaning making on a systemic level is made possible for the learners. For example, from the textbook "We usually symbolise the field with field lines. Field lines show how a positive test charge would move if placed in the field. Where the electric field is strong the field lines will be closely spaced".

Mathematics refers to mathematical concepts, theorems, representations, mathematical reasoning and methods. Words used in a purely everyday manner such as "most" and "part of" are not included.

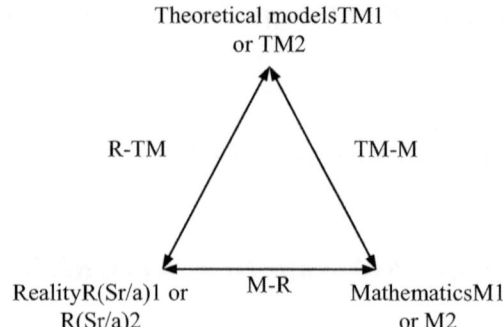

Fig. 13.2 Depicts the analysis framework for relations between *Reality/Reality School*, *Theoretical models* and *Mathematics* in physics teaching. It is adapted from Hansson et al. (2015) and developed.

13.3.7 Technical Use of Mathematics (M1)

Technical use of mathematics indicates that mathematics is viewed as a "calculation tool" with an instrumental use of mathematics as a consequence (Karam 2014; Pietrocola 2008; Uhden et al. 2012). M1 is often but not always coupled to TM1.

13.3.8 Structural Use of Mathematics (M2)

Structural use of mathematics means that mathematics is used as a "reasoning instrument" and that there is an emphasis on interpretations or consequences and on using logical reasoning (Karam 2014; Pietrocola 2008; Uhden et al. 2012).

In Fig. 13.2 the developed framework is depicted through a schematic diagram showing the links with the distinctions described above within each of the three entities in the ternary perspective on physics teaching.

13.4 Results

Results from the analysis of two sections of a physics textbook and the associated physics lessons are presented below.

13.4.1 Example 1: The Energy Principle, Work and Potential and Kinetic Energy

The textbook covers the energy principle, work, potential energy and kinetic energy in five pages, including tasks, and this is the start of a chapter called energy (chapter in the textbook). Previously Hansson et al. (2015) showed how the teacher and the students communicated around the content of these book pages. In this chapter, we will put those results in relation to the results from a similar analysis of the textbook. The focus will be on relations between Reality, Theoretical model and Mathematics (Fig. 13.2).

13.4.2   Introduction to the Energy Concept and the Energy Principle

13.4.2.1   The Textbook

The chapter in the textbook is introduced with a qualitative description of the energy concept, the energy transformation and the energy principle. Mainly relations between the *Theoretical model* and *Reality* are communicated in this introduction. Relations between Theoretical models and Reality are communicated, for example, when the model description of the energy principle (TM2) is linked to the description of experience from the real world (R1) (see Table 13.1).

After the introduction there is a section called "Energy types" with deeper explanations of the introduced concepts and phenomena and where the energy principle is covered. The energy principle (Theoretical model) is qualitatively described without direct links to neither Reality nor Mathematics. The reader also gets to know that different units can be used such as joule (expressed as the most important unit abbreviated J), newton metres, calories or kilowatt hours, but the relations between them are not discussed. The energy principle is written with words within a frame, which marks its importance.

Table 13.1 Communicated relations between Theoretical models (TM), Mathematics (M) and Reality (R) in the introduction of the chapter

Textbook	Categories
Energy is a basic concept in all of science. At each event energy is transformed, not only when the lights are on or cars collide but also when thoughts are thought and buds burst. Energy is constantly converted between different forms. The total amount of energy cannot be changed but only moved from one place to another. Most of the energy on earth comes from the sun. The sun's energy-rich rays heat our planet to make it habitable. Since different parts of the earth are warming to different extents, winds and ocean currents also occur. A small portion of the sunlight is captured by the green plants that store the energy in chemical form. When we and other animals eat the plants or their fruits, a part of the energy is transferred to us so we can live.	R1, TM2

13.4.2.2 Teachers' Lecture

In the teacher's lecture, she starts by telling the students that the lesson is about energy and the conservation of energy, without explicitly mentioning the energy principle but talking about transformation of energy in, for example, eating and running (TM2). A student mentions the energy principle explicitly, and some others associate energy with wind, the sun and nuclear power (R1). The teacher does not make explicit use of the students' examples in her explanations of the energy principle or of energy transformation, for example, in a windmill. The relation communicated by the teacher between the energy principle and Reality is mainly implicit. The teachers' introduction where she presents the content presented so far has a duration of 5 min, and the corresponding text in the textbook is two thirds of a page in total.

13.4.2.3 The Textbook vs Teachers' Lecture

The text is analysed as mainly communicating relations between R1 and TM2 for this part. The teacher, on the other hand, is only linking R1 and TM2 implicitly as in the example mentioned.

13.4.3 High-Quality and Low-Quality Energy

High-quality and low-quality energy is the next section of the textbook. The concept exergy is introduced and defined as the useful part of the energy and as the amount of kinetic energy that is possible to extract from the energy form at hand (TM2, R1). There is a description of how the usefulness often is measured in percent where the usefulness is the relation between the exergy and the total amount of energy. A relation between the Theoretical model and Mathematics is communicated (TM2, M2), and it is shown how mathematics can be used structurally to reason about high-quality and low-quality energy forms. It is taken for granted that the students know what "the relation between" means mathematically and that they understand the concept of percentage ("For high-quality energy forms, the part of exergy is near a hundred percent", TM2, M2). No part of the text in the textbook is highlighted in a box or otherwise. The teacher does not mention high-quality and low-quality energy or exergy in her lecture.

In summary in this section, the textbook communicates relations between TM2 and R1 and relations between TM2 and M2. The teacher, as said above, does not deal with the content of this section.

13.4.4   Work

In the following sections of the textbook, the concepts "work", "potential energy" and "kinetic energy" are dealt with in this order. Here the teacher departs from the order in the book and deals with potential energy first, followed by a routine problem; after that the concept "work" is dealt with followed by kinetic energy. When all formulas are written on the whiteboard (approximately 3 min have been used introducing the formulas), they are used to solve different kinds of problems. The following presentation takes a starting point in the textbook, but we also do some comparisons with communication during the teacher's lesson where these concepts were introduced.

13.4.4.1   The Textbook

The section of the book dealing with the concept "work" starts out with a definition of work (Table 13.2). Here a relation between Theoretical model (TM2) and Reality School (objects that are moved, RSr1) is communicated (TM2, RSr1 in Table 13.2). However, the emphasis is on Theoretical model/concept (TM2). In a box (marked with * in Table 13.2), to the right of the written definition, the word "Work" is used as a heading, and then the correlation between work, force and distance follows in a formula. Here a relation between Theoretical model and Mathematics is communicated (TM1, M1 in Table 13.2). Mathematics is used in a technical way – the reader is told how to calculate the work, but the text does not include reasons for this. Also, there are no units mentioned for F and s, only for W. If all units were mentioned, the logic of the formula would have been more apparent.

The book then continues with two examples of solved problems. In the first problem, Helen helps Jonas to push start the car. She pushes the car with 400 N for 12 metres, and then the car starts (RSa2). The questions asked are how much work Helen uses to push the car and what becomes of the performed work. The solution of the problem starts out stating the formula to use (the one written in a box on the previous page) (R2, TM1), it is said that the distance (12 m) and the force

Table 13.2   Description of work in the textbook used

Textbook	Categories
Work	TM2, RSr1
If a force moves an object a certain distance, the force will transfer a certain amount of energy. We say that the force is doing work. Work is denoted by Wa	
Work*	TM1, M1
$Wa = F \cdot s$	
where F is the force and s is the distance that the force operates. F and s should always be measured in the same direction. The unit of work is Newton metre abbreviated Nm. 1 Nm = 1 J	

(400 N) are known, and then the work is calculated using the formula. This we argue is an activity where the relation between Theoretical model and Mathematics is in focus, more specifically with a focus on a technical use of mathematics (TM1, M1). It is shown how the given values are put into the formula. Then it is said that the work generates kinetic energy for the car. The next assignment is similar to the first, with the difference that the person pulls a sledge diagonally upwards. Thus, the component in the direction of the movement needs to be calculated. It is shown in the book that the force needed to calculate the work is given by multiplying the force in the cord with cosine for the angle, and this is further explained by a figure with force arrows (M2).

In both examples, it can be said that there is a relation communicated between Theoretical model (TM1) and Reality (RSa2). The examples are descriptions of events that could happen in reality, but it is still an "altered reality", a reality where people are depicted to know forces in amounts of Newton needed to push a car or to pull a sledge. And a reality where people, of an unknown reason (reasons are not given in the assignment in the book), are interested in knowing how much work (in a physics sense) they perform. In the solution to the assignment in the book, it is not communicated how you could know whether the reached result is reasonable or what the result could be used for. What is communicated to the reader, however, is the central importance of *being able to* determine the force, i.e. an instrumental approach to theoretical models.

In sum in the introduction to work, TM2 is emphasised in the textbook. However, in the solved examples, the relations focused by the book in this section are between TM1, M1/M2 and RSa2 (altered).

13.4.4.2   The Lesson

As said above the teacher deals with potential energy before she deals with work. She does the transition through an example where she is calculating the potential energy of a 5 kg weight and then lifts the weight from the floor to her counter and says that work is done when you change the energy (TM1, RSr1). Then the concept work is introduced by her writing a definition and a formula on the whiteboard and telling the students that F and s need to have the same direction (TM1). She gives the classical example that no work is done, in a physics meaning, when you are walking around carrying a weight, even though you are getting tired doing so. However, she does not work further on the link to *Reality* by, for example, discussing why you are getting tired or how you can justify the choice of definition (TM1, R1). Instead the teacher turns to showing how to do a decomposition of a vector (M1). Then she turns to define kinetic energy as described below, after which she demonstrates an example where a car is braking and the work done is calculated (TM1, RSa2). This is another example of an altered reality. The large picture is that the focus of the classroom communication is on the relationship between Theoretical models (TM1) and Mathematics (M1). Mathematics is used in a technical way (*how* the work is calculated and *how* you decompose a vector in components).

13.4.4.3 The Textbook vs the Teachers' Lecture

In summary the classroom communication has the same focus as the book (a formula in a box emphasises how you are supposed to perform the calculations, not why this works or how the results from the calculations are related to the reality). The relations focused by the teacher for this section is between TM1, M1 and R1/RSa2. Similar to the book, she communicates with an instrumental use of models applied in real-world situations. However, the focus on TM2 in the beginning of the section in the textbook is not present in the lesson.

13.4.5 Potential and Kinetic Energy

13.4.5.1 Textbook

After the two examples, a section about potential energy follows in the book. In this section, a relation between Reality (R1) and Theoretical model is communicated at first, but very fast the text leaves this relation and instead turns to a focus on the relation between Theoretical model and Mathematics. With the help of mathematics, a formula is derived in a structural manner (TM2-M2). Also "zero-level" is introduced. The formula is finally placed (similar to what was described above) in a box, a procedure that serves as a way to communicate the importance of the mathematical shape of the Theoretical model. The goal of the text seems to be formulas in boxes (TM1). The section about kinetic energy is structured in a similar way. The formula for kinetic energy (TM2) is derived from earlier relations (M2). And then also this formula ends up in a box. After that three solved example assignments follow.

In the first example, the energy principle is used in a simple situation of adding kinetic and potential energy of a falling stone in an altered realistic context (RSa2-TM2). In the second example, Mats is swinging. It is told that he starts stationary from 2.0 m above the lowest point of the ground level of the swing. The problem is about what velocity Mats has when passing the lowest point. Several relations are communicated in the problem and the associated solution. At first a relation between an altered Reality (RSa2) (a person swinging) and Theoretical model (TM2-M2) (the energy principle and the concepts potential and kinetic energy) is communicated. This is done by starting out with the statement, referring to the law of energy, i.e. "the total amount of energy is always the same". It is stated that: "We choose to put the zero-level in the lowest point of the swing", and it is also stated the values for h "before" and "after" and v "before". Then the book states that the values are to be inserted in the relation:

$$mgh_{before} + \frac{mv^2_{before}}{2} = mgh_{after} + \frac{mv^2_{after}}{2}$$

and by division eliminate m and finally determine v_{after}.

Here to some extent, a structural use of mathematics (M2) is communicated through reasoning about the assignment using mathematics. However also a technical focus is present, showing how values are inserted into the formula and, e.g. how you should eliminate m (M1). Also in the third example, several relations are communicated. In the beginning of the solution, a relation between reality and model is communicated, followed by a focus on the model where it is also related to mathematics as a way of reasoning (structural use of mathematics). At last the relation between model and reality is returned to. After these solved examples, the book continues with assignments for students to work with.

13.4.5.2   Lesson

In the teachers' introduction of potential energy, she introduces the concept zero-level but does not show how the formula ($W = mgh$) could be derived (as was the case in the textbook). Instead she only writes the formula on the whiteboard without giving any justification (TM1). Thus, she does not take the chance for structural use of mathematics that the book invites to (though rather shortly). Through this way of teaching, the teacher is, even more than the book, emphasising an instrumental approach to theoretical models and a technical use of mathematics. The latter is emphasised by an example (routine problem) where she puts numbers into a formula, but due to the object being at the zero-level (RSr2), no calculation is really necessary which was actually pointed out by a student saying that the answer is zero due to the zero-level. The teacher writes $5 \cdot 9.82 \cdot 0 = 0$ anyway.

In a similar way, the formula for kinetic energy is written on the whiteboard without justifying it (TM1), while the book is deriving it. The only comment to the formula given by the teacher is about how v^2 should be understood (M1). An example is given of how the formula can be used (TM1). The example is about a car with the mass 1000 kg that drives with a velocity of 90 km/h. To some extent a relation about reality (cars that drive) and the model (kinetic energy) is communicated but only in passing. The focus is not on what the model describes or why it could be useful to know the car's kinetic energy but instead on how the kinetic energy can be calculated by the use of the formula (R2, TM1, M1). That the relation between model and reality was secondary to the relation between model and mathematics (technical use) during the lesson was underlined by the teacher giving an answer with four digits of accuracy. She ended the lesson with asking the students if they had airbags in their cars and had a short discussion (about 1 min) about force distribution depending on distance. Here a relation between Theoretical model (TM2) and Reality (R2) is communicated.

13.4.5.3   The Textbook vs the Teachers' Lecture

In summary, the textbook has a stronger emphasis on relational and structural aspects of the models (TM2) and mathematics (M2), respectively, than the teacher has. She mainly focuses on technical use of mathematics (M1) in an instrumental approach to models (TM1). Systematic descriptions of reality (R2), is communicated in both contexts, but also altered reality (RSa2) in the textbook.

13.4.6   Example 2: Electromagnetism

This section is about textbook coverage and teaching of electric fields during two lessons in a physics class during the third year of the Science programme. The textbook is dealt with at two occasions in a classroom lecture (40 min) and a problem-solving session (80 min). A discussion on links made between Mathematics, Theoretical models and Reality in the communication during these lessons can be found in Hansson et al. (2015). Here the focus is a comparative analysis of textbook and teaching using the extended model presented above.

13.4.7   Electromagnetism and Electric Fields

13.4.7.1   The Textbook

"Electric fields" is the first section in the textbook of a chapter on Electromagnetism. In the introduction to the chapter, the importance of electromagnetism in concurrent societies is stressed – "Many are unaware of the importance of Electromagnetism in society". The page is dominated by a picture of a singer using a microphone and a chain of technical apparatus utilising electromagnetism, from microphone through CD and hard disks to loudspeakers. Generators and transformers in connection to production and transportation of electricity are discussed. The introduction exclusively contains references to Reality (R) and Reality School-reduced (RSr) and Theoretical models (TM1). The links to reality comprise sound from a loudspeaker (R1), electromagnets transforming an electric signal to sound (RSr1). Theoretical models are used instrumentally (TM1): concepts like electric and potential energy, magnetic fields and electricity are mentioned.

The first section after the introduction is "Electric Fields" (3 pages). It starts with an introductory text followed by a solved exemplary problem. A representation from a theoretical model is depicted at the top of the page. It shows examples of charges with electric field lines (TM1). It is analysed as TM2 and RSr1 for the display of unidentifiable objects called "plates" and "ring". Electric fields are related to transformer stations (RSr1) with shielded walls, but no explanations relating to the effects of shielding in reality. The text discusses appliances where electric charges are accelerated and directed through the use of electric fields (TM2). It is also stated that the course covers only homogenous fields and cases where a charged particle has a velocity component perpendicular to the field (TM2).

In the solved example, the problem is about electrons in an electron tube, hence a connection to Reality School, but in a systematic way (RSr2) in the formulation of the problem. The solution to the problem on the other hand does not specify the theoretical model explicitly (TM1), and the mathematical treatment becomes technical (M1), with no variables explained, and a bare mentioning of real components (RSr1). However, a qualitative reasoning comparing forces generated by gravitation and electric field strength including kinetic energy is classified as TM2.

13.4.7.2   The Lesson

The teacher starts the lesson with a demonstration of electric fields in a parallel capacitor. The students were shown a parallel-plate capacitor charged by a hand-powered generator. The teacher asked the students why the spark appears, and they concluded that the capacitor wants to equalise the charges. The voltage required for a spark was shown and that the spark occurred more often and with less required voltage when the distance between the plates decreased, "Is this logical?" asked the teacher, thus directing attention to relationships between Reality School (RSr2) and structural use of mathematics (M2). The teacher connected implicitly to theoretical models (TM1) by calculating the field strength required for the capacitor in the demonstration, using formula $E = U/d$ in a structural way (M2). The analysis of this part concludes that links were made between RSr2, TM1, and M2.

The teacher drew a parallel-plate capacitor on the whiteboard that emphasised a theoretical model-based reasoning (TM2) (see Fig. 13.3) and asked "Will it [the electron] move at a constant speed or accelerate? And why?" The students were uncertain, and the teacher pointed out that there is an electric force acting downwards but no force acting upwards and explained that there is a net force, which will make the electron accelerate. The teacher and students' reasoning was now focused on the relation between Reality School (RSr) and Theoretical models with an emphasis on a relational use of Theoretical models (TM2). The teacher draws attention to that they now can use the formulas $F = ma$ and ($E = F/Q$ and

Fig. 13.3 Figure drawn of an electron in an electric field on the whiteboard by the teacher

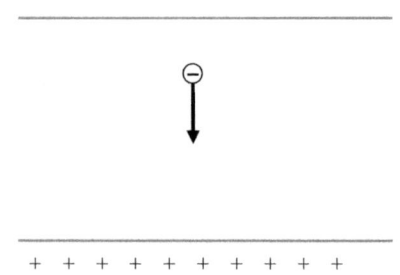

write) $F = EQ$, noticing, "F is the force that the electron senses in the electric field" indicating the force vector in Fig. 13.3.

13.4.7.3   The Textbook vs Teachers' Lecture

The teacher diverged from the introduction of the section in the textbook by starting directly with a demonstration of a capacitor, whereas the textbook begins with students' everyday experiences of electromagnetism (R).

The textbook illustrates field lines and makes links to both *Reality* and *Reality School-reduced*, i.e. between electrons and charged objects (TM2-RSr1), and in real-world applications like inkjet printers (TM1-R1). The classroom demonstration links the theoretical constructs only to *Reality School-reduced* in encompassing phenomena observed through the use of objects and measuring equipment that is not part of the students' everyday lives, but it treats Reality School in a systematic way (TM, RSr2).

13.4.8   Projectile Motion in Electric Fields

13.4.8.1   Textbook

The following text is about projectile motion in electric fields (see Table 13.3). An introductory text describes how an electric charge is influenced by an electric field (TM2). The text relates to reality (RSr1) in comparing to the effects from a gravitational field on an object in projectile motion. The text states that both cases result in a path that can be described by a parabola; the analogy is not discussed or explained (TM1, M1).

The page is dominated by a schematic figure depicting an inkjet printer (R1) with four named objects: ink patron, charged electrode, deflecting plates and paper. Except for the paper, the three objects are categorised as RSr2. The figure caption explains how droplets of ink are accelerated towards the paper and directed. The book relates to a familiar object for the students, inkjet printers (R1), and the figure depicts through a schematic diagram how droplets can be controlled by electric fields, represented by field lines (RSr2, TM1).

Thereafter the text describes three central components in an oscilloscope: an electron canon, a guiding system (deflecting plates) and a fluorescent screen. The function of the components is described in a text with references to Reality School and Theoretical models (see Table 13.3).

The following page contains a solved problem. The problem is related to Theoretical Models (TM2), but notice that the RSr relations above are not mentioned in the problem. In contrast to the above, the problem has no figures or references to reality explaining how the sought deflection and direction should be understood. The solution uses relations not explicitly present in the text that allows space for

Table 13.3 Text in the textbook about projectile motion in electric fields

Textbook	Categories
When an electric charge is in an electric field, it will be accelerated in the direction of the field. Perpendicular to the field the velocity will be unaffected	TM2, M2
This is similar to the situation for a projectile motion where an object is accelerated towards the ground by gravitation but keeps its velocity parallel to the ground. Just like the projectile motion, the path for the charge in an electric field will be a parabola	RSr1, TM1
Figure caption	
In an inkjet printer, small charged droplets are accelerated in the direction of the paper. On the way, they pass an electric field deflecting them vertically. The computer manages the potential between the plates so that each drop arrives where it is supposed to	RSr2, TM1
An oscilloscope has three different components, an electron canon, a guiding system and a fluorescent screen. The electron canon accelerates the electrons so that they have a high velocity straightforward. The guiding system decides how much the electrons should be deflected using the electric fields. One directs the electrons' motion sidewise, the other heightwise. The electron beam sweeps the whole time from right to left since the voltage over the electric field is varied. On the oscilloscope, you can adjust the sweep speed. At the same time, the voltage over the other field will decide how much up or down the beam should hit the screen. When the electrons reach the fluorescent screen, the hit point lights up.	RSr2, TM1

a qualitative reasoning about the theoretical model (TM2) and structural use of mathematics (M2). However, the textbook's way of presenting this formula seems primarily to entice technical use of mathematics and formulas.

13.4.8.2 The Lesson

After the demonstration the teacher discussed projectile motions in electric fields and solved a problem directly related to the textbook example during most of the first lecture (see Table 13.4). The task formulated by the teacher was:

> In an electric field the field strength is 200 V/m. An electron is fired into the field moving at the velocity 2 Mm/s. The field is 15 cm long.
>
> (a) Determine the deflection in y-direction.
> (b) At what angle does it exit [out of the field] in the y-direction?

Before formulating the task, the teacher drew a figure in which an electron is fired into an electric field. She asked the students how the electron will move when it enters the field. The students suggested that the electron will travel in a projectile motion, which the teacher sketched and continued (Fig. 13.4):

> Teacher: The electron will move in a curve or projectile motion, along a parabolic trajectory we might say. Compare with Chap. 1, was it? So, we must apply the parabolic trajectory again but in electric fields. And what was important when you calculate projectile trajectories? (TM1)
> Student: Velocity.

Table 13.4 Analysis of textbook example on electrons in an electric field

Textbook	Categories
An electron enters a 15-cm-long homogenous electric field with the velocity 3.5 Mm/s. The electric field strength is 140 V/m directed perpendicular to the direction of the electron	TM2, M2
a) How much is the electron deflected by the field?	
b) What direction does the electron have when it leaves the field?	
a) We start by calculating how long the electron will be in the field. We calculate only perpendicularly to the field and denote this with index v. $s_v = v_v \cdot t \Rightarrow t = \frac{s_v}{v_v} = \frac{0.15}{3,5 \cdot 10^6} s \approx 4,29 \cdot 10^{-8}\ s$ To be able to calculate the deflection, we need to calculate the acceleration $a = \frac{F}{m} = \frac{q \cdot E}{m} = \frac{1,602 \cdot 10^{-19} \cdot 140}{9,1 \cdot 10^{-31}}\ m/s^2 \approx 2,46 \cdot 10^{13}\ m/s^2$ Then we calculate the deflection in the direction of the field according to $s = \frac{at^2}{2} = \frac{2,46 \cdot 10^{15} \cdot \left(4,29 \cdot 10^{-8}\right)^2}{2}\ m \approx 0,023\ m$ **Answer:** The electron is deflected 2.3 cm b) The velocity perpendicular to the field is unchanged 3.5 Mm/s. In the direction of the field, we get v = a·t = 2.46·1013 · 4.29· 10^{-8} m/s = 1,055 Mm/s The direction we get as $\tan \alpha = \frac{v}{v_v} = \frac{1,055 \cdot 10^6}{3,5 \cdot 10^6} \Rightarrow \alpha \approx 16,8^0$ **Answer:** The electron is deflected 17° in the electric field	TM1, M1

Fig. 13.4 An electron in an electric field. (Figure drawn by the teacher)

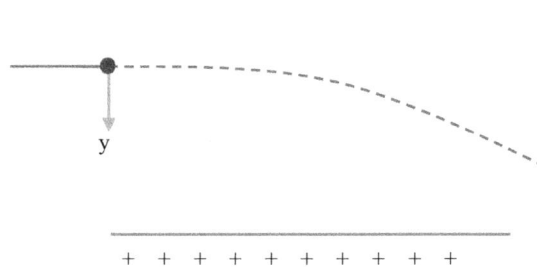

Teacher: Velocity. Expand!

Student: In different directions.

Teacher: That's right, x-direction and y-direction for themselves (TM1). We have already established that it accelerated in the y-direction. How was it in the x-direction then?

Student: It is constant.

Teacher: Yes, good!

The teacher continued to write the formulas $E = U/d$ and $F = Eq$ before she formulated the first part (a) of the task. Thus, appropriate formulas were given before the task (TM1), in contrast to the book's introduction of the task. The figure the teacher drew illustrates the plate capacitor in the demonstration and thus links RSr2 to TM1, in contrast to the textbook example that emphasises TM2 and M2 explicitly including the electric field.

The teacher conducted a dialogue with the students and invited them to reflect on the subject. The emphasis was on reasoning linked to theoretical models (TM2-RSr2) with technical use of formulas (M1). Then the teacher continues with the next part (b) of the task:

> T: We will now determine the angle of it when it exits (the teacher marks the angle, see Fig. 13.5.) If we compare with the beginning when it ran in parallel [with the plates], what is the angle, then, when it bends?
> S: When it leaves the field?
> T: Yes.

The teacher's reasoning was directed by an instrumental use of the theoretical model (TM1), which is limited in relations between velocity concept and tangent vector (TM1, M1) to the parabolic trajectory. The total velocity was not talked about as a tangent vector at different points of the parabolic path. The focus was on components in the "x- and y-direction".

Another student stated that they knew the velocity. The teacher agreed and drew (Fig. 13.6) and continued her reasoning:

> T: If we can find out the velocity of the [electron] in the x-direction and y-direction when it just gets out of the field, then we can with a little trigonometry to find out the angle too; we may call it alpha. How do we do that?

The teacher clarified their reasoning about velocity and the marked angle with an illustration (Fig. 13.6). The teacher argued that the angle α can be determined by the formation of a triangle where a technical representation of the problem is in focus (M1) – instead of describing how the components of forces act on the electron using a structural approach connected to the theoretical model.

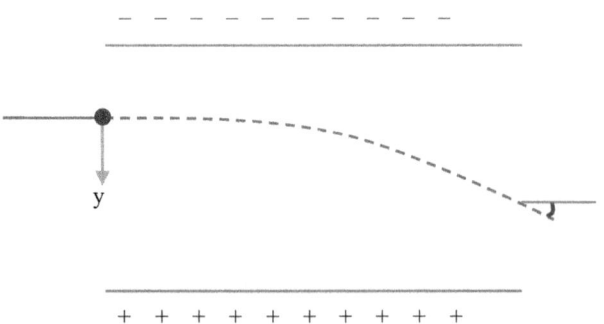

Fig. 13.5 An electron in an electric field. (Adjusted figure drawn by the teacher)

Fig. 13.6 The teacher's drawing of velocity vector components

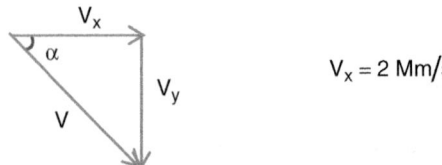

$V_x = 2\ Mm/s$

Fig. 13.7 The teacher's
corrected drawing of the
velocity vector components

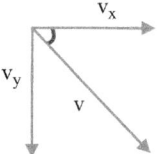

One student then suggested that they should use "acceleration times time". The teacher calculated $v_y = v_0 + at = 0 + 3.5 \cdot 10^{13} \cdot 7.5 \cdot 10^{-8}$ m/s $\approx 2{,}637{,}750$ m/s ≈ 2.64 Mm/s.

> T: Did you all get that? So, from the beginning when it came into the field, it had in the y-direction the velocity 0 m/s. Then it accelerates downwards. Then, just when it comes out, the velocity is instead 2.64 m/s.

The teacher replaced Fig. 13.6 with a parallelogram of forces (see Fig. 13.7). Commenting "it should look like this if you're picky". The reasoning now related to TM and illustrated the components of the resultant force, but in an instrumental way (TM1).

The teacher pointed out that they can use the Pythagorean theorem to calculate the total velocity. One student stated that they could use the tangent and the teacher stayed with a technical approach and wrote $v_y/v_x = \tan \alpha$, $9.8/15 = \tan \alpha$ and $\alpha = 22.8$ degrees (M1).

When the task is completed, the teacher asked: "What use could we have for all this? Can you think of anything, or is it hard?" The students related to their reality (R1) and mentioned "old" TVs and inkjet printers, both of which are mentioned in the textbook, so it is not totally clear if they are independently thinking of their everyday lives. The teacher confirmed the mentioned applications and commented on how electric fields are used in these two cases.

13.4.8.3   The Textbook vs Teachers' Lecture

One may notice that there are clear differences between the textbook and the teacher's reasoning in relation to the first described problem. The argumentation in the textbook is based on the existence of a field direction; the electron moves perpendicular to the field direction, and, to make this clear, the textbook uses an index v on the variables s_v (speed) and v_v (velocity). While the teacher communicated the context in terms of x-direction and y-direction and made use of variables v_x and v_y and emphasised that "the electron moves parallel to the plates" (RSr2) not explicitly relating the movement to the field direction, the teacher conducted reasoning related to a technical mathematical context (M1) and instrumental use of Theoretical Models (TM1) in more or less directly connecting M1 to Reality School (RSr2). Also, in the textbook's formulation of the task, you are asked to calculate the "direction" of the electron when it leaves the field. Thus, it makes an implicit connection to the vector concept but leaves the interpretation of the concept of "direction" to the reader, enabling a more structural reasoning. In the teacher's phrasing, it has a more technical framing (M1). It is stated that there is an angle to calculate with reference to the plates (see Fig. 13.5).

13.5   Discussion

In this chapter, we have described a framework for analysing relations between *Theoretical models*, *Reality* and *Mathematics*. The framework has been developed from an earlier version presented in Hansson et al. (2015) and Redfors et al. (2016). The developed framework has been tested on textbook sections and on associated physics lessons. The focus on relations between Theoretical models, Reality and Mathematics in classrooms as well as textbook analysis has been rare in the research literature. This is the case even though the communication of these relations is of central importance for the teaching of physics and more specifically the possibilities and constrains for a meaningful learning for students. The here presented fine-tuned framework aims at contributing to the research field in this respect.

Using the analysis framework makes it possible, as has previously been shown, to shed light on what and how different relations are communicated during different kinds of physics lessons and as shown in this chapter also what and how the same relations are communicated in physics textbooks. From this kind of analysis, we can get information, for example, about the amount of time spent at communicating different relations during a physics lesson. We have earlier shown how the relation between Theoretical models and Mathematics was more frequently communicated than the relation between Theoretical models and Reality (cf. Hansson et al. (2015)). This could be viewed to be a problem in relation to students' meaning making. We also found in that study that the teacher frequently communicated a technical use of mathematics (M1); this was the case not only in problem-solving situations (which has been shown in previous research as well) but also during labwork and during teacher-led lessons (Hansson et al. 2015). The analysis described in this chapter showed that the textbook has a more pronounced structural use of mathematics (M2) compared to the teacher. The major difference here is that the book deduces and proves formulas, whereas the teacher mainly uses them.

In this chapter, the analysis framework has been further developed to include possibilities to distinguish between instrumental (TM1) and relational approaches (TM2) to theoretical models. The results presented in this chapter show that the investigated textbook more often communicates relational approaches to theoretical models (TM2) compared to the teacher. The teacher frequently communicated an instrumental approach to theoretical models (TM1). This is the case when she introduced formulas as a kind of recipe to be used to calculate quantities or solve problems, without focusing why the formula is valid or how it is connected to other concepts, relations or the overall theoretical model.

However, we argue the textbook invites teachers and students to such an instrumental approach to models through, for example, its way of highlighting formulas. Formulas are accentuated in specific squares on the pages communicating their importance and the formulas as the goal of the text – an instrumental emphasis (TM1). Hence, in highlighting formulas and not the underlying theoretical models, the textbook invites to an instrumental approach to theoretical models, and the teacher in instances follows this path. An example is that one section of the text (about energy and exergy), which does not result in a formula, was excluded by the teacher in her presentation.

Furthermore, in the theoretical framework presented here, we have included possibilities to distinguish between Reality and Reality School (including reduced and altered reality) and between whether reality was only mentioned (recall experience) and whether the real-world phenomenon was referred to in detail and developed upon (systematic description). The analysis gives that when some form of reality is referred to, it is most often a reduced reality (RSr) or an altered reality (RSa). While the reduced reality refers to idealised labwork situations, an altered reality refers to a reality in which people ask questions and/or know things not normally asked for or known in an everyday situation. This is important in relation to possibilities and constrains in relation to students meaning making. Of course, referring to known situations of interest for students could be one way to increase interest and the perceived importance of science. But, when the reality referred to is an altered reality, these possibilities run the risk of being lost and could instead have negative consequences for students. For example, in a study of students' perspectives on PISA science assignments, Serder and Jakobsson (2015) show how "students' positioning themselves as being different from and opposed to the fictional pictured students who appear in the backstories of the test" (p. 833). We argue that many of the reality situations discussed in this chapter (from the textbook and from the lessons) run the same risk.

The references to Reality School differ for the two investigated courses. It is most often Reality School-altered (RSa) for the first course (first-year class), while in the second course (third-year class), references to Reality School-reduced (RSr) dominate. Notice that the Reality School referred to in the second course, often, is a systematic description (RSr2). This also applies to the teacher, who in this sense often seems to follow the book. In the case of the introduction of electric fields in the second course, the textbook relates to everyday reality (R), i.e. inkjet printers. But the inkjet printer is described by a component diagram, categorised as Reality School-reduced with a systematic description (RSr2), and the teacher takes it from there and relates mainly to RSr2 in her effort to communicate the abstract content of the theoretical model and link it to something more tangible, i.e. in using parallel plates to demonstrate electric discharges, but without further reference to reality (R).

In working to introduce the electric field chapter, the teacher invites the students to discuss relations between the introduced concepts. She interacts with the students and in doing so induces a general discussion. However, in this she relates mostly instrumentally to Theoretical models (TM1) and uses Mathematics technically (M1) helping the students to find the correct mathematics and formulas. She is also, together with the textbook, leading the students to a foremost technical use of mathematics and instrumental approach to theoretical models for their learning process. An interesting example is the presentation of the case of a moving electron in an electric field, where the textbook refers to a relational approach to theoretical models (TM2) by focusing on the path of the electron in relation to the direction of the electric field lines in an effort to connect a structural use of mathematics (M2) to a relational use of theoretical models (TM2). Here, the teacher instead focuses on the context of the problem and talks about the path of the electron in relation to the alignment of the plates, hence linking a systematic approach to a reduced Reality School (RSr2) directly to structural use of mathematics (M2), placing the communication in a more instrumental way concerning the theoretical model. This transfers the problem to a mathematical context that the students might more easily master (cf. chapter by Planinić et al. (2019)) but does not facilitate their understanding of the physics in context.

The developed framework has been found useful when it comes to classroom as well as textbook analysis. The fine-tuning of the Theoretical model and the Reality entities makes it possible to focusing different relations of interest in the specific situation. In the case described here, we can see that the teacher has an overall instrumental approach to theoretical models, and the book also to some extent invites to such an approach. This is an example of how the framework could be used for comparisons of different presentations of physics content knowledge. In the case presented here, textbook-teacher comparisons were made, but also teacher-teacher or textbook-textbook comparisons are of course possible. It is of course also possible to in a specific analysis exclude the differentiations done here concerning an entity or focus solely on relations made between the two of them. In such a way, our intention is that the framework should enable analysis with different (but related) focuses in the future.

13.6 Implications for Teaching

The specific results presented here could point to interesting issues to take into consideration for textbook authors and publishers. The case described here shows how invitations to an instrumental approach to theoretical models in textbooks could be a path followed (and even strengthened) by teachers. In the same way, the results presented here strengthen previous research in other contexts (e.g. Serder and Jakobsson 2015) on the necessity to discuss consequences for students of the everyday contexts related to in the teaching of science. In the case described here, we mostly see reality references in respect of Reality School (reduced or altered), which underlines the importance of this discussion also in the context of textbooks and teachers' presentations during physics lessons.

Even though this case study has to be followed by studies of other textbooks and teachers, the results presented here raise the issue of how textbook authors could strengthen the invitations to more relational approaches to theoretical models. In the same way, teachers could be made aware of their choices and uses of textbooks. Analysis of textbooks as well as teachers' own practice, with the starting point in the suggested framework, could be a tool for teachers to shed light on when relations between the three entities (Theoretical models, Reality and Mathematics) are communicated and not communicated. As pointed out above, it is possible for the teacher to focus on any combination of the three entities and stress differentiation when that is of interest for a specific situation.

The framework makes possible comparisons between textbook and teacher as described here but also between textbook-textbook and teacher-teacher. The case study described here as well as in Hansson et al. (2015) could be used as example analyses. These cases could be a starting point for preservice and in-service teachers' own analysis of relevant textbooks and/or lessons. It makes it possible to analyse consequences of categorisation of statements (R1-2, RS1-2, M1-2, TM1-2) by teachers and in textbooks for class communication. The focus of such analyses can be on all aspects of the triangle of analysis in Fig. 13.2 or on any combination of the eight entities described above. This kind of activities, as part of physics teacher education and professional development, could be a way for teachers to raise their awareness of relations communicated between Theoretical models, Reality and Mathematics. The here presented model could be used to study the implementation of teaching recommendations. For instance, recommendations are described in the chapter by Planinić et al. (2019) in this book. Different ways of addressing and linking the three corners of the triangle of analysis are likely to promote communication and learning in differing ways for the students. Such an awareness could be an important starting point for teachers to act to change her/his traditional teaching focus in desirable directions.

Acknowledgements We acknowledge the financial support of Swedish Research Council (2015-01643) and the efforts of the GIREP organisation and of Gesche Pospiech as the editor of this book.

References

Adúriz-Bravo, A. (2012). A 'semantic' view of scientific models for science education. *Science & Education, 22*(7), 1593–1611.

Angell, C., Lie, S., & Rohatgi, A. (2011). TIMSS Advanced 2008: Fall i fysikk-kompetanse i Norge og Sverige. *NorDiNa, 7*(1), 17–31.

Bachmann, K. E. (2005). *Læreplanens differens: Formidling av læreplanen til skolepraksis.* Trondheim: NTU.

Bliss, J., Monk, M., & Ogborn, J. (1983). *Qualitative data analysis for educational research.* London: Croom Helm.

Erduran, S., & Dagher, R. (2014). *Reconceptualizing the nature of science for science education: Scientific knowledge, practices and other family categories* (Contemporary trends and issues in science education) (Vol. 43). Dordrecht: Springer Verlag.

Frejd, P. (2012). Teachers' conceptions of mathematical modelling at Swedish Upper Secondary school. *Journal of Mathematical Modelling and Application, 17*(1), 17–40.

Giere, R. N. (1988). *Explaining science: A cognitive approach.* Minneapolis: University of Minnesota Press.

Hanson, N. R. (1958). *Patterns of discovery.* Cambridge: Cambridge University Press.

Hansson, L., Hansson, Ö., Juter, K., & Redfors, A. (2015). Reality – Theoretical models – Mathematics: A ternary perspective on physics lessons in upper-secondary school. *Science & Education, 24*(5–6), 615–644. https://doi.org/10.1007/s11191-015-9750-1.

Johansson, A., Andersson, S., Salminen-Karlsson, M., et al. (2016). *Cultural Studies of Science Education*. https://doi.org/10.1007/s11422-016-9742-8.

Karam, R. (2014). Framing the structural role of mathematics in physics lectures: A case study on electromagnetism. *Physical Review Special Topics – PER, 10*, 010119-1-010119-23.

Karam, R, Uhden, O., & Höttecke, D. (2019). The "math as prerequisite" illusion: Historical considerations and implications for physics teaching. *Chapter in this book.*

Koponen, I. T. (2007). Models and modelling in physics education: A critical re-analysis of philosophical underpinnings and suggestions for revisions. *Science & Education, 16*(7–8), 751–773.

Krey, O. (2019). A mathematics sensitive look at what is taught and learned in physics. Chapter in this book.

Lederman, N. G. (2007). Nature of science: Past, present, and future. In S. K. Abell & N. G. Lederman (Eds.), *Handbook of research on science education* (pp. 831–879). Mahwah: Erlbaum.

Nelson, J. (2006). Hur används läroboken av lärare och elever? *Nordic Studies in Science Education, 2*(2), 16–27.

Pask, C. (2003). Mathematics and the science of analogies. *American Journal of Physics, 71*(6), 526–534.

Pietrocola, M. (2008). Mathematics as structural language of physical thought. In M. Vicentini & E. Sassi (Eds.), *Connecting research in physics education with teacher education* (International Commission on Physics Education) (Vol. 2). ICPE.

Planinić, M., Sušac, A., Ivanjek, L., & Milin-Šipuš, Ž. (2019). Comparing student understanding of graphs in physics and mathematics. *Chapter in this book.*

Redfors, A., Hansson, L., Hansson, Ö., & Juter, K. (2016). A framework to explore the role of mathematics during physics lessons in upper-secondary school. In N. Papadouris, A. Hadjigeorgiou, & C. Constantinou (Eds.), *Insights from research in science teaching and learning* (pp. 139–151). New York: Springer International Publishing.

Sánchez, G., & Valcárcel, M. V. (1999). Science teachers' views and practices in planning for teaching. *Journal of Research in Science Teaching, 36*(4), 493–513.

Serder, M., & Jakobsson, A. (2015). "Why bother so incredibly much?": Student perspectives on PISA science assignments. *Cultural Studies of Science Education, 10*(3), 833–853.

Skemp, R. R. (1976). Relational understanding and instrumental understanding. *Mathematics Teaching, 77*, 20–26.

Skolverket. (2012). *Curriculum for the upper secondary school*. Stockholm: Fritzes. [http://www.skolverket.se/publikationer?id=2975].

Skolverket. (2016). *Uppföljning av gymnasieskolan*. [http://www.skolverket.se/publikationer?id=3642].

Triantafillou, C., Spiliotopoulou, V., & Potari, D. (2016). The nature of argumentation in school mathematics and physics texts: The case of periodicity. *International Journal of Science and Mathematics Education, 14*(4), 681–699.

Uhden, O., Karam, R., Pietrocola, M., & Pospiech, G. (2012). Modelling mathematical reasoning in physics education. *Science & Education, 21*(4), 485–506.

Chapter 14
Starting with Physics: A Problem-Solving Activity for High-School Students Connecting Physics and Mathematics

E. Bagno, H. Berger, E. Magen, C. Polingher, Y. Lehavi, and B. Eylon

14.1 Introduction

The interrelations between physics and mathematics in the learning of high-school physics are manifested in several aspects of physics teaching (Sherin 2001; Bing and Redish 2009; Uhden et al. 2012; Karam 2014; Redish and Kuo 2015). These interrelations, used by teachers, have been conceptualized into four "phys-math patterns," each of which addresses different teaching goals (Pospiech and Oese 2014; Lehavi et al. 2015, 2017; Pospiech and Geyer 2016). The phys-math patterns reflect how teachers "travel" in their teaching between the two domains and within each of them, always starting from the physics domain. One of these patterns, the "application pattern," describes how teachers employ the phys-math interrelations in problem-solving – an endeavor that occupies much of high-school physics teachers' time and attention. Here we focus mainly on this pattern.

E. Bagno (✉) · H. Berger · B. Eylon
The Weizmann Institute of Science, Rehovot, Israel
e-mail: esther.bagno@weizmann.ac.il; Bat-sheva.Eylon@weizmann.ac.il

E. Magen
The Weizmann Institute of Science, Rehovot, Israel

Ostrovsky High School, Raanana, Israel

C. Polingher
The Weizmann Institute of Science, Rehovot, Israel

Hemda Schwartz-Reisman Science Education Center, Tel Aviv, Israel

Y. Lehavi
The Weizmann Institute of Science, Rehovot, Israel

The David Yellin Academic College of Education, Jerusalem, Israel

© Springer Nature Switzerland AG 2019
G. Pospiech et al. (eds.), *Mathematics in Physics Education*,
https://doi.org/10.1007/978-3-030-04627-9_14

Research on the problem-solving habits of high-school physics students shows that students often start solving problems by using mathematical manipulations or by looking for seemingly relevant formulae (Mason and Singh 2010; Kim and Pak 2002; Van Heuvelen 1991; Byun and Lee 2014; Heller and Heller 2010). Research also indicates that a technical use of formulae may decrement the development of students' understanding of physics (Bagno et al. 2008; Karam 2014).

Chi and her collaborators (Chi et al. 1981) report on a fundamental difference between how experts and novices address problem-solving. For example, they found that novices tend to sort problems according to their surface features (e.g., blocks on inclined planes), whereas experts sort them according to the underlying physics principles. Chi et al. claim that "experts in physics problem solving, engage in qualitative analysis of the problem prior to working with the appropriate equations . . . and . . . this method of solution for the experts occurs because the early phase of problem solving (the qualitative analysis) involves the activation and confirmation of an appropriate principle-oriented knowledge structure, a "schema." Apparently, qualitative analysis of a problem by using physics terms and principles is an essential skill that can assist high-school physics students in narrowing the gap between how they approach a problem in physics and how experts do it.

Here we describe the "Starting with Physics" activity, which attempts to activate the "principle-oriented knowledge structures" mentioned above. Students are asked to carry out an activity by focusing on the use of appropriate physics concepts and principles together with their mathematical manifestations (e.g., graphs and their descriptions) and to delay the use of formulas and other technical mathematical manipulations. Our goal is to stress, in the context of problem-solving, the power of a concise set of physics principles for explaining a phenomenon described in a problem, before using mathematical manipulations and techniques.

Another important goal that guided us in the design of this activity is an attempt to build a "learner-centered activity" supporting students' learning. In this regard we used the knowledge integration (KI) perspective on learning, (Linn and Eylon 2006), according to which learners build their knowledge when teachers stimulate four learning processes:

1. Eliciting prior knowledge: learners become aware of their preexisting knowledge
2. Adding new ideas: learners are introduced to ideas that are new to them. These ideas may originate from various sources such as a teacher, a textbook, a peer, or the Internet.
3. Developing criteria to evaluate ideas: questions and tests that the learners use to determine whether they consider the ideas as acceptable. Examples of such criteria are whether the origin of the new ideas is reliable (i.e., based on scientific principles) and whether contradictions exist within the ideas acquired or between them and the ideas that are already known to the learner.
4. Sorting out and reflecting: this is a metacognitive learning process in which learners reflect on and differentiate between their preexisting ideas and the newly acquired ones based on specific criteria.

The four processes do not necessarily appear one after another and not always in the same order. These learning processes formed the basis for designing the procedure through which our activity was carried out. However, many other teaching methods can promote these learning processes (e.g., peer instruction, context-rich problems).

The research literature reports on a large number of empirical studies, investigating the relationships between designs of such teaching methods that attempt to promote KI and learning outcomes (Linn and Eylon 2011).

We carried out a study in the context of implementing the activity in high-school physics classes. The research in this study was aimed at examining students' reasoning throughout the activity and their reflections regarding how the activity contributed to their learning. In addition, we investigated teachers' views regarding the activity and its contribution to physics learning.

The following sections describe the activity, the study, and teachers' views.

14.2   The "Starting with Physics" Activity

One of the main goals of physics instruction is to promote students' "physics understanding," as manifested in their ability to describe a phenomenon qualitatively and explain it by using physics concepts and principles. However, the usual structure of a standard physics problem allows students to have an "escape route" from this important goal. A problem in physics often consists of a paragraph describing a phenomenon, followed by a set of questions. Both experienced physics teachers and physics education researchers agree that students tend not to thoroughly read the introductory paragraph nor try to understand the problem. Instead, they turn to formulas and mathematical manipulations that seem relevant to them, without examining whether they are valid in explaining the phenomenon under consideration.

The "Starting with Physics" activity was designed as follows:

(a) Students receive only the first part of the problem consisting of a textual description of the phenomenon and the relevant mathematical information, without any subsequent questions. Thus, they are prompted to address the problem conceptually first with nothing to calculate.
(b) At the beginning of the activity, students are asked to divide the phenomenon into events and to describe and explain each event by using physical concepts and principles without using equations.
(c) Then, students are asked to list the physical concepts and principles on which they based each event's description and explanation.

Figure 14.1 shows an example of the "Starting with Physics" activity in the context of electrostatics. Based on the KI perspective, we implemented the activity in a four-phase learning cycle. The cycle consists of "individual work" in which each student fills in the table in Fig. 14.1. In order to save class time, this phase may be carried out as homework. This is followed by a "group work" phase that usually takes place in class. The students work in small groups on the same activity, evaluate their individual work, add new ideas, and reach a consensus (or

The motion of a charged particle between charged plates

1. Individual work

Many electric systems (for example, a particle acceleration system) contain charged plates similar to the system shown below.

The system contains three charged plates A, B and C parallel to each other. The distance between plates A and B is different from the distance between plates B and C. There is a small hole in the center of plate B (see the illustration, but assume that the plates are much larger than the distances between them).

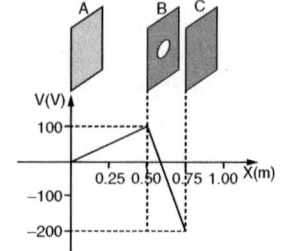

The attached graph describes the electric potential between the plates.

Consider the following phenomenon:
A negatively charged particle is released from rest at the center of plate A and it starts moving.
Fill in the table below according to the following:
- a. If possible, divide the phenomenon into events that differ from each other regarding the nature of the moving particles, the acting forces, and more. For each event indicate its starting and ending points. Use as much as possible diagrams, graphs, or illustrations. If needed, add rows to the table.
- b. The "physics" of each event must include a description of the event and its explanation using physical concepts and principles (do not use equations).
- c. List, in a separate column, the physical concepts and principles on which you based the event's description and explanation.

Events			"Starting with Physics"		
1	2	3	4	5	6
	Start	End	Describe and explain the event by using physical concepts and principles	List the physical concepts and principles	Diagrams
Event I					
Event II					

2. Group work
Discuss your individual work with your friends. If necessary, modify your table.
3. Whole-class discussion
Group work is discussed under the teacher's guidance.
4. Individual reflection
If you were helped by the activity, describe how.

Fig. 14.1 The "Starting with Physics" activity

have a disagreement). The next phase is a "whole-class discussion" in which a representative of each group presents to the plenum the group's consensus as well as any disagreements; all the issues raised in the group work are discussed, under the teacher's guidance, and the class formulates a summary. The activity culminates in "individual reflection" in which each student individually accounts for what he or she has learned during the activity.

As can be seen, the design of this activity balances "problematizing" and "structuring" two complementary mechanisms of scaffolding problem-solving: (1) *Structuring* a task refers to reducing its complexity and limiting the choices of the problem-solver. (2) *Problematizing* directs one's attention to aspects that one might otherwise overlook. Instruction should be balanced between structuring and problematizing so that tasks will be manageable to learners yet challenging and engaging (Reiser 2004; Yerushalmi and Eylon 2016). In our study, this activity (see Fig. 14.1) was carried out by two experienced 12th grade teachers with 31 students.

14.3 Research on Students' Use of Physical Concepts and Principles in Performing the Activity

14.3.1 Research Questions

We studied students' answers in the table, focusing on the following questions:

1. How did students in this activity use the *physical concepts and principles* in describing and explaining the events in a phenomenon? (From column 4 in the table)
2. How did students list the *physical concepts and principles* on which they based each event's description and explanation? (From column 5 in the table)

14.3.2 Methodology

The phenomenon in the activity exemplifies two apparently different events that share the same underlying physical principles. The two events are not identical, since in the first event the electric charge moves from a low potential to a high potential, whereas in the second event it moves from a high potential to a low potential. This information is conveyed by a graph (see Fig. 14.1) and leads to differences between the description and explanation of the two events in the direction of the electric field, the electric force, the acceleration, and the velocity of the charge (column 4 in Fig. 14.1).

However, the list of the physical concepts and principles should be the same (column 5 in Fig. 14.1).

Our considerations in the content analysis of the students' answers in the table were based on the answers written by a top-level student (see Fig. 14.2). This student focused mainly on the following two aspects:

1. The connections between physical concepts and principles
2. The connections between mathematics and physics

We will indicate below how the above two aspects were manifested in the paragraph written by this top-level student.

(a) *Graph V(x) has a constant slope, and therefore, the electric field is uniform.*
(b) *The force exerted on the particle is constant because the electric field is constant.*
(c) *The electric field is directed to the left due to the higher electric potential at plate B.*
(d) *Since the particle is negatively charged, the electric force acting on it is directed toward plate B.*
(e) *Due to this force, the particle moves with constant acceleration, and its speed increases.*

Manifestations of the Two Aspects in the Paragraph Statement *(a)* in this paragraph – *Graph V(x) has a constant slope* – is a mathematical statement leading to a physical conclusion – *. . . the electric field is uniform*. This conclusion is followed in statement *(b)* by a sequence of physical concepts, starting with the relationship between the field and the force and then the electric charge – *the force exerted on the particle is constant because the electric field is constant.*

In statement *(c)* the direction of the electric field is determined by referring back to the graph (a mathematical representation) – *the electric field is directed to the left due to the higher electric potential at plate B*. Next, in statement *(d)* an important relationship exists between three central physical concepts (field, force, and charge) – *since the particle is negatively charged, the electric force acting on it is directed toward plate B*. Finally, in statement *(e)*, the student relates to dynamics and kinematic concepts and concepts within kinematics – *due to this force, the particle moves with constant acceleration, and its speed increases.*

14.3.3 Findings on Research Question 1

How did students in this activity use *physical concepts and principles* in describing and explaining events in a phenomenon?

The findings are based on all students' answers in column 4 of the table in Fig. 14.1.

Events			"Starting with Physics"		
1	2	3	4	5	6
	Start	End	Describe and explain the event by using physical concepts and principles	List the physical concepts and principles	Diagrams
Event I	Particle at plate A	Particle at plate B	a) Graph $V(x)$ has a constant slope and therefore the electric field is uniform. b) The force exerted on the particle is constant because the electric field is constant. c) The electric field is directed to the left due to the higher electric potential at plate B. d) Since the particle is negatively charged, the electric force acting on it is directed towards plate B. e) Due to this force, the particle moves with constant acceleration and its speed increases.	a) Existence of the field, the gradient of the potential b) The magnitude of the force is proportional to the strength of the field. c) The force on a negative charge points in an opposite direction to that of the field. d) Newton's 2nd law e) The acceleration is in the direction of the field.	
Event II	Particle at plate B	Particle at plate C (or before C)	a) All the events have the same arguments; however, the directions are opposite. b) The slope of $V(x)$ is negative, but the field is positive. c) The force is to the left, but the field is to the right.	The slope (the gradient) $\downarrow$ Field $\downarrow$ Force $\downarrow$ Acceleration • Newton's 2nd law • The acceleration is in the opposite direction of the velocity. 	

Fig. 14.2 The filled-in table of a top-level student

1. **Most of the students' statements dealt with the two aspects mentioned above: the connections between physical concepts and the connections between mathematics and physics.**

Students took advantage of the mathematics in the connections between physics and mathematics to enhance their understanding of physics; they formed connections between concepts or ideas within physics (either **within a domain** such as electrostatics or **between domains** such as kinematics and dynamics). In this respect, they employed what we termed "a phys-math exploration pattern," characterized by beginning with a certain physical phenomenon or system; then a mathematical representation is studied, and finally, the ramifications of the mathematical analysis for the case in hand are discussed with new physical insights (Lehavi et al. 2015).

2. **Students realized physics-related similarities between seemingly different events.**

This was reflected by the fact that most students used the same concepts and principles in describing the two events. Moreover, the findings regarding their individual reflections, described below, indicate that they were cognizant of this.

3. **There was a progression from the description of the first event to that of the second one.**

About 70% of the students described and explained the second event, in a more general manner than the first event. This finding was more frequent among top-level students: *The events are similar; however, the directions of the forces are opposite.*

14.3.4   Findings on Research Question 2

How did students list *physical concepts and principles* on which they based each event's description and explanation?

The findings are based on all students' answers in column 5 of the table in Fig. 14.1

1. **Most of the students used the same principles for the two apparently different events.**

In most of the students' tables, the list of the physical concepts and principles was the same for the two events. Some of the students did not even bother to write the same concepts and principles again for the second event. Some left the relevant box in the table empty and noted that it should be the same. Further support for this finding comes from the "whole-class discussion" in one of the classes. When the classroom summary was formulated under the teacher's guidance, the students suggested leaving the box of the concepts and principles for the second event empty, since it is identical to that of the first event.

2. **Some students summarized, in the second event, the whole sequence of reasoning by a concept map representing the connections between underlying physical concepts and principles** (typical of top-level students).

This is exemplified in column 5 of Fig. 14.2. In addition, this particular student also described the acceleration by representing it in a graph.

14.4   Research on Students' Views on the "Starting with Physics" Activity

14.4.1   Research Questions

1. How did students, in their individual reflections, refer to the *goals of the activity and its contribution* to their learning?
2. What congruence can be found between the students' individual reflections and their use of concepts and principles in performing the activity?

14.4.2   Methodology

The data for this analysis originates from students' individual reflections on the activity.

Whereas the first part of the activity involved team learning and **whole-class discussions, the individual reflection required students to report on what** they had learned from the activity. In order to enable the students to come up with a variety of ideas, the individual reflection was phrased in an open manner: "If you were helped by the activity, describe how."

We started the analysis by dividing students' reflections into statements. All together, we identified 50 statements. In the analysis, we looked for congruency between students' reflective statements and their answers in the tables. Accordingly, the analysis was guided in a top-down manner and referred to the following:

1. Reflections about connections and their congruence with the ones students wrote in column 4 of the tables
2. Reflections about physical relationships and their congruence with the ones students wrote in column 5 of the tables
3. Other ideas that students brought up

Note that some of the statements provided information on more than one of the three foci of reflection.

14.4.3 Findings on Research Questions 1 and 2

1. **Reflections about connections**: Of the 50 reflective statements, about 50% dealt with connections. The different types of connections that were found in column 4 of the tables were also found in students' reflections.

   The following are some examples:

- **Connections between physics and mathematics**: About 25% of the statements dealt with the ways by which the students understood the physical meaning of the mathematical representations.

    - *I understood that the slope of the graph can also indicate whether the field is constant.*
    - *I understood that the gradient is the derivative of the potential.*
    - *The activity helped me understand the meaning of the formula: $E = -\Delta V/\Delta x$.*
    - *The activity helped me mainly in better understanding graphs and in relating to and connecting between a graph and an event.*

- **Connections between concepts or ideas within a physics domain and/or between physics domains**: About 20% of the statements dealt with different aspects of physical connections.

    - *It helped me in better understanding the relationship between distance, potential, and the field.*
    - *I understood that the field is the slope of the potential.*
    - *It helped me understand how the potential affects the forces acting on a charged particle.*
    - *It clarified for me that a relationship exists between the potential, the field, the force, the acceleration, and the velocity.*

2. **Reflections about physical principles**: About 20% of the statements dealt with physical principles resembling those we found in column 5 of the table.

    - *It helped me to better understand the motion of a charged particle in an electric field.*
    - *It clarified for me that a relationship exists between the potential, the field, the force, the acceleration, and the velocity.*

3. **Reflections dealing with metacognitive issues:** About 50% of the statements dealt with different types of metacognitive issues:

    - Understanding the goals of the activity and how they are promoted by its structure:

        - *I was helped by the activity. I now better understood the material that we learned and how one can describe and analyze better an exercise before starting to solve it.*

- *Yes, I better understood the theory as well as interpreting and understanding a graph.*
- *Yes, the activity enhanced my understanding of how many events and parts are in a problem and what happens to the particle in each part.*

• Promotion of various learning capabilities:

- *It underscored the rule that one should always check the given information in order to verify what really occurred.*
- *How to analyze a situation according to a graph of V vs. x and what consequences can be derived from this graph.*

• Realizing the relationships between the studied topics:

- *Yes, the activity summarized for me the materials studied and connected all the relevant topics.*

14.5 Summary of Research on Students

In studying students' use of concepts and principles in the activity, we found that the activity achieved its goal: students indeed engaged in physics during the activity rather than "jumping" to formulae and technical mathematical manipulations. Most of their statements actually dealt with various types of connections: the connections between physical concepts and the connections between mathematics and physics. We also found that students managed to describe the two apparently different events similarly and some of them even provided a more comprehensive and general description in the transition from the first event to the second one.

In studying students' views concerning the activity, we found that students, in their individual reflections, mentioned explicitly the formation of connections between physics and mathematics, between concepts or ideas within a physics domain, and/or between physics domains. They also referred to the important role of physics principles in describing events. Interestingly, we also found in students' reflections different types of metacognitive issues such as how the activity contributed to their learning capabilities.

14.6 Teachers' Reports on Using the Activity in Their Practice

Two important questions are to what extent and how is this activity useful for teachers in their practice and what did teachers think about its contribution to learning physics. We had an opportunity to examine these questions in the context of professional development programs for teachers in which they were introduced to several "learner-centered" activities. They implemented the activities in their classes and brought materials from their classes (such as students' answers in responding to questions on the activities) for collaborative reflections with their peers. This "evidence-based" approach is a powerful method for teachers' learning and impacts teachers' practice (Berger et al. 2008; Harrison et al. 2008; Eylon et al. 2008).

We audiotaped the discussions and also interviewed several teachers. Most teachers rated the "Starting with Physics" activity as the highest one. The teachers referred to both the physics learning aspect and to various phases of the activity.

A common finding is that the teachers found that the activity contributes to their practice and to students' learning of physics. They also reported the importance of carrying out the different phases. In addition, their reports indicate that this activity can be used in various formats and in different physics domains (e.g., mechanics, electrostatics), and therefore, it provides ample opportunities for teachers to use it in their practice on a regular basis. The following are some examples from reports of three teachers who participated in these professional development programs: Ella, Ziva, and Tibi (all pseudonyms).

Ella became convinced that the activity has a real impact on her students' ability to relate physical principles to the events in a problem. She also pointed out that each of the phases has its own importance. In her words: "In the individual work, each student is forced to expose his or her own knowledge, whereas in the group work, they learn from each other; in the class discussion, the teacher helps them to correct mistakes that are found during the activity." Ella also reported that, "Decomposing a complex situation into several events and dealing with each of them separately simplifies the activity for most of the students."

Ziva was very enthusiastic about this activity as well. She uses it in her classes on a regular basis. In order to save class time, she usually asks her students to perform the individual phase at home. In an interview held with Ziva, she said: "This activity, which I am so attached to, no doubt caused a new language to develop in my classes. This language includes, for example, the term 'event'. This word is now familiar to my students in the context of problem solving. I find myself solving with my students complex problems by decomposing them into their events. I even started to include tasks such as 'decompose the problem into its events and give the event an appropriate title' in my exams. Ninety percent of the exams are better organized now. I think that this organization has to do with my explicit request to relate to each event separately." Ziva claimed that the activity enables her to emphasize the common underlying physical principles of apparently different problems: "Usually I spend a whole lesson solving each of the very similar problems I gave for homework. My students insist on it. With this activity, they leave me alone, since they realize that you can solve many problems by using the same ideas; and it serves as a supporting framework for problem solving in physics."

Another teacher Tibi reported that in analyzing his students' worksheets he found that the group discussions had greatly contributed to students' understanding. He also said that in the "whole-class discussion" phase, his students easily realized the similarity between this electrostatics problem and other problems from mechanics, having the same underlying principles. He suggested that it is necessary to carry it out with students several times in order to bring about its habitual use.

Indeed, in an interview held with Tibi, several years after the professional development program, he said that he uses the activity regularly in his classes in the following format: he invites students to write on the blackboard their descriptions of the events, and he encourages others to justify the descriptions. Tibi also said that he encourages his students to reflect on the activity and to express explicitly what ideas they have learned during the activity and what still remains unclear. In his words: "Describing the underlying physics of the problem, before they start with the formulas and going back to the physics after they have finished with the formulas, is so important. This resembles debugging."

We also found other teachers' views that were similar to those illustrated here. Teachers used the activity in a wide range of formats that they found were feasible and useful for their students.

14.7   Discussion and Implications

In this paper we described an activity that aims to promote students' ability to describe and explain a phenomenon qualitatively by using physical concepts and principles rather than engaging in technical mathematical manipulations. The "Starting with Physics" activity was very effective in activating "principle-oriented knowledge structures" (Chi et al. 1981). Instead of technically misusing the phys-math relationships, students focused on physics concepts and principles and their relations to mathematical aspects.

Research on implementing the activity in physics high-school classes indicated that in carrying out the activity, most of the students managed to describe the two apparently different events similarly. They referred to various types of connections between physics concepts and principles and connections between physics and mathematics. Furthermore, some of the students' responses may indicate that they use mathematical ideas (e.g., the slope of the electric potential as an indicator of the electric field) rather than technics when analyzing a physical event. Such a perspective (an exploration phys-math pattern rather than an application one) was found to characterize more expert teachers (Lehavi et al. 2015). This positive finding may encourage further research on examining in detail the above described activity and models for its implementation in frameworks such as professional learning communities of teachers.

In their reflections the students explicitly mentioned different types of connections as well as the role of physical principles in describing events. They also referred to metacognitive issues. In particular, students mentioned the rationale underlying the activity's design and its important contribution to their learning. In a more detailed analysis of students' actual work in the table and their reflections (not reported here), we found congruency between their answers and their views. Some students even suggested that activities of this kind should be encouraged by giving them extra credit.

This activity was highly appreciated by physics teachers. They claimed that it emphasizes the common underlying physical principles of apparently different problems and supports problem-solving in physics. However, it is necessary to carry it out with students several times in order to bring about its habitual use. Since this activity is generic, it is suitable for many standard A level physics problems. We already have a large pool of problems in the format of this activity filled out by teachers and tried out by many students.

Several directions can be explored in future research: What can be learned from the data that students bring from the individual work to the peer discussion and from the discourse that follows? How do students evolve in their ability to fill in the tables in the activity correctly and exhaustively (i.e., use properly all the relevant concepts and their interrelations)? What impact may such an activity have on low grades students?

Such studies can enable one to better understand the underlying mechanisms leading to student and teacher learning in the context of this activity.

References

Bagno, E., Berger, H., & Eylon, B. S. (2008). Meeting the challenge of students' understanding of formulae in high-school physics: A learning tool. *Physics Education, 43*(1), 75–82.

Berger, H., Eylon, B., & Bagno, E. (2008). Professional development of physics teachers in an evidence-based blended learning program. *Journal of Science Education and Technology, 17*(4), 399–409.

Bing, T. J., & Redish, E. F. (2009). Analyzing problem solving using math in physics: Epistemological framing via warrants. *Physical Review Special Topics – Physics Education Research, 5,* 020108.

Byun, T., & Lee, G. (2014). Why students still can't solve physics problems after solving over 2000 problems. *American Journal of Physics, 82,* 906.

Chi, M. T. H., Feltovich, P. J., & Glaser, R. (1981). Categorization and representation of physics problems by experts and novices. *Cognitive Science, 5*(2), 121–152.

Eylon, B., Berger, H., & Bagno, E. (2008). An evidence-based continuous professional development program on knowledge integration in physics: A study of teachers' collective discourse. *International Journal of Science Education, 30,* 619–641.

Harrison, C., Hofstein, A., Eylon, B., & Simon, S. (2008). Evidence-based professional development of teachers in two countries. *International Journal of Research in Science Education, 30,* 577–591.

Heller, K., & Heller, P. (2010). *Cooperative problem solving in physics – A User's manual: Why? What? How?* Alexandria: The National Science Foundation.

Karam, R. (2014). Framing the structural role of mathematics in physics lectures: A case study on electromagnetism. *Physical Review Special Topics – Physics Education Research, 10,* 010119.

Kim, E., & Pak, S.-J. (2002). Students do not overcome conceptual difficulties after solving 1000 traditional problems. *American Journal of Physics, 70,* 759.

Lehavi, Y., Bagno, E., Eylon, B. S., Mualem, R., Pospiech, G., & Böhm, U. (2015). Towards a PCK of physics and mathematics interplay. In The *GIREP MPTL 2014 Conference Proceedings.*

Lehavi, Y., Bagno, E., Eylon, B. S., Mualem, R., Pospiech, G., Böhm, U., Krey, O., & Karam, R. (2017). Classroom evidence of teachers' PCK of the interplay of physics and mathematics. In T. Greczyło & E. Dębowska (Eds.), *Key competences in physics teaching and learning* (Springer proceedings in physics) (Vol. 190, pp. 95–104). Cham: Springer.

Linn, M. C., & Eylon, B. S. (2006). Science education: Integrating views of learning and instruction. In P. A. Alexander & P. H. Winne (Eds.), *Handbook of Educational Psychology* (2nd ed., pp. 511–544). Mahwah: Lawrence Erlbaum Associates.

Linn, M. C., & Eylon, B. S. (2011). *Science learning and instruction: Taking advantage of technology to promote knowledge integration*. New York: Routledge.

Mason, A., & Singh, C. (2010). Helping students learn effective problem solving strategies by reflecting with peers. *American Journal of Physics, 78*, 748.

Pospiech, G., & Geyer, M.-A. (2016). Physical – Mathematical modelling in physics teaching. In *Electronic proceedings – Key competences in physics teaching and learning* (pp. 38–44). Wroclaw: Institute of Experimental Physics, University of Wrocław. Abgerufen von. http://girep2015.ifd.uni.wroc.pl/.

Pospiech, G., & Oese, E. (2014). Use of mathematical elements in physics – Grade 8. In *Active learning – In a changing world of new technologies* (pp. S. 199–S. 206). Prag: Charles University in Prague, MATFYZPRESS Publisher.

Redish, E. F., & Kuo, E. (2015). Language of physics, language of math: Disciplinary culture and dynamic epistemology. *Science & Education, 24*, 561. https://doi.org/10.1007/s11191-015-9749-7.

Reiser, B. J. (2004). Scaffolding complex learning: The mechanisms of structuring and problematizing student work. *Journal of the Learning Science, 13*, 273–304. https://doi.org/10.1207/s15327809jls1303_2.

Sherin, B. (2001). How students understand physics equations. *Cognition & Instruction, 19*, 479.

Uhden, O., Karam, R., Pospiech, G., & Pietrocola, M. (2012). Modelling mathematical reasoning in physics education. *Science & Education, 20*(4), 485. https://doi.org/10.1007/s11191-011-9396-6.

Van Heuvelen, A. (1991). Learning to think like a physicist: A review of research-based instructional strategies. *American Journal of Physics, 59*(10), 891–897.

Yerushalmi, E., & Eylon, B. (2016). Problem solving in science learning. In R. Gunstone (Ed.), *Encyclopedia of science education*. Heidelberg: Springer.

Part IV
Facilitating Mathematization by Visual Means

Chapter 15
Taking the Phys-Math Interplay from Research into Practice

Yaron Lehavi, Roni Mualem, Esther Bagno, Bat-Sheva Eylon, and Gesche Pospiech

15.1 Introduction

15.1.1 The "Phys-Math" Interplay

Although physics and mathematics can be regarded as autonomous distinct disciplines, physics, since its modern evolution, has been considered to be heavily interrelated with mathematics. Hence, as students' learning of physics evolves with age, they become increasingly more acquainted with this Phys-Math interplay and its many facets and thus encounter its complexity and unique features. This, as was shown from research, presents difficulties for many students at various levels. Current research indicates that learners, at different ages and levels, lack the ability to construct the mathematical models of physical processes or to describe the physical meaning of mathematical constructs. Bagno et al. (2007) found that high school students face difficulties in describing the physical meaning of formulae. Interestingly, difficulties in constructing equations to match situations described in words were also found at higher learning levels (engineering majors)

Y. Lehavi (✉)
The Weizmann Institute of Science, Rehovot, Israel

The David Yellin Academic College of Education, Jerusalem, Israel

R. Mualem · E. Bagno · B.-S. Eylon
The Weizmann Institute of Science, Rehovot, Israel
e-mail: esther.bagno@weizmann.ac.il; bat-sheva.eylon@weizmann.ac.il

G. Pospiech
Technische Universität Dresden, Dresden, Germany
e-mail: gesche.pospiech@tu-dresden.de

© Springer Nature Switzerland AG 2019   335
G. Pospiech et al. (eds.), *Mathematics in Physics Education*,
https://doi.org/10.1007/978-3-030-04627-9_15

(Clement et al. 1981). Rebmann and Viennot (1994) discussed the difficulty of many university physics students in applying and interpreting algebraic sign conventions consistently in a variety of topics.

Research in the context of DC circuits showed that skills in correctly using mathematics in problem-solving do not guarantee proper qualitative reasoning by both high school students and their teachers (Cohen et al. 1983). Baumert et al. (2010) have shown that teachers' mathematical knowledge highly reflects the quality of their explanations.

In the past, mathematics within a physics education context was mainly examined within the context of problem-solving (Bagno et al. 2007; Redish and Smith 2008). However, the above-mentioned findings may indicate that there is more to it and that the context of physics teaching involves an interplay between physics and mathematics, which is worthy of being the subject of research in the context of physics education. Indeed, some researchers pointed out that there is a blending of conceptual and formal mathematical reasoning during the mathematical processing stage (Kuo et al. 2013; Hull et al. 2013).

In this respect, our previous studies on expert high school physics teachers' views with regard to the "Phys-Math" interplay and the ways by which they implement it revealed that they employ specific paths in their teaching strategies when navigating back and forth between the two domains. Our findings indicate that teachers practice the use of Phys-Math interplay in order to foster different teaching goals. When following such practice, the teachers employ different specific "patterns" that follow different "steps" between physics and mathematics and within each domain. Each of these patterns serves different teaching goals in the general PCK framework (Lehavi et al. 2013, 2015, 2016a, b). We have identified four such patterns, each of which serves different teaching goals as well as practices (Table 15.1).

Here we will focus on the construction pattern and we will exemplify two Phys-Math teaching strategies that are aligned with this pattern:

Table 15.1 Phys-Math patterns, teaching goals, and teaching practices

Pattern	The teaching goal	The teaching practices
A. Exploration	To demonstrate how Phys-Math is used to explore the behavior of physical systems	Exploring, within mathematics, ramifications for the physical system: borders (of validity, of approximation), extreme cases, among others
B. Construction	To demonstrate how Phys-Math is used in constructing a model for physical systems	Constructing and developing (from experiments or from first principles) mathematical tools to describe and analyze physical phenomena
C. Broadening	To demonstrate how Phys-Math can be used in broadening the scope of a physical context	Adopting a bird's-eye view and employing general laws of physics, symmetries, similarities, and analogies
D. Application	To demonstrate how Phys-Math aids in problem-solving	Employing already known laws and mathematical representations in problem-solving

(A) A strategy according to which students employ visual representations of processes and arrive step-by-step from a description of a phenomenon to its mathematical representation. This strategy, called Visual Mathematics (VM), will be demonstrated here in two contexts: mechanics, where it was already shown to be successful (Mualem and Eylon 2010), and in the context of teaching energy.

(B) A strategy by which students construct their mathematical representations based on empirical results. We will demonstrate how formulae describing a change in energy (its decrease or increase) in a specific process that follows the principles laid down by Joule, in his famous mechanical equivalent of the heat experiment, can be constructed from experiments.

15.1.2 Construction Pattern Strategy I: Visual Mathematics

The Phys-Math interplay, previously described, can be regarded as the bridge between qualitative thinking and mathematical quantitative thinking with regard to a given physical situation. Students first encounter this challenge at the junior high school (JHS) level when they are asked to employ Newton's laws in addressing real-life situations. Here we will describe a strategy designed to guide the students in predicting and explaining everyday phenomena and situations using tools of visual representations and Visual Mathematics (in the form of vectors and free-body force diagrams).

The need to develop a teaching strategy in order to assist students in developing mathematical representations of physical situations is based on various studies indicating students' difficulties in this aspect (Hake 1998; Minstrell 1983; Redish 1999; Halloun and Hestenes 1985; McDermott 1984).

It was suggested that mathematics can even inhibit students' qualitative understanding in these domains. It was demonstrated that many high school students use a "Plug and Chug" technique: they read the question, they search for the correct mathematical formula and perform the needed manipulation, and finally, they check the solution in the back of the textbook.

Consequently, many physics education researchers emphasize the importance of acquiring some qualitative understanding of basic concepts in physics as early as during middle school and suggest that these concepts be taught within familiar everyday contexts (Pugh 2004).

15.1.2.1 The Conceptual Framework

The Visual Mathematics (VM) strategy in the context of introductory mechanics at the junior high school (JHS) level was developed in order to enable students to analyze everyday situations by using physical terms and applying a qualitative understanding of Newton's laws. It also aims at changing students' interest in physics and their views regarding its importance. In order to apply the VM strategy, we chose a systemic approach to teach mechanics which focuses on the concepts of interaction and systems.

15.1.2.2   The Systemic Approach Used to Teach Newton's Laws

The systemic approach begins with the concept of "interaction," recommended by several physics educators (Reif 1995; Karplus 2003), and emphasizes the concept of a system. Later, the concept of force and Newton's third law are introduced, arriving, finally, at the laws of motion (Newton's first and second laws). At the first stage, the students are encouraged to analyze the interactions between components of the entire system before focusing on a specific object. This approach is especially useful in analyzing complex situations, as well as ill-defined problems that characterize authentic situations familiar to students.

The VM strategy, corresponding to the systemic approach, provides a method to subdivide a problem-solving procedure into simple successive steps that are easy to follow. By following these steps, the students strengthen their understanding of how Newton's laws should be applied to analyze a physical situation such as that of a donkey pulling a wagon and how such an application leads to constructing a mathematical model of a mechanical phenomenon.

Next, we will describe and exemplify the VM strategy as it was used in teaching Newton's laws and then demonstrate how it can also be implemented in teaching energy.

15.1.2.3   Applying the VM Strategy in Teaching Newtonian Mechanics

The VM strategy consists of the following sub-step instructions:

(A) *System characterization:*

   A1. Illustrate the situation by a block diagram in which the blocks represent the components of the system.
   A2. Represent (via a table) all of the interactions between objects within the system.

(B) *From systems to forces*:

   B1. Mark all the pairs of forces and the direction of each force in the block diagram using the table of interactions.
   B2. Select an object and all the forces (represented by arrows) that act *on it* using the block diagram. Indicate for each force the object that exerts it.

(C) *Forces and motion (applied repeatedly to each object chosen in B.2)*:

   C.1 Assign a magnitude value to each force (depending on the arrow's length) according to the given situation and Newton's first or second law.

C.2 Write down the proper equations for the given situation.

Let us demonstrate the VM strategy for a situation in which a car pulls a wagon but they are not moving. The students are asked to explain why the car is not moving. The steps of the VM strategy are illustrated below (Fig. 15.1).

In the first step – a physics step – the students are asked to represent a real-life situation by a block diagram and then describe it in an interactions table. The sub-step B.1 emphasizes Newton's third law (N3) using schematic vector representations of the block diagram – a blending of physics and mathematics. When proceeding to step B.2, the students are asked to ensure that all the forces that act on the selected object (car) appear and they must indicate, by marking them,

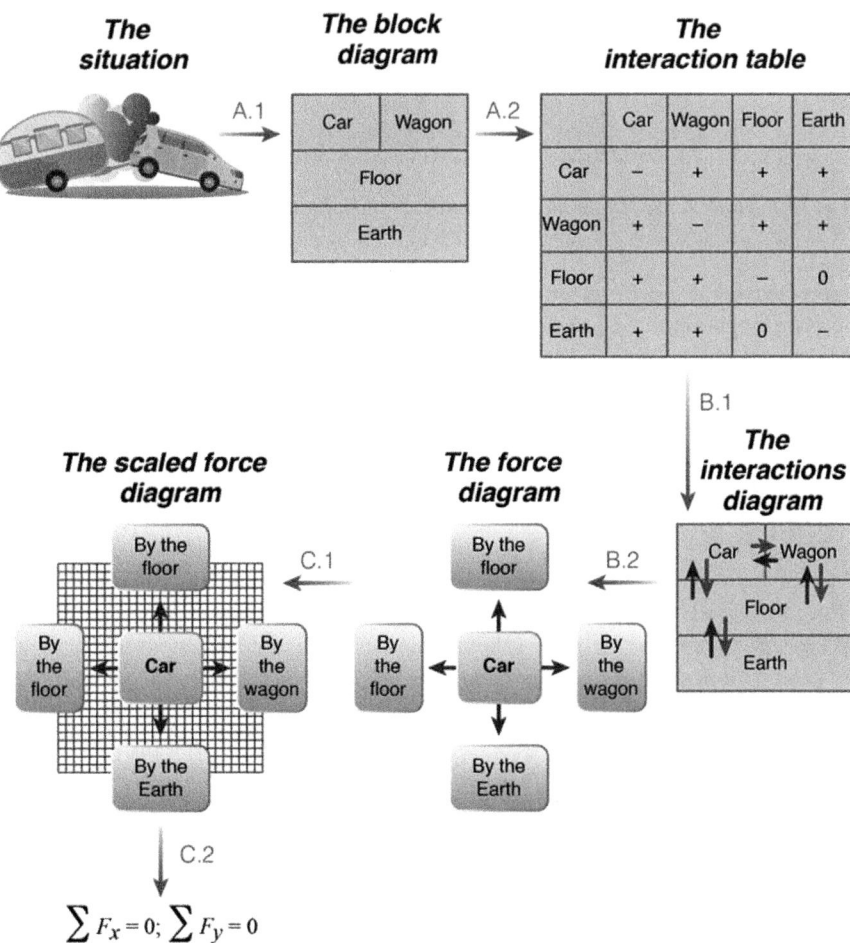

Fig. 15.1 The full Visual Mathematics strategy. Note that the order of the blocks in the block diagram corresponds to a real situation

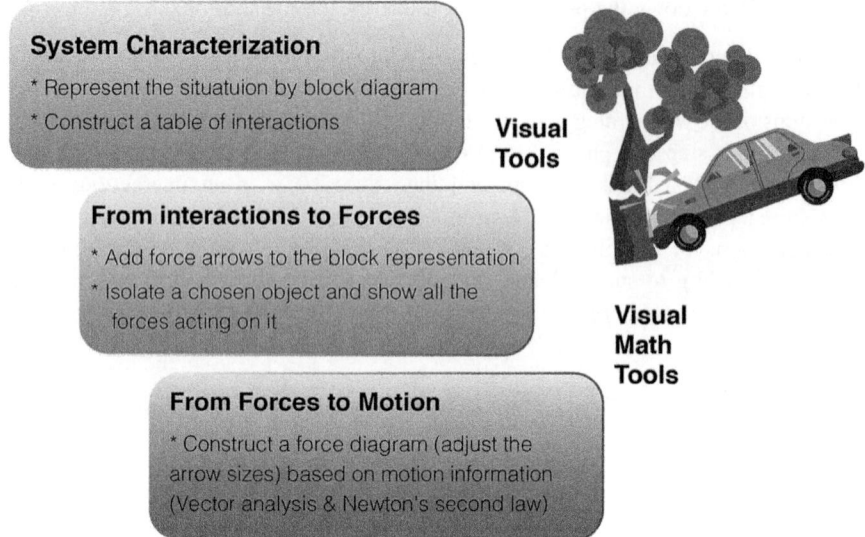

System Characterization

* Represent the situatuion by block diagram
* Construct a table of interactions

Visual Tools

From interactions to Forces

* Add force arrows to the block representation
* Isolate a chosen object and show all the forces acting on it

Visual Math Tools

From Forces to Motion

* Construct a force diagram (adjust the arrow sizes) based on motion information (Vector analysis & Newton's second law)

Fig. 15.2 The VM strategy for mechanical problems

the objects that exert each of these forces – again a Phys-Math combination. The relative magnitudes of the forces that act on the object from the different interactions are considered in step C.1, thus reinforcing the quantitative nature of the Phys-Math interplay. Finally, in step C.2 the students arrive at the corresponding formal representation of Newton's laws of motion.

This strategy, which encourages students to distinguish between forces and motion, enables the students to link, via Newton's second and first laws, between the two. It also fosters students' mathematical formulation of complex situations by (a) deducing forces from motion conditions as described above; (b) deducing motion characteristics from a force diagram; and (c) predicting, by using Newton's laws, what will happen in a situation, by observing the outcome and explaining it (POE – Predict, Observe, and Explain). In our example, the wagon motion condition is given (it does not move) → hence, the net force along each axis should equal to zero → Newton's first law dictates that the arrows along each axis should be of equal length in opposite directions → $\sum F = 0$.[1] One can see how the VM strategy leads students to go, step-by-step, from the phenomenon's description to its mathematical formulation. Each of these steps demonstrates how physics and mathematics are interrelated and how an understanding of mathematical concepts can enhance students' understanding of physics.

Figure 15.2 summarizes the VM strategy employed for using Newton's laws in solving mechanical problems.

---

[1] Vector notation was left out due to the student's age.

15.1.2.4 The Teaching Approach

The VM strategy requires a specific teaching sequence consisting of presenting the conceptual framework and the qualitative strategy in a combined manner. During the teaching process, several selected situations, illustrated as caricatures (see Fig. 15.1), are analyzed several times. The students carry out an analysis corresponding to the conceptual level that they have reached, until they can perform the complete analysis and can employ all the concepts learned in the program (a spiral analysis).

Some cases are chosen to motivate the introduction of additional concepts necessary to analyze the situations, using the strategy as a platform for introducing new concepts (e.g., friction).

The following example illustrates how this Phys-Math interplay process enables students to gain meaningful learning even when they have not yet studied all the principles and basic concepts: Consider a situation described below (Fig. 15.3). Before introducing friction, the force diagram is constructed as described in (a). This diagram leads to the conclusion that the dog should move to the left, which is inconsistent with the actual situation. This apparent inconsistency creates the need to introduce friction (b).

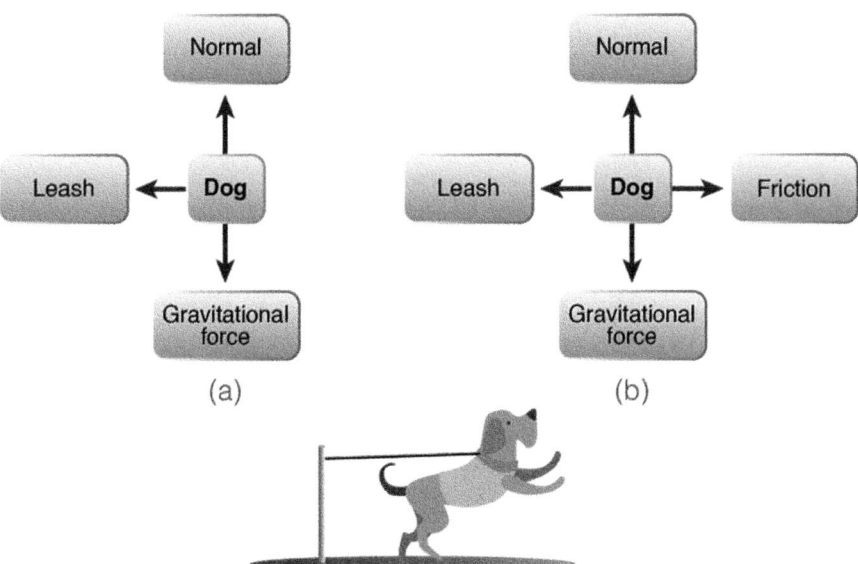

Fig. 15.3 A partial force diagram leads to inconsistency with the actual situation, hence forcing the addition of friction

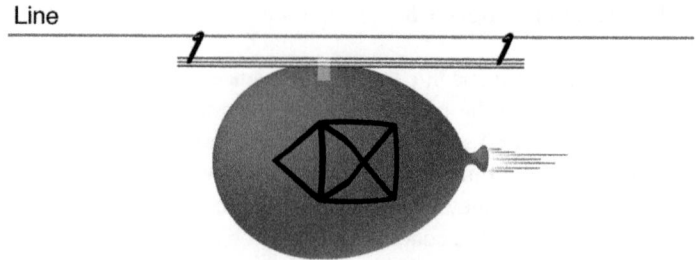

Fig. 15.4 A small balloon is allowed to move along a straight line when the air filling it is released. Explain why a half-filled balloon is not moving, whereas a fully filled balloon does

15.1.2.5   Visual Mathematics and Problem-Solving

In a previous study, the teaching approach described above was found to foster JHS students' problem-solving performances (Mualem and Eylon 2010). This study examined how the teaching method contributed to students' ability to explain and predict phenomena by employing Newton's laws. The study encompassed pre- and post-knowledge questionnaires, administered to junior high school (JHS) (n = 460) and pre- and post-interviews with students (n = 69).

The findings indicate that there was a significant improvement in students' performances in providing explanations and predictions with regard to given phenomena. Moreover, the gain in students' performances in solving Force Concept Inventory (FCI) items sometimes exceeded that of older and more experienced students.[2]

When reexamining the research findings from a Visual Mathematics' point of view, one may gain an additional insight for new possible research questions that might rise from the new VM point of view. Consider, for example, the following transcript, extracted from a pre- and post-instruction interview with a typical ninth grader, who tackled the following question: Predict and explain what will happen to the balloon in Fig. 15.4 when the air is released from it.

In the interview before the instruction, the student answered: "The balloon will go this way (left) because the air that is released is pushing it." Then, the student was asked to explain why a half-filled balloon will not move. The student's answer was "There is not enough power now so the balloon is not moving."

In the interview after the instruction, the student used physics terms and force diagrams to explain the situation: "Because there is an interaction...the air exerts a force on the balloon this way (points to the correct direction). It will move if the pushing force is **greater** than the friction force . . . ."

And when the student was asked "Suppose you release the balloon but it doesn't move?" His answer was "the friction force can exert a force up to a certain

---

[2]This was reported by Redish et al. for university-level students.

magnitude and when you have **a larger magnitude** the object will move ... but here this didn't happen ... so the balloon didn't move."

Whereas in the pre-interviews the student exhibited only intuitive reasoning, in the post-interviews he also used physical terms and force diagrams, namely, a physical language. This student, like many others in the research, exhibited a more expert-like performance.

In addition to this improvement, the student also exhibited an ability to provide a quantitative judgment of the situation by comparing the magnitudes of the forces and canceling out their total sum. This example clearly demonstrated that the student acquired a new proficiency: an ability to examine phenomena by employing an intuitive mathematical point of view – a clear Phys-Math skill. Such a gain encouraged us to adapt Visual Mathematics for a new context: teaching energy.[3]

15.1.3 Applying a Visual Mathematics Strategy for Teaching Energy

In this section we will demonstrate how the Visual Mathematics strategy can be applied to contexts other than mechanics by showing how it could be adapted in the context of teaching energy.

Since there are many approaches to teaching energy, we will first describe the one for which we adapted the VM strategy – a common approach used in our schools. We will then outline how the VM strategy can be used in order to arrive at a mathematical formulation of energy considerations with regard to certain phenomena.

15.1.3.1 Background: The Energy Change Approach

Energy poses a great challenge for curriculum designers, who attempt to arrive at coherent and consistent teaching, since its meaning and special language are far from being agreed upon (Bevilacqua 2014; Lehavi and Eylon 2018). This is mostly manifested by the lack of a consensus with regard to what is energy and what is meant by energy types/forms, conversion/transformation, transfer, and conservation. From a scientific perspective, science emphasizes the changes in the value of energy in analyzing processes which, unlike the value of energy itself, can be absolutely determined.[4]

Thus, a change in the quantity of energy is measurable and thus, it is of physical importance (Reif 1967, p. 202; Reif 1965, p. 129; Quinn 2014, p. 18).

---

[3]This adaptation was not yet tested.

[4]The value of energy does not appear in the first law of thermodynamics – only its quantitative changes.

This equivocal nature of energy led us to construct an approach that focuses on processes and the quantitative change in the energy values that can be attributed to them, rather than on the concept of energy as it describes a static state (Eylon and Lehavi 2010; Lehavi et al. 2014; Lehavi and Eylon 2018). This approach enabled us to develop a spiral, coherent, and consistent curriculum for teaching energy from the seventh to ninth grade in Israel. The resulting curriculum employs only a quantitative description of changes in energy – its increase or decrease – in providing meaning to the traditional energy vocabulary (energy "types/forms," "transfer," "transformations," and conservation). According to this interpretation, the usual terms height energy, elastic energy, and kinetic energy, among others, do not represent different "energies" but, rather, provide labels that can remind one of the different types of *processes* through which the energy of an object has been increased or decreased. The quantitative nature of the energy change approach motivated us to try to integrate the VM strategy within it.

15.1.3.2   The Didactics of the Energy Change Approach

From a didactic perspective, the energy change approach aims at drawing students' attention to the following[5]:

I. Systems and various processes occurring within them and between them.
II. The fact that some systems have a unique feature: changes within their borders seem to be never correlated with changes outside their borders (isolated systems).
III. The fact that processes can be characterized *quantitatively* by changes in certain observable/measurable variables.
IV. The fact that changes in the variable values occur simultaneously and, for some variables, in an opposite direction.
V. The fact that temperature change accompanies all processes and therefore it can be used (jointly with a standard body) to define an entity (energy change). Thus, the measured change in energy attributed to different processes enables one to compare quantitatively between them.[6]
VI. The fact that an energy increase/decrease can provide the traditional vocabulary (energy "types," "transfer," "transformations," and conservation) with a quantitative interpretation.

These aspects of energy, and especially the last one, suggest that Visual Mathematics has the potential to significantly improve the "energy change" approach for teaching energy.

---

[5]This order is not compulsory.

[6]This follows Joule's approach, according to which one can attribute a measurable quantity for different processes by the same operation: measuring the *maximal* change in the temperature of a standard body that each process can cause.

15.1.3.3 Employing Visual Mathematics in Teaching Energy Change

The Visual Mathematics strategy is based on the idea that one of the tools used to correlate physics and mathematics is a symbolic visualization of physical concepts. In VM this idea is used in such a way that the visualization also has quantitative features and therefore lends itself to mathematical manipulations. A good example, which was demonstrated above, is the arrow's representation of force.

With regard to teaching energy concepts, the VM strategy aims at systematically structuring energy change as a quantitative, measurable concept that lends itself to mathematical investigation. Similar to the case of mechanics, the energy change approach emphasizes the concepts of system and interactions. However, an energy description of phenomena differs from the Newtonian description in a few important aspects: (A) The interactions do not necessarily involve forces; (B) the effect of the interactions on *all* the interacting objects cannot be separated (if energy conservation is considered); (C) the energy description focuses on *processes* rather than on *situations*; (D) it may involve the surrounding of objects (e.g., air or even empty space); and (E) the spatial arrangement (up-down, left-right, among others) is not relevant (energy is not a vector). These differences require some adaptations in employing the VM strategy and its symbols. For example, following aspect (B) the interaction arrows here are double headed. We also added symbols for processes and energy increase/decrease.

The VM strategy adapted for energy teaching will adhere to the following sequence of steps:

1. Tell the "story" of actual events.
2. Choose the system of interacting objects.
3. Mark the arrows that indicate the interaction(s), and add to them the observable/measurable changes.
4. Describe the process by the changing variables.
5. Add arrows of energy increase/decrease.
6. Relate energy quantitative changes (based on empirical results such as those described in the next section) to the observable characterizing variables.[7]
7. Add formulae to each energy increase/decrease (use the plus/minus signs).
8. If the system is isolated, sum the formulae to zero.

Figure 15.5 illustrates how such a strategy can be applied to the process of light absorption.

As one can see, some of the representations are directly borrowed from the Newtonian VM strategy, whereas others have been adapted or added. For example, objects are represented by blocks in both cases; the length of an arrow as an indicator of force magnitude is replaced here by the length of a wide arrow that indicates the magnitude of the energy change (its increase or decrease). The zero net force

---

[7]This means that for each change in a specific variable, the students will be able to relate to the corresponding change in the amount of energy. No formula is required at this stage.

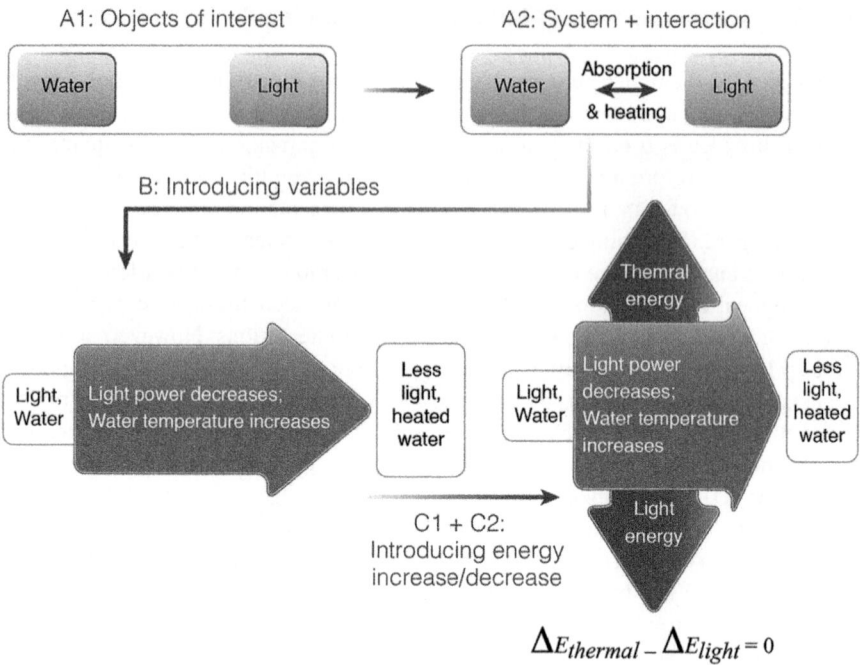

Fig. 15.5 A full description of a derivation of a mathematical representation of energy conservation

required by Newton's first law is manifested by the zero net change in energy dictated by the energy conservation law.

The idea to begin from a phenomenon and construct step-by-step the mathematical formulation is maintained here similar to the Newtonian case. Here the idea of a system and its boundaries are put forward, keeping in mind the energy conservation. After arriving at the law of energy conservation, it can be used, similar to Newton's laws of motion, as a monitoring tool: whenever this law is doubted, the students are encouraged to rethink their analysis of the phenomenon at hand and check whether they left out some process and its corresponding change in energy. The full energy change diagram may help students arrive at the energy conservation formula.[8]

---

[8]Bear in mind that, according to the teaching approach, the value of the energy change corresponding to each process is experimentally predetermined.

15.2 Construction Pattern Strategy II: Constructing Formulae from Experiments

15.2.1 The Phys-Math Didactics of Joule's Experiment

As is well known, James Prescott Joule's experiments enabled him to find quantitative relationships between temperature change and other phenomena: electrical, chemical, and mechanical (Joule 1850). These experiments allowed him not only to generate a conceptual unification of various phenomena, otherwise considered to be disconnected, but also to provide a formal mathematical description of the relations he discovered. These principles of Joule's approach, which provided the grounds for our understanding of the concept of energy change as a measure of processes of different types, can assist teachers in justifying the use of the language of energy in analyzing phenomena that belong to different domains of science. The great importance of Joule's conclusion with regard to the generality of his standard measure of different phenomena and the corresponding mathematical representation renders reproducing his main empirical conclusions in a teaching context highly desirable.[9]

From a science education perspective, the inclusion of Joule-like experiments follows the call to present energy as a measurable quantity (Millar 2014, p. 196). Thus, the energy change approach described above follows Joule's spirit and uses the heating phenomenon as a standard against which one can compare the results of measuring different processes related to disparate phenomena. Specifically, performing an experiment similar to Joule's famous paddle wheel experiment can demonstrate that (A) a mechanical process – a height decrease – could be regarded as a heating phenomenon and thus equivalent to other such phenomena and (B) measuring the temperature rise in such an experiment can result in mathematical formula.[10] A very simple device enabled us to relate quantitatively, through the process of heating a "standard" body, a change in energy to changes in the variables that characterize the change in height. The results of such an experiment were used to arrive experimentally at the known mathematical relations between the change in the amount of energy and the changes in such variables (Lehavi et al. 2016a, b).

As described above, the Phys-Math didactics follow certain tracks that involve making leaps between physics and mathematics. We will present here how such a process was presented to junior high as well as high school teachers participating in several workshops (N = 105 in four workshops). In constructing these workshops,

---

[9]Surprisingly, performing Joule-like experiments, which are crucial for quantifying energy change in different phenomena and hence for laying the ground for energy conservation, was excluded from many school physics curricula (Bécu-Robinault and Tiberghien 1998). This occurred despite the recognized importance of Joule's experiments for teaching the subject of thermodynamics (Sichau 2000).

[10]This equivalence between mechanical and nonmechanical processes cannot be deduced from mechanical laws (Arons 1999).

we took into consideration the finding from our previous research that all Phys-Math teaching patterns began from physics and hence, the initial challenge for the teachers participating in our workshops was to find and test as many ways as they could come up with to change the reading of a thermometer. We then discussed with the teachers the possibility of using heating (not heat) as a means of comparing *quantitatively* different processes, and we suggested naming the entity thus measured as an energy change (an operational definition).[11] We discussed with them whether the existence of a *linear relationship* between ΔT and an energy change is a reasonable assumption, since one would expect that carrying out the same process twice will result in double the temperature rise. We suggested interpreting this as "the change in energy in the second process is twice as much as that in the first one." With regard to Joule's paddle wheel experiment, the linear relationship between ΔT and energy change means that dropping the same load twice from the same height (as Joule did) will result in double the temperature rise.[12] We then introduced the device (the "mini-Joule meter") we developed for measuring the heating caused by a descending small plastic bottle of water used as a weight and asked the participants to conduct the experiment.[13]

The teachers performed the experiment in small groups by first dropping a constant weight from different heights and then dropping different weights from a constant height. We specifically refrained from providing the participants with any standard measuring instrument such as a standard ruler or a beam balance. As a result, each group invented its own standards of measuring height and mass. The teachers were then asked to summarize their results in graphical representations. The results of such an experiment are presented in Fig. 15.6.

The teachers were also asked to arrive at a formula from the graphical representations.

Several Phys-Math-related topics were then discussed:

 I. How should the dots representing the empirical results be connected?
 II. What is the mathematical meaning of the linear fit?
 III. What are the physical ramifications of the linear fit?
 IV. What are the physical ramifications of the units' insensitivity of the linear fit?
 V. What is the difference between the variable Δh, which changes during each measurement, and the total Δh and m, which change across sequential measurements?
 VI. How can we combine two linear relationships into one formula?

---

[11]Interestingly, no one (according to our experience so far) suggested falling as one of these processes.

[12]Of course this requires that the heated standard body be well isolated.

[13]The heart of our device lies in using a wine bottle cork with a digital thermometer inserted in it. When the cork revolves around the thermometer, friction heats up the metallic probe of the thermometer and the sensor within it. The "standard" object is the thermometer probe (instead of a fixed amount of water in Joule's original experiment), and the cork plays the same role as the paddles. It also serves as a very good insulator.

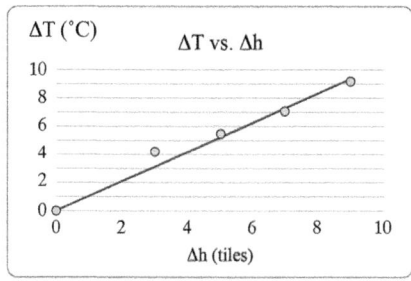 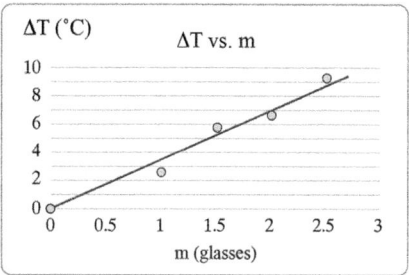

Fig. 15.6 Measuring energy change via ΔT in a Joule-like experiment. Note the choice of arbitrary units

In addition, the following observations were suggested to explain the interplay between the two domains of physics and mathematics:

(A) The idea that the temperature change in a standard body can be agreed upon as a measure of the change in the quantity of energy when this body is heated by some process and that this belongs entirely to the physics domain.

(B) Experiments, which also belong to the physics domain, will always provide a finite number of results. Hence, "connecting the dots" by a linear (or any other) fit provides a "leap" to the domain of mathematics. This leap enables one to perform mathematical manipulations suited for continuous variables.

(C) In mathematics, a linear graphic representation (or any other functional representation) has general features that are not sensitive to the chosen units. Therefore, obtaining a good linear fit to the experimental results by groups that employed different units may allow one to conclude that these results could be regarded as a general law of nature.

(D) The following conclusions belong to the Phys-Math domain:

(a) A change in temperature (a physical quantity) was directly proportional (a mathematical statement) to both the change in height and the mass of the falling body (physics).

(b) The conclusion that $\Delta E = W \cdot \Delta h$ (where W denotes the weight of the descending object) is a valid mathematical statement about our physical understanding of nature.

15.2.2 Ramifications for Teachers' PCK

The two construction strategies might be considered in the context of physics teachers' PCK with regard to the unique domain within physics education – the Phys-Math domain. The Phys-Math PCK framework, adapted from Magnusson et al. (1999) and Etkina's (2010) frameworks (see Fig. 15.7) may assist teachers in:

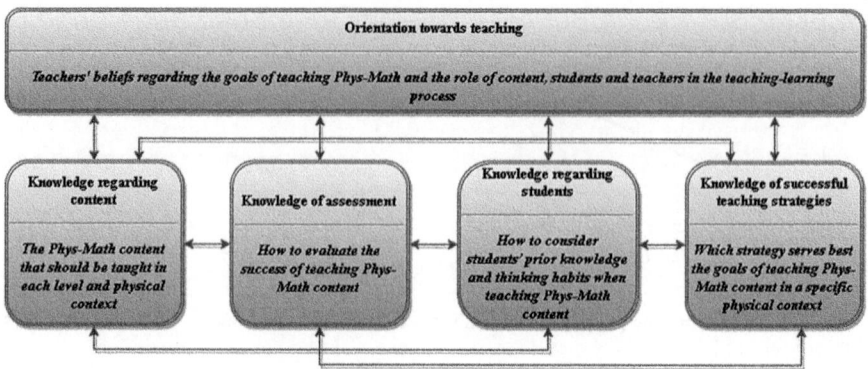

Fig. 15.7 The adapted Phys-Math PCK framework

(a) Helping students develop "science process" skills.
(b) Representing a particular body of knowledge.
(c) Transmitting the facts of science.
(d) Facilitating the development of scientific knowledge by confronting students with contexts to explain that challenge with regard to students' naïve concepts.
(e) Involving students in investigating solutions to problems.
(f) Representing science as an inquiry process.

The two examples described above, of constructing, via a Phys-Math interplay that follows a certain pattern, a mathematical representation of physical phenomena, carry with it some important values for teachers' PCK. These examples belong to different parts of the adapted Phys-Math PCK framework (Lehavi et al. 2017). Whereas the Visual Mathematics strategy may be related to knowledge regarding students and successful teaching strategies, the empirical-based formula construction can also be related to the teachers' knowledge regarding content (see Fig. 15.7).

The positive results of using the Visual Mathematics strategy in the context of mechanics and the experience we attained with the formula construction procedure may assist us in fostering a systematic development of teachers' Phys-Math PCK (Pospiech et al. 2015). This may make teachers more aware of the role that the Phys-Math interplay plays beyond being merely a tool for solving problems in physics. We therefore highly recommend incorporating similar activities into teachers' training.

15.3 Summary

As mentioned before, many students use the "Plug and Chug" approach and actually "bypass" many of the teaching objectives relevant to the Phys-Math interplay. The next step is to apply this approach with high school students (see Fig. 15.8).

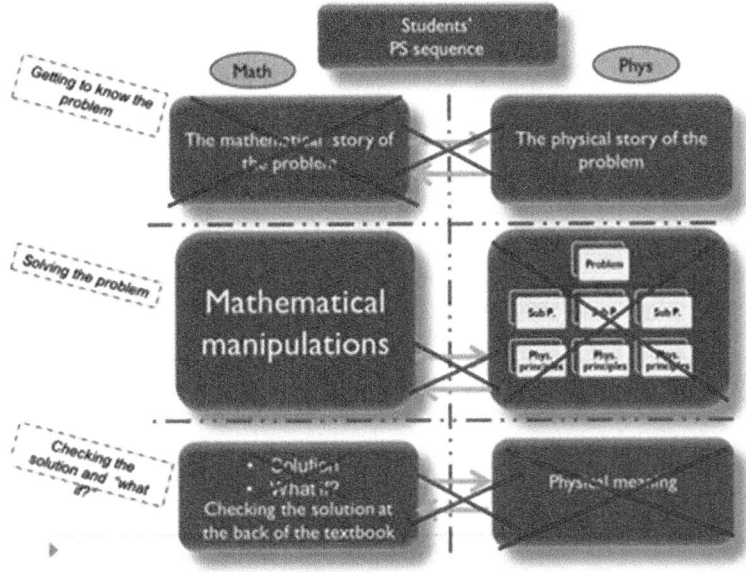

Fig. 15.8 Students' "Plug and Chug" problem-solving strategy as an incomplete Phys-Math interplay

However, the students may miss learning the relationships between the physical and mathematical themes of a certain problem, and they may have difficulties in subdividing a problem into subproblems, each of which may be handled using different physical laws and their corresponding mathematical representations. Without a good knowledge of the various ways to navigate between the two disciplines (the "patterns") and the different goals they serve, students may have difficulties in monitoring their solution for a certain problem and in attaining the physical meaning of its solution.

If we want high school students to achieve meaningful learning, we must develop their Phys-Math interplay abilities, as described in Fig. 15.8.

References

Arons, A. B. (1999). Development of energy concepts in introductory physics courses. *American Journal of Physics, 67*(12), 1063–1067.

Bagno, E., Eylon, B., & Berger, H. (2007). Meeting the challenge of students' understanding formulas in high-school physics – A learning tool. *Physics Education, 43*(1), 75–82.

Baumert, J., Kunter, M., Blum, W., Brunner, M., Voss, T., Jordan, A., Klusmann, U., Krauss, S., Neubrand, M., & Tsai, Y.-M. (2010). Teachers' mathematical knowledge, cognitive activation in the classroom, and student progress. *American Educational Research Journal, 47*(1), 133–180.

Bevilacqua, F. (2014, June). Energy: Learning from the past. *Science & Education, 23*(6), 1231–1243.

Bécu-Robinault, K., & Tiberghien, A. (1998). Integrating experiments into the teaching of energy. *International Journal of Science Education, 20*(1), 99–114.

Clement, J., Lochhead, J., & Monk, G. (1981). Translation difficulties in learning mathematics. *American Mathematical Monthly, 88*(4), 286–290.

Cohen, R., Eylon, B., & Ganiel, U. (1983). Potential difference and current in simple electric circuits: A study of students' concepts. *American Journal of Physics, 51*(5), 407–412.

Etkina, E. (2010). Pedagogical content knowledge and preparation of high school physics teachers. *Physical Review Special Topics – Physics Education Research, 6*(2). Retrieved from http://www.univ-reims.fr/site/evenement/girep-icpe-mptl-2010-reims-international-conference/gallery_files/site/1/90/4401/22908/29702/30688.pdf

Eylon, B. S., & Lehavi, Y. (2010). Position paper: Energy as the language of changes. In P. R. L. Heron, M. Michelini, B. S. Eylon, Y. Lehavi, & A. Stefanel (Co-organizers), *Teaching about energy. Which concepts should be taught at which educational level?* A workshop held within: GIREP – ICPE – MPTL 2010, University of Reims, France.

Hake, R. R. (1998). Interactive-engagement versus traditional methods: A six-thousand-student survey of mechanics test data for introductory physics courses. *American Journal of Physics, 66*(1), 64–74.

Halloun, I. A., & Hestenes, D. (1985). The initial knowledge state of college physics students. *American Journal of Physics, 53*(11), 1043–1055.

Hull, M., Kuo, E., Gupta, A., & Elby, A. (2013). Problem-solving rubrics revisited: Attending to the blending of informal conceptual and formal mathematical reasoning. *Physical Review Special Topics – Physics Education Research, 9*(1), 010105.

Joule, J. P. (1850). On the mechanical equivalent of heat. *Philosophical Transactions of the Royal Society of London Series A, 140*, 61–82. Retrieved from http://links.jstor.org/sici?sici=0261-0523%281850%29140%3C61%3AOTMEOH%3E2.0.CO%3B2-M

Karplus, R. (2003). *Introductory physics – A model approach.* Captains Engineering Services, Inc.

Kuo, E., Hull, M. M., Gupta, A., & Elby, A. (2013). How students blend conceptual and formal mathematical reasoning in solving physics problems. *Science Education, 97*, 32.

Lehavi, Y., Bagno, E., Eylon, B. S., & Cohen, E. (2013). *Can math for physics teachers impact their conceptual knowledge of physics.*

Lehavi, Y., Eylon, B. S., Hazan, A., Bamberger, Y., & Weizman, A. (2014). Focusing on changes in teaching about energy. In M. F. Taşar & G. Üniversitesi (Eds.), *Proceedings of the World Conference on Physics Education* 2012 (pp. 491–498), Istanbul, Turkey.

Lehavi, Y., Bagno, E., Eylon, B. S., Mualem, R., Pospiech, G., Böhm, U., et al. (2015). Towards a pck of physics and mathematics interplay. In C. Fazio & R. Sperandeo-Mineo (Eds.), *The GIREP MPTL 2014 conference proceedings.* Dipartimento di Fisica e Chimica, Università degli Studi di Palermo.

Lehavi, Y., Amit Yosovich, A., & Barak, S. (2016a). Bringing Joule back to school. *School Science Review, 97*(361), 9–14.

Lehavi, Y., Bagno, E., Eylon, B., Mualem, R., Pospiech, G., Böhm, U., Krey, O., & Karam, R. (2016b). Classroom evidence of teachers' PCK of the interplay of physics and mathematics. In T. Greczyło & E. Dębowska (Eds.), *Selected contributions from the International Conference GIREP EPEC 2015, Wrocław Poland, 6–10 July 2015* (pp. 95–104). https://doi.org/10.1007/978-3-319-44887-9_8

Lehavi, Y., et al. (2017). Classroom evidence of teachers' PCK of the interplay of physics and mathematics. In T. Greczyło & E. Dębowska (Eds.), *Key competences in physics teaching and learning. Springer proceedings in physics* (Vol. 190). Cham: Springer.

Lehavi, Y., & Eylon, B. S. (2018). Integrating science education research and history and philosophy of science in developing an energy curriculum. In *History, philosophy and science teaching* (pp. 235–260). Cham: Springer.

Magnusson, S., Krajcik, J., & Borko, H. (1999). Nature, sources and development of pedagogical content knowledge for science teaching. In J. Gess-Newsome & N. G. Lederman (Eds.), *Examining pedagogical content knowledge: The construct and its implications for science education* (pp. 95–133). Dordrecht: Kluwer Academic Publishers.

McDermott, L. C. (1984, July). Research on conceptual understanding in mechanics. *Physics Today, 37*(7), 24–32.

Minstrell, J. (1983). Getting the facts straight. *Science Teacher, 50*(1), 52–54.

Millar, R. (2014). Towards a research-informed teaching sequence for energy. In R. F. Chen, A. Eisenkraft, D. Fortus, J. Krajcik, K. Neumann, & J. C. Nordine (Eds.), *Teaching and learning of energy in K-12 education*. New York: Springer.

Mualem, R., & Eylon, B. S. (2010). Junior high school physics: Using a qualitative strategy for successful problem solving. *Journal of Research in Science Teaching, 47*(9), 1094–1115. https://doi.org/10.1002/tea.20369.

Pospiech, G., Eylon, B. S., Bagno, E., Lehavi, Y., & Geyer, M. A. (2015). The role of mathematics for physics teaching and understanding. In *The GIREP MPTL 2015 Conference Proceedings.* Italian Physical Society. https://doi.org/10.1393/ncc/i2015-15110-6.

Pugh, K. (2004). Newton's laws beyond the classroom walls. *Science Education, 88*(2), 182–195.

Quinn, H. (2014). A physicist's musings on teaching about energy. In R. Chen, A. Eisenkraft, D. Fortus, J. Krajcik, J. Nordine, & A. Scheff (Eds.), *Teaching and learning of energy in K-12 education* (pp. 15–36). Cham, Heidelberg, New York, Dordrecht, London: Springer.

Rebmann, G., & Viennot, L. (1994). Teaching algebraic coding: Stakes, difficulties, and suggestions. *American Journal of Physics, 62*(8), 723–727.

Redish, E. F. (1999). Millikan lecture 1998: Building a science of teaching physics. *American Journal of Physics, 67*(7), 562–573.

Redish, E. F., & Smith, K. A. (2008). Looking beyond content: Skill development for engineers. *Journal of Engineering Education, 97*(3), 295–307.

Reif, F. (1965). *Fundamentals of statistical and thermal physics McGraw-Hill series in fundamentals of physics*. New York: Mcgraw-Hill Book Company.

Reif, F. (1967). *Statistical physics: Berkeley physics course* (Vol. 5). New York: Mcgraw-Hill Book Company.

Reif, F. (1995, December). Understanding and teaching important scientific thought processes. *Journal of Science Education and Technology, 4*(4), 261–282.

Rozier, S., & Viennot, L. (1991). Students' reasoning in thermodynamics. *International Journal of Science Education, 13*, 159–170.

Sichau, C. (2000). Practicing helps: Thermodynamics, history, and experiment. *Science & Education, 9*(4), 389–398.

Chapter 16
Algodoo as a Microworld: Informally Linking Mathematics and Physics

Elias Euler and Bor Gregorcic

16.1 Introduction

This chapter uses two case studies of high school and undergraduate students interacting with a two-dimensional sandbox modelling software, *Algodoo*, to show how physics students can make use of the mathematical representations offered by the software in unconventional yet meaningful ways. We show how affordances of the technology-supported learning environment allow the emergence of student creative engagement at the intersection of mathematics and physics. In terms of learning, the activities studied here are relevant in two central ways: (1) they open up alternative conceptual learning pathways for students by allowing them to access and engage with the content in original, self-directed and creative ways; (2) in doing this, the studied activities carry significant potential to motivate students and support their intrinsic interests.

Much of the existing research focused on digital learning environments in physics education comprises investigations of *simulation* software, such as the studies which examine PhET simulations (Perkins et al. 2006; Wieman et al. 2008), GeoGebra simulations (Arnone et al. 2017) or Physlets (Dancy et al. 2002). For this chapter, we consider simulations as those digital learning environments which allow students to interact with pre-built models of real or hypothesized situations (National Research Council 2011). As such, simulations are typically designed around a specific phenomenon or set of phenomena so as to provide students with access to particular disciplinary concepts. Much of the research into the use of simulations has produced strong support for their benefit to learning in many different contexts (for a comprehensive review, see Plass and Schwartz 2014).

E. Euler (✉) · B. Gregorcic
Disciplinary Domain of Science and Technology, Physics, Uppsala University, Uppsala, Sweden
e-mail: elias.euler@physics.uu.se; bor.gregorcic@physics.uu.se

© Springer Nature Switzerland AG 2019   355
G. Pospiech et al. (eds.), *Mathematics in Physics Education*,
https://doi.org/10.1007/978-3-030-04627-9_16

By way of contrast, we choose in this chapter to investigate the unique learning opportunities afforded by a less phenomenon-specific digital learning environment, *Algodoo*. In doing so, we make use of the notion of Papertian *microworlds* (1980), a term which refers to digital environments which offer more opportunities for creativity and invention than what is typically offered by simulation software (Plass and Schwartz 2014). While simulations tend to allow users to explore the effects of a set of parameters within the given phenomenon, microworlds provide users with the freedom to build their own environments and phenomena, making possible a wider range of scenarios within the same software. As Laurillard (2002, p. 162) explains, people who use simulations are 'controlling a system that someone else has built', while those using microworlds are 'building their own runnable system'.

In our investigation of the learning afforded by software such as *Algodoo*, we examine two cases of students using *Algodoo* on an interactive whiteboard (IWB) to carry out physics tasks. In particular, we examine how the students in both of these cases make creative use of the mathematical representations available within Algodoo as they reason about physics phenomena. We assert that it is precisely the way in which *Algodoo* seems to function as a microworld, which we refer to as its 'microworldiness', which enables the students to utilize mathematical representations in their playful[1] yet meaningful exploration of physics phenomena. By this we mean that the interactional and representational affordances of *Algodoo* which align with the characteristics of a microworld seem to allow the students in our cases to create and manipulate mathematical representations in ways that are both unconventional and also productive from a physics education perspective.

With the two cases presented in this chapter, we also examine how *Algodoo* and similar digital learning environments might be a useful way for mathematical representations to become interesting and meaningful for students as they engage with physics. Our analysis shows that, while using *Algodoo*, students can interact with mathematical representations in spontaneous, playful ways. Digital learning environments like *Algodoo* may, thereby, provide potentially motivating alternatives for students to make connections between the physical world and the formalisms we use to describe it.

We begin by reflecting on what it means to informally learn physics, followed by a brief review of the instructional philosophy of microworlds advocated by Seymour Papert in his book, *Mindstorms* (1980). Thereafter, we detail some of the features of *Algodoo*, highlighting some of the options the software offers for generating mathematical representations. Finally, we present the two cases of students using *Algodoo* to show that, when used in an appropriate manner, *Algodoo* appears to function as a microworld by supporting students in their creative implementation of mathematical representations.

---

[1] *Playful* is used in this chapter to mean voluntary, intrinsically motivating (pleasurable for its own sake) and/or creativity-driven (inspired by Rieber 1996).

16.2   Learning Formal Ideas in Informal Ways

By mastering the many mathematical representations used in physics (Van Heuvelen 1991), physicists can employ a diverse range mathematical tools such as force vectors, motion diagrams and graphs to conceptualize phenomena in terms of formal physics models and to appropriately solve problems (Hestenes 1992). Through their commitment to internalizing how nature is described by their discipline, physicists cultivate, among other things, a mathematically enhanced perspective towards the phenomena that they encounter. However, and perhaps not unexpectedly, this is not necessarily the case for most students while they learn physics. For students who are not adequately familiar with – or at least not confident in – the formal, mathematically intensive concepts of physics, the techniques used by physicists to describe the world are often not readily compatible with the students' daily experience of phenomena. There exists for such students a significant difference between how they perceive the world and the way in which physics canonically represents it using formal mathematics. Indeed, students' difficulties with navigating this difference are a common interest for physics education researchers, as evidenced by this very book or, found for example, in McDermott et al.'s (1987) famous discussion of students' difficulties when attempting to interpret kinematics graphs and relating them to their real-world counterparts.

In response to the sometimes-unnavigable disparity between the physical world and the mathematics which physicists use to describe it, many students gain access to the implications of formal physics concepts by other means than an explicit application of mathematics. This can be seen in students' informal cultural exposure to speed and speedometers from cars. Today, the notion of a speedometer can be called upon by physics students as they make sense of velocity and acceleration, something which was impossible for either Galileo or Newton to do in the time before speedometers were invented. Students who grow up in a culture where the enforcement of speed limits is a common occurrence, where a car's top speed is listed in advertisements and where they can ride in a car with an omnipresent visual display of their speed have a corpus of informal experiences which they can and, certainly do, involve in their reasoning with physics concepts such as velocity and acceleration.

In his book *Mindstorms*, Papert (1980) argued that the informal learning culture surrounding students is what provides them with the necessary *materials* with which they can construct their understanding of the world and incorporate them into their understanding of formal physics models. Thus, when the topic of velocity is discussed in a physics context, students from a speedometer-rich culture need not first conceptualize the idea of 'speed-in-general' to begin to become familiar with the concept in the formal physics sense. Such students are able to come to the physics classroom already equipped with the materials from their culture (in this example, their experiences around speedometers) with which they can build new understanding. Surely it should be noted that, as with any previously constructed understanding that students bring to a physics classroom, an everyday experience with speedometers neither certifies that students will automatically intuit physics, nor does it ensure that students will contextualize their understanding of kinematic quantities in the manner consistent with the discipline of physics (Trowbridge and McDermott 1980, 1981).

Nonetheless, in this chapter we explore how, as an environment rich in mathematical representations, *Algodoo* can provide resources to students which might act in a similar manner to the speedometer, providing them with access to materials which they can recruit in the construction of their own understanding of physics. We suggest that when combined with appropriate instructional approaches, *Algodoo* can not only expose students to mathematical ideas as they are used in physics but can also provide an environment for students where they are able to engage in playful inquiry and draw on mathematical representations in a spontaneous and nonthreatening way. Similar to how speedometers can be used as materials for conceptualizing velocity and acceleration in a physics context, the carefully crafted mathematical representations provided within *Algodoo* can be spontaneously recruited as rich materials in students' inquiry into physical phenomena.

16.3   Papert and Microworlds

After observing how young students tended to struggle with reasoning in terms of systematic procedures (necessary for tasks such as ordering beads in all possible combinations along a string), Seymour Papert argued in favour of creating environments rich in the necessary materials for students 'to build powerful, concrete ways to think about problems involving systematicity' (1980, p. 22). In *Mindstorms*, he presented a family of computer languages called LOGO systems (typically involving small programmable robots) as an example of an educational programming language that could enrich the learning environment to promote logical and systematic thinking skills in young students. Papert argued that LOGO systems could provide students with a sufficiently enticing environment for them to develop, in a relatively intuitive and spontaneous way, a mathematical language to communicate with computers. Just as learners of French might immerse themselves in the French language by visiting France, he suggests learners of mathematics could immerse themselves in the 'Mathland' (p. 6) cultivated in the LOGO systems.

Papert intended to provide students with an arena where they could explore formal topics in informal ways. By including what he characterized as *microworlds*, Papert aimed to make computer programming and even the formalisms of Newtonian mechanics accessible to students. In contrast to what he considered the often ineffective and ingenuine approaches taken by much of traditional education, Papert believed that the use of microworlds would result in 'Piagetian learning' or informal 'learning without being taught' (p. 7). He believed that this could be done by providing arenas which were rich in the building blocks needed for students to explore, create and experience formal concepts for themselves. In order to motivate and facilitate the students' learning process, Papert argued that microworlds needed to allow students to become active *builders* in the environment and support them in taking the initiative to engage creatively with the provided materials. The role of a microworld was thus twofold: it needed to (1) offer the correct *materials* for students to recruit and (2) provide a *space* where students could be inspired to create with these materials.

In arguing for builder-focused microworlds, Papert developed what is referred to as a *constructionist* perspective on learning. This constructionist approach places explicit emphasis on the students' act of building – or *constructing* – as a means of learning. In this way, the constructionist perspective can be seen as a special case of the broader, more commonly known perspective of constructivism.

In the time since *Mindstorms*, a body of research has amassed examining the function of microworlds. Abelson and diSessa (1980) quickly adopted the LOGO systems in the teaching of advanced mathematics in the Logo Group of the MIT Artificial Intelligence Laboratory, and the term 'microworld' has persisted in the education research community in the many years since (e.g. diSessa 1988; Jimoyiannis and Komis 2001; Mayer et al. 2003; Miller et al. 1999). However, somewhat contrary to Papert's optimistic view of microworlds, many of the modern reports suggest that more is needed than an environment that simply provides opportunities for exploratory learning if achievement of specific learning goals is desired. Many researchers claim there is a need for some imposed structure of activities or curriculum around a microworld for the environment to become educationally useful (Rieber 2005; White 1984). For example, research has shown that, while using LOGO systems, many students do not spontaneously generate the powerful ideas that Papert had intended unless the microworld is used within a context that is 'well engineered and targeted at well-defined learning objectives' (Miller et al. 1999; referring to work such as Pea and Kurland 1984; Clements 1986, 1990; Klahr and Carver 1988; Lehrer et al. 1989). Even the definition of a microworld, and how the concept of microworlds compares to *simulations* or *games*, is not without contention in the literature. Rieber (1996) discusses the classification of microworlds, suggesting that a learning environment can be regarded as a microworld if it acts as such for a particular learner:

> In a sense, then, it is the learner who determines whether a learning environment should be considered a microworld since successful microworlds rely and build on an individual's own natural tendencies toward learning. It is possible for a learning environment to be a microworld for one person but not for another. (p. 46)

For Rieber, a learning environment should be considered as a microworld in its specific use within a particular context. It is precisely this user-subjective perspective on microworlds that we use in this chapter: whether or not *Algodoo* can be unanimously identified as a microworld for all students, we illustrate how the software *acts* as a microworld for certain students as they use it on an IWB, particularly when dealing with mathematical concepts in a physics context.

In this chapter we present two cases where, aligned with Papert's criteria for a digital microworld, *Algodoo* seems to (1) offer a diversity of mathematical materials – especially in the form of dynamic mathematical representations – and (2) provide an arena within which students are inspired to explore and create with these materials as they engage with physics phenomena.

16.4 Algodoo

Algodoo (www.algodoo.com) is a two-dimensional sandbox software which was inspired, at least in part, by Papert's constructionist approach to learning (Gregorcic and Bodin 2017). At first glance, *Algodoo* resembles other digital drawing software such as *Microsoft Paint, Corel Draw* or *Adobe Illustrator* in that it contains various toolbars for creating objects of different geometrical shapes, colours and sizes. However, unlike these other digital drawing platforms, *Algodoo* allows users to press play and has the user-drawn objects dynamically interact. These objects will bounce off each other, roll around, swing from ropes, etc. Thus, users are able to create *scenes* by using a diverse set of available construction elements within *Algodoo*, which include physics-relevant elements such as springs, axles, motors, thrusters, ropes and fastening tools. These scenes typically contain constructions ranging from simple systems (e.g. spring-mass pendula, balls rolling down the slopes or two-body gravitational systems) to more elaborate ones (e.g. suspension bridges, cars and engine transmission systems). When users create systems of objects within *Algodoo* and press the play button, the scenes they have built then evolve in accordance with Newtonian mechanics in two dimensions.

Unlike mathematics modelling tools such as *Modellus* and *Matlab,* which feature an exposed, editable architecture, *Algodoo* is not designed for users to easily change every aspect of the rules governing the virtual world. For example, while users can turn gravity or air resistance off, the underlying mechanics of object interaction cannot be altered from a two-dimensional Newtonian system. Indeed, some researchers might see this algorithmic opacity as a hindrance to students' learning of how to model (Hestenes 1995); however, others argue that *Algodoo* retains a level of *semi-transparency* for students that allows them the opportunity to create and manipulate a virtual world without requiring the prior knowledge of programming (Gregorcic and Bodin 2017). In doing so, *Algodoo* can facilitate new and potentially beneficial experiences in a digital learning environment for those users without fluency in coding languages (Euler and Gregorcic 2018; Gregorcic 2016). In fact, *Algodoo* appears to be an intuitive program for students at both high schools and universities, so much so that these students can, in a matter of minutes, start engaging in creative activities even when they use the software for the first time (Gregorcic et al. 2017a). *Algodoo*'s ease-of-access is a key component in our consideration of it functioning as a microworld.

The other important characteristic of *Algodoo* is that it provides, through visual and interactive means, a range of dynamic representations which have been shown by research to contribute to effective physics learning (e.g. Rosengrant et al. 2009). In what follows, we discuss *Algodoo*'s capability for representing mathematical concepts. Specifically, we emphasize the utility of combining *Algodoo* and an IWB, which can provide students with a collaborative space for engaging with mathematical and physics concepts.

16.4.1 Representations Afforded by Algodoo

Algodoo, like any other computer-based model of phenomena or modelling system, is built up from formal mathematical relationships in its source code. The software can track the motion of the objects created within it, therefore allowing it to display quantities such as momentum, force, velocity and position. This is due to the fact that these quantities are part of the internal structure that manifests in the external user interface (Plass and Schwartz 2014).

Algodoo dynamically updates visual representations in real time (i.e. while the microworld runs), which allows users to access and manipulate physical quantities describing virtual objects in ways that would be impossible to achieve in a traditional physics laboratory, in a classroom or in everyday life. Nonetheless, while including these mathematical aspects, the *Algodoo* environment still retains many characteristics of the world students experience every day. In the software, users can grab, move and even throw virtual objects, which can then be observed to bounce off each other, slide, tumble and generally behave in ways that most people can relate to their everyday experiences with real-world objects.

By including mathematical representations (e.g. Fig. 16.1) of quantities like dynamic vector arrows (e.g. velocity, momentum and force), numbers and sliders representing values of physics-relevant quantities (e.g. density, restitution, coeffi-

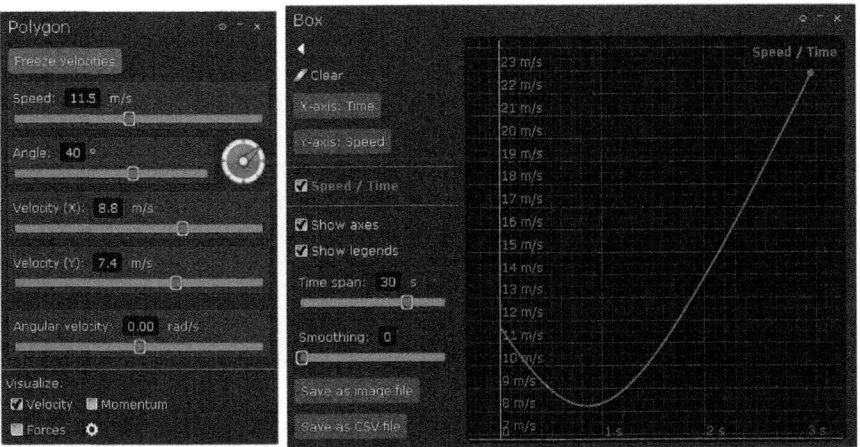

Fig. 16.1 Two examples of the representations provided by *Algodoo*, namely, the 'Velocities' tab (left) and a graph from the 'Show Plot' function (right) (In the Velocity tab, sliders for changing the speed, angle, velocity (x), velocity (y) and angular velocity are provided along with a wheel which displays the angle of velocity and checkboxes for displaying vectors (for velocity, momentum and forces) on the selected object(s). In the graph, various quantities can be assigned to the axes, and the options are provided to display the title ('speed/time' in this case), the axes and the legends. The slider-labelled 'time span' allows the user to select the length of time to include (from the most recent 'run' of the simulation), while the slider-labelled 'Smoothing' allows the user to smooth the graph of the data.)

cient of friction) and plots of quantities (e.g. kinetic energy vs. time, x-position vs. y-position) alongside the visually accessible virtual world, *Algodoo* superimposes formal physics and mathematical ideas onto a more familiar world of physical, albeit simulated, interactions. *Algodoo* provides opportunities for students to explore and engage in open-ended and creative tasks where they can experience physics-relevant, mathematical ideas *in action* and interact with physics content in new pedagogical ways which are not typically available. For example, students can observe the forces acting within a suspension bridge, which they may have built themselves, by selecting to display *Algodoo*'s overlay of dynamically changing force vectors on top of the bridge itself.

The close interplay of the mathematical representations within an intuitively manipulable virtual world gives students and instructors access to a rich collection of meaning-making resources. These resources can be employed to help students develop a better understanding of the meanings embedded in mathematical representations that are used in physics and may even encourage them to make use of these representations in their communication of physics ideas.

The creative potential of *Algodoo* appears to be significantly enhanced when used in combination with a large touch screen, such as an interactive whiteboard (IWB) (Gregorcic 2015a). The IWB provides students with common perceptual ground which they can visually appreciate in small groups (Roth and Lawless 2002) and which they can refer to using environmentally coupled hand gestures (Goodwin 2007; Gregorcic et al. 2017b). This allows students to engage with *Algodoo* in collaborative exploration and communication (Mellingsæter and Bungum 2015). The affordances of the *Algodoo*-IWB setup for multimodal communication allow students to address conceptually interesting ideas even when their knowledge of corresponding vocabulary is limited. Where they struggle to find words to express meaning, they can resort to gestures, such as pointing to patterns and values on the screen (Gregorcic et al. 2017b). The pronounced gestural and interactional components of student communication in front of the IWB can also provide researchers with a better insight into students' meaning-making than paying attention to their speech alone (Euler and Gregorcic 2018; Gregorcic et al. 2017b).

16.5 The Cases

In order to illustrate how the *Algodoo*-IWB setup can provide new opportunities for students to learn how to appropriately use mathematical representations in playful (yet useful) ways, we present two cases of open-ended physics activities which utilized the *Algodoo*-IWB setup: (1) one where students threw planets into orbits and (2) another where students rolled an object down a ramp. The data for both of these cases was video recorded in a small room with researchers (referred to as instructors) present to act as the facilitators of the activity, to push the students to further clarify their thinking out loud, and/or to aid the students with any technical difficulties with *Algodoo* or the IWB.

In order to present the data in a manner which captures both the speech of the students and also their gestural activity, we include a multimodal transcript (Bezemer and Mavers 2011) comprising written excerpts of talk[2] and line illustrations drawn from frames of the video data (which are occasionally augmented by close-ups of the relevant *Algodoo* menus). Each line of the transcript is numbered and labelled with the speaker or actor responsible for the speech or action contained in the line ('S1' to 'S5' for Student 1 to Student 5, respectively; and 'In' for the instructors). Actions such as gestures or manipulations of the IWB are included as *italicized* text in [brackets] and represented visually by illustration when useful. In the section that follows, each excerpt of transcript is followed by a summary of the what was said and done by the students to make explicit the things we wish to highlight from the students' interactions.

While the physics content varies between the two data sets presented here, we will show how in both instances, the presence of representational options within *Algodoo* led students to coordinate their discussion and creative inputs around complex mathematical representations in ways which we can appreciate as appropriate for the learning of physics. While exploiting the open, microworld-like nature of *Algodoo,* students were able to creatively link mathematics and physics through their informal use of mathematical representations.

16.5.1 Case 1: Vector-Sense in Orbital Motion with the 'Velocity' Tab

Our first case comes from a data set collected as part of a previous study on the use of IWBs in astronomy instruction (Gregorcic 2015a) where small groups of high school students used the *Algodoo*-IWB setup to explore celestial motion. The students were presented a scene in *Algodoo* which involved a central circular body with an attractive potential – representing a star or planet in an astronomical system. The students used the *Algodoo*-IWB setup to qualitatively investigate Kepler's laws of planetary motion (Gregorcic et al. 2015, 2017b), specifically following the prompt to explore how relatively smaller bodies behave in the vicinity of the larger central massive body. The students drew planet-like or moon-like objects and, by swiping on the IWB, threw these objects into orbit around the star-like object located in the centre of the scene. It was also possible for the students to send objects into orbit by pausing the simulation, placing the object at the desired radius away from the central circle, assigning a velocity to the object and then running the simulation. Some groups chose to display the force vectors or velocity vectors of the objects as these objects orbited the central object (referred to hereafter as the 'Sun'). Gregorcic designed the Kepler's laws scene in *Algodoo* to provide students with 'hands-on

---

[2]The data collection session for Case 1 originally took place in Slovenian, but we have translated the speech into English for the purposes of this chapter.

[access] to [the] otherwise experimentally inaccessible topic' of orbital mechanics (2015a, p. 515).

Analysis of the data collected in these sessions has previously focused on social and embodied aspects of students' learning of physics concepts with the IWB-*Algodoo* setup (Gregorcic 2015a, b; Gregorcic et al. 2017b). For the analysis presented in this chapter, we instead focus on the students' engagement with mathematical representations of velocity.

We begin by highlighting an excerpt from a session where a group of three students – who we refer to as Student 1 (S1), Student 2 (S2) and Student 3 (S3), along with the Instructor (In) – try to send an object into orbit by setting its initial velocity within the 'Velocities' tab in the drop-down menu (while the simulation is paused). They estimate the initial conditions (radius and velocity) necessary to send the object into orbit by comparing them to that of an already orbiting object from before. They press the play button and then watch as the newly launched object collides with another object that was already orbiting the Sun. The collision sends the new object out of the frame of view and pushes the original object into a new orbit around the Sun. While the new object is sent out of the frame of view, its Velocity menu remains open in *Algodoo*. We include sections of the transcript to illustrate the informal exploration that took place after the students first observe the collision.

1	**S2:**	Okay…
2	**S1:**	Aha!
3	**In:**	What happened now?
4	**S1:**	This one's trajectory changed, but it remained constant.
5	**S1:**	And it's losing speed.
6	**S2:**	No, it's not losing speed.
7	**S1:**	[*points to the slider for speed*] (Fig. 16.2)
8	**S1:**	One of them is losing speed.
9	**S2:**	Yeah, yeah. That one.
10	**S1:**	Yeah, that one, yeah. That one that is going away.
11	**In:**	Ah, now you're looking at that one!

Excerpt Summary In this exchange, we see the students make sense of the behaviour of the two objects after the collision. They notice how the originally orbiting object has been pushed into a new, stable orbit – which Student 1 refers to as being 'constant' (line 4) and which we take to mean stable in time (self-repeating on a closed trajectory). Noticing how the Velocity tab is displaying a decreasing speed, the students quickly come to realize that the Velocity tab is still showing data for the runaway object, which is now out of sight, past the edge of the view in *Algodoo*.

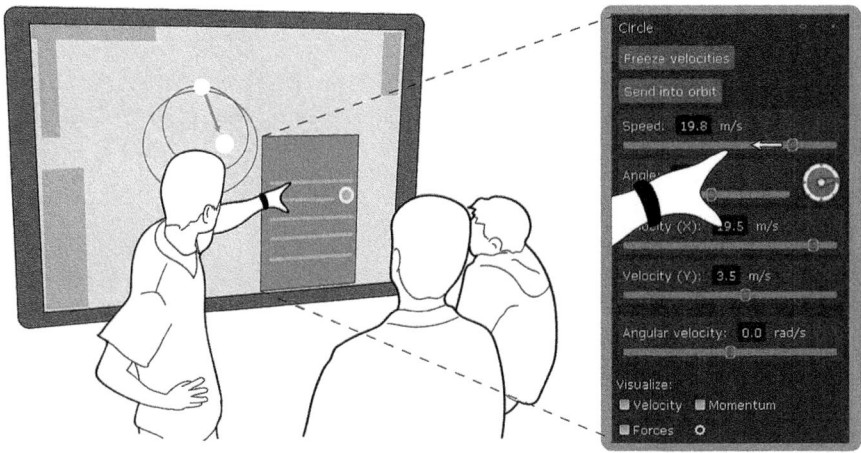

Fig. 16.2 Student 1 (left, with Student 2, middle, and Student 3, right) points to the moving slider labelled Speed within the Velocity tab as he emphasizes that one of the objects is 'losing speed' (line 6) (It should be noted that the values for Speed, Angle, etc. in the Velocity tab are an approximate recreation and do not necessarily reflect the exact values seen by the students during the session. These values are 'unrealistic' for objects on planetary scales, but their usefulness holds in their proportions to one another and their qualitative changes over time.)

(continued from above)

12	**S3:**	Turn its angle, so it will come back.
13	**S1:**	[*laughs*]
14	**S2:**	[*starts dragging the Angle slider to the right, changing the angle at which the runaway planet is travelling*]
15	**In:**	You can also turn the little wheel if you want to turn the angle. There, on the right side
16	**S1:**	And let's add some speed . . . Or not. It's already coming back! [*performs a U-turn gesture in front of the IWB*] (Fig. 16.3)
17	**In:**	So, you noticed something interesting.
18	**S1:**	So, now it's slowly coming back into orbit. Because it's becoming faster. [*points to the speed slider, where the value is increasing*]
19	**S3:**	Yes.
20	**S2:**	Yes.

Excerpt Summary   Here, Student 3 suggests that they 'turn [the planet's] angle' (line 12) in order to bring it back into sight. Student 2 then drags the Angle slider to the right to change the angle at which the planet is travelling, and the instructor suggests that he can also use the Wheel to change the angle. After Student 2 changes the angle, Student 1 initially wants to alter the object's speed as well but changes his mind as he watches the angle spontaneously rotate with the motion of the planet. He interprets the changing angle as the planet reversing direction and he gestures with his hand in a U-turn motion. He also notices that the Speed slider is moving

366   E. Euler and B. Gregorcic

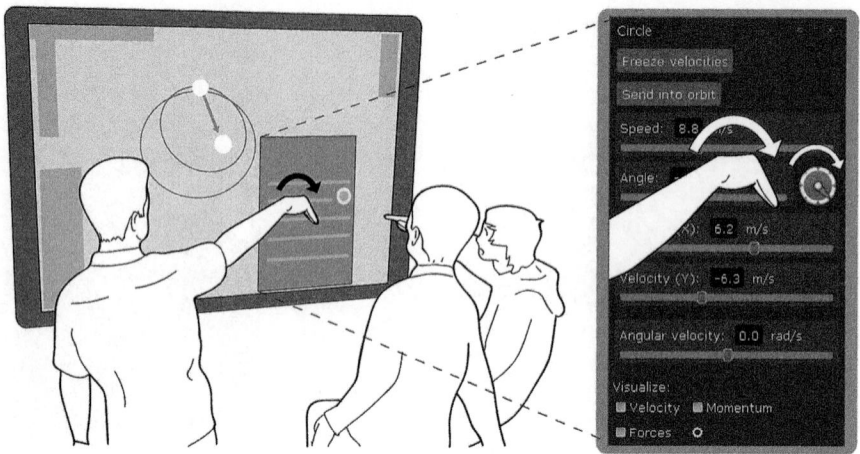

Fig. 16.3 Student 1 gestures in front of the IWB with a U-turn gesture (downwards) as he vocalizes that the runaway planet is 'already coming back' (line 16). Student 3 points towards the wheel of the velocity menu as it turns with the changing trajectory of the planet

to the right, which he interprets as meaning that the object's speed is increasing. He explains this as the object 'slowly coming back into orbit' (line 18), and the other two students agree.

(continued from above)

21	**In:**	Coming into orbit, what does that mean?
22	**S3:**	Closer...
23	**S1:**	Closer to the [Sun].
24	**S2:**	Actually, it is already kind of in orbit, unless it will crash into it. Because it... because it is attracting it. It means it will... [*starts gesturing a large curve in the air*]
25	**S1:**	Just a moment. Considering it was travelling away from this object and it was losing speed...
26	**S2:**	Yes, it was.
27	**S1:**	And there was no resistance...
28	**S2:**	It was in orbit from the beginning, but...
29	**In:**	Okay. Okay. Interesting observation. It was flying away. It was losing speed.
30	**S2:**	It was losing speed and it had no resistance.
31	**S2:**	Yes, but that's normal. If you have a body out here and a gravitational force between them, and there is no other force, and you don't accelerate [the body out there], its speed will get smaller until it will turn around and travel the other way. [*mimics the motion of a planet moving away from the Sun and then back towards it with his hand*] (Fig. 16.4)

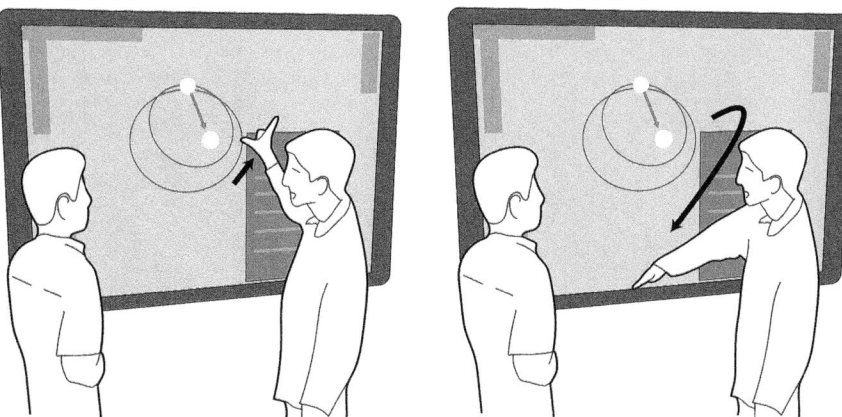

Fig. 16.4 Student 2 gestures to show the movement of a planet as it is accelerated by the Sun. We can interpret this explanation as one that uses a Newtonian model of Sun-planet interaction

32 **S2:** Which is interesting, but . . . I mean, it's interesting . . .
33 **S1:** Yeah, I get it.

Excerpt Summary Here, the students engage in a discussion about orbital motion and the underlying mechanisms that govern the changes in an object's velocity. While Student 1 first has an issue with the slowing down of a planet in a frictionless environment, Student 2 is able to explain how the object's behaviour makes sense in a system with gravitational force (line 31, which we interpret as a Newtonian perspective). Student 2 supports his argumentation with environmentally coupled hand gestures, symbolizing the movement of the planet and the direction of forces (Fig. 16.4).

(continued from above)

34 **S1:** Aha, okay, now its angle started changing, which means . . . [*starts repositioning himself in front of the IWB, pointing to the Velocity tab*] (Fig. 16.5)
35 **In:** Oh, yes, now you are observing that body just through [the Velocity tab].
36 **S2:** Yeah, um . . . Good point.
37 **S1:** [*laughs*]
38 **S2:** [*uses the Zoom tool to zoom out, revealing more of the space around the Sun*]
39 **S1:** Here it is. [*notices the runaway planet on the left side of the Sun, close to the edge of the screen*]

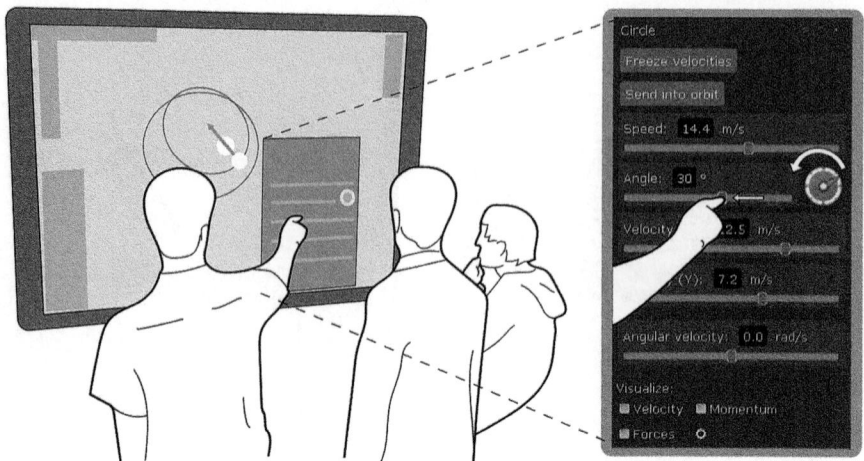

Fig. 16.5 Student 1 notices the changing velocity of the runaway object in the Velocity tab. He repositions himself in front of the IWB and points to the changing Angle slider

40 **S2:** It's here. [*pointing to the runaway planet*]
41 **S1:** Let's do it by hand.
42 **S2:** Let's zoom out more. Can we zoom out more?
43 **S1:** No.
44 **In:** This is the most zoomed out it can be.
45 **S1:** Quickly. [*turns the angle wheel CW, in the direction towards the Sun*]
46 **S2:** But now we are changing its things again.
47 **S1:** [*drags the speed slider to the right and the planet starts travelling faster towards the Sun*]
48 **S2:** It is going to crash directly into it.
49 **S1:** [*adjusting the direction using the angle wheel*] So now it is already growing. [*watches as the speed slider spontaneously moves to the right*]

Excerpt Summary Again, Student 1 notices an increased rate of change in the object's angle of velocity by watching the Velocity tab, all while the planet remains outside the field of view in the scene. The instructor points out that the students are interpreting the motion of the planet by just looking at the values in Velocity tab, to which the students respond by zooming out to find the object (now on the left side of the Sun) just as it is about to fly out of the field of view. Student 1 quickly manipulates the object's velocity by changing the angle (turning the wheel CW towards the Sun) and then increasing its speed (by dragging the Speed slider to the right). Finally, he watches the object and its Velocity tab simultaneously and notices that the Speed slider continues to move to the right as the object accelerates towards the Sun.

In the excerpts of transcript presented above, we see that, although the students originally speculated that the runaway object was lost after the collision, they noticed that the velocity of the runaway object changed in a way that suggested it would return if they kept waiting (meaning that the runaway object was in some type of orbit). Despite the object being absent from the frame of view in *Algodoo*, the students were able to track the motion of the object through the Velocity tab still open from before the 'play' button was pressed. They watched the Speed slider move and the Angle wheel rotate, interpreting them to understand that the runaway object was slowing down and turning back towards the Sun. The students were then able to propose explanations (which we identify as consistent with a formal, Newtonian model) for the patterns of motion seen in the Velocity tab. In the end, they located the runaway object in a zoomed-out field of view and manipulated its velocity so that it started moving back directly towards the Sun.

16.5.1.1 Analysis of Case 1

The case included above is an example of how a group of students creatively used one of the representations within *Algodoo*, namely, the Velocity tab, in a playful, unconventional way, which we can see was still meaningful from a physics learning perspective. From this case, we discern two functions for which students used the Velocity tab: (1) as a *tool* for manipulating (or setting) the velocity of an object and (2) as a *representation* which was recruited in making sense of the motion of an object.

The first function of the Velocity tab, as a tool for manipulating the velocity of an object, can be seen initially when the students used the Velocity tab to put a newly created object into motion (giving the object an initial velocity before hitting play). Then, once the collision had sent the object far away from the Sun, the students used the Velocity tab to manipulate the object's motion dynamically (with *Algodoo* running). This manipulation appeared in two instances: first as Student 1 changed the angle of the object's velocity (line 14) and again when the same student redirected the object towards the Sun (lines 45–49). In all of these instances, the presence of *Algodoo*'s Velocity tab, which allowed the students to set and manipulate the velocity of the object with sliders and a wheel, provided an opportunity for the students to engage with the orbital task creatively. More traditional approaches to the learning of orbital motion often do not provide such a means for interacting with objects' velocities as they relate to orbits. In this case, the students were able to test their own ideas of orbital mechanics, giving them ownership of the result, all while they utilized a mathematically rich interface. The manner in which the Velocity tab was used as a dynamic tool for the manipulation of velocity showcases our first concrete example of *Algodoo*'s microworldiness: the software seems to have provided the students with mathematically rich materials while also allowing the students to be creative and self-directed in their activities.

The second role that the Velocity tab played in the presented case was that of a representation recruited in making sense of the motion of an object. During most of the episode, the Velocity tab served as a monitoring device for the orbiting object outside the field of view of the scene. Formally, the velocity vector of an object in two dimensions can be expressed in terms of speed and angle (magnitude and radial direction) or as the sum of the *x*- and *y*-components of the velocity. Interestingly, in the *Algodoo* environment, the students sent an object into motion and *observed* its components, interpreting the motion of the runaway object intuitively as they tracked the changes in the angle and speed. Thus, even without being prompted to discuss vector magnitudes or components, the students were able to demonstrate fluency (at least, partially) of vector sense in relation to two-dimensional motion. The presence of the Velocity tab allowed the students to spontaneously move between a familiar, informal experience of motion (the visual movement of the object on the IWB surface) and a formal mathematized representation of motion (within the sliders and wheel of the Velocity tab). Indeed, the limited field of vision in *Algodoo*, which made the students unable to watch the object's motion as they would normally, along with the persistence of the Velocity tab, which provided them with a dynamically updated rendition of the runaway object's velocity data, encouraged the students to interpret and make creative use of the mathematical representation made available by the software.

Though the significance of the dynamically changing information on the Velocity tab was not initially appreciated by the students, as they began to make sense of what was happening, they were able to interpret the motion of the runaway planet from the controls in the tab, translating the information of the sliders and wheel into more familiar, everyday language of gesture and speech. We see this when the students noticed one of the objects 'losing speed' (line 5), after which Student 1 started making sense of the changing angle and slowly increasing velocity of the runaway planet with an explanatory gesture (line 16). Student 1 reinterpreted the information within the Velocity tab with a gesture, transforming the meaning carried in the software into a dynamic mode of expression.[3] He then engaged with the Velocity tab as a source of information about the motion of the runaway object until he is able to demonstrate his interpretation of what is going on in a more conceptual way (see Fig. 16.3).

Beyond functioning in the two ways described above, the *Algodoo*-IWB learning environment was successful in encouraging students to spontaneously produce an explanatory model for the patterns of motion. This can be seen when Student 1 questioned the motion of the runaway object (line 25). Student 2 responded by proposing an explanation for the patterns of motion consistent with a Newtonian model of orbital motion (line 31). Student 2's interpretation of the patterns seen on the Velocity tab gave rise to explanatory talk and gesture about the behaviour of orbiting objects in general. In this way, the Velocity tab within *Algodoo* appears to have behaved as a point of departure for further inquiry, providing some

---

[3]This process of transforming meaning from one mode of expression to another is sometimes referred to as *transduction* in multimodality circles (Jewitt et al. 2016). For a discussion of how transduction may be a key concept in physics learning, see Volkwyn et al. (2018).

mathematical materials which students were compelled to observe and explain in a science-like discussion (Etkina 2015; Gregorcic et al. 2017b).

This can be taken to demonstrate, in a slightly different manner, how *Algodoo* can act as a microworld for students. That is, the students were inspired by the setup and the activity to not only explore and create within the mathematically rich environment but to also begin taking science-like approaches to solving the problems they encountered (Gregorcic et al. 2017b). Consequently, a case might be made for how microworlds like *Algodoo* can offer alternative ways to promote both the learning of nuanced content knowledge at the intersection of mathematics and physics and also the adoption of the behavioural patterns used by scientists, all while promoting active engagement and creativity.

We see from Case 1 that, when using *Algodoo,* students can use mathematical representations in a creative way, therein becoming inspired to discover how a physical system works. The students' use of the mathematical materials provided by *Algodoo* was both playful – due to *Algodoo*'s open-ended, creativity-driven structure – and meaningful for their understanding of the physics formalisms that underpinned the activity. It is precisely this richness of the digital environment, the way in which *Algodoo* is an explorable sandbox populated by mathematically rigorous representations, which seems to have made possible the unique, meaningful interaction presented above.

Indeed, in the case presented here, the particular affordances of *Algodoo* that resulted in students' meaningful use of mathematical representations were paired with an instructional strategy of open-ended – but task-based – inquiry and exploration with some guidance from an instructor. Throughout the activity, the instructor engaged with the students to help direct them in their exploration. If the students had simply been given the Kepler's law scene without any instruction or guiding activity, it is unlikely that they would consistently end up manipulating the velocity in such fruitful ways. Nonetheless, the above case shows an instance where the microworldiness of *Algodoo* contributed to a group of students' creative inquiry while at the same time engaging them with formal representations of motion.

16.5.2   Case 2: Graphical Representations in Kinematics with 'Show Plot'

We now present the second case to illustrate the potential for *Algodoo* to promote creative and meaningful use of mathematical representations. This case focuses on pairs of students that used *Algodoo* in an activity alongside a physical ramp and a hockey puck on a table (hereafter referred to as the ramp-puck setup, see Fig. 16.6). The data collection comprised six students, all of whom were selected on a volunteer basis and observed pairwise on separate occasions (i.e. three separate groups of two students). Like the session in Case 1, the participating students were provided with an IWB running *Algodoo*; however, to foster a direct link between

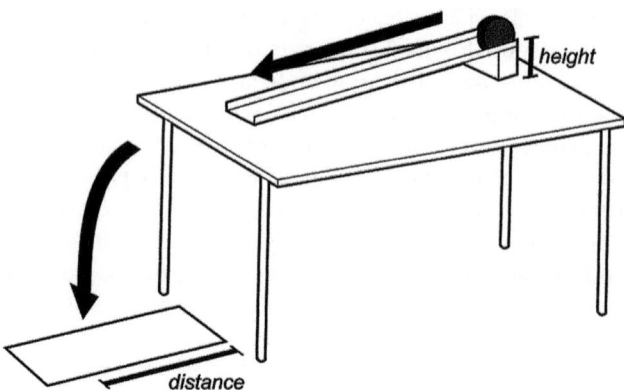

Fig. 16.6 The ramp-puck setup used by the students in Case 2 alongside *Algodoo* running on an IWB. The 'height' (above the table) and the 'distance' (horizontally along the floor from the edge of the table) are labelled

digital and physical learning environments, the students were asked to use *Algodoo* in parallel with the physical ramp-puck setup while answering a specific physics prompt. Specifically, the students were asked to convince the researchers of the relationship between (1) the height above the table from which the puck is released on the ramp and (2) the horizontal distance from the edge of the table which the puck travels before hitting the ground.

Each of the pairs of students were part of a larger, three-part session where they (1) familiarized themselves with the functions of the *Algodoo* (a duration of approximately 1 h), (2) completed the ramp activity (1 h) and then (3) concluded with a short discussion about their impression of *Algodoo* and the activity as a whole (30 min). As in Case 1, these sessions were all video recorded and transcribed for analysis.[4]

The original aim of the study was to examine how students use *Algodoo* in combination with real experiments when faced with a physics task (Euler and Gregorcic 2018). However, as in Case 1, we have found examples within the data of students engaging with a variety of mathematical representations. We can see some of the students coordinating physical observations and mathematical ideas within *Algodoo* in a manner that suggests the digital environment encourages the meaningful use of mathematical representations. The particular excerpt that we present here shows how one pair of students, referred to as Student 4 (S4) and Student 5 (S5), used the 'Show Plot' tool to quantify aspects of the puck's motion in a virtual model of the ramp-puck setup they had created. The excerpt we present here illustrates how the students can recruit and interpret graphical representations in *Algodoo*, as they attempt to quantify a physics phenomenon.

---

[4]For Case 2, the sessions were conducted in English, though the native language of both of the students was Swedish.

Fig. 16.7 Student 5 (right, with Student 4, left) rotating the ramp portion of the ramp-puck model to the desired angle. Here, the horizontal rectangle functions as a virtual table, the tilted rectangle functions as a virtual ramp and the circle functions as a virtual puck. In this scene, as opposed to the scene in Case 1, the ground is represented by a horizontal plane towards the bottom of the screen and gravity acts vertically downwards

Case 2 begins as the students finish setting up the virtual model of the ramp-puck experiment in *Algodoo*. They place two rectangular objects (representing the ramp and the table) and the circular object (the puck) in such a spatial arrangement that when they press the play button, the puck rolls down the ramp, continues off the table and then hits the ground below (Fig. 16.7). The students then try to address the prompt by finding a way in which they can measure the distance the puck travels horizontally from the edge of the table before hitting the ground.

After constructing the virtual ramp-puck setup, the students run the scene to check the function of their model. The circle successfully rolls down and off the rectangles before hitting the ground. The students immediately wish to measure the distance that the puck travels from the edge of the horizontal rectangle, but *Algodoo* does not include a purpose-built distance measuring tool. Student 4 stumbles upon the Show Plot tool. He opens the Show Plot tool and explores its possibilities for representing plots of various physical quantities for the selected object (the virtual puck in this case) in the form of a two-dimensional graph. He discovers that *Algodoo* allows you to plot different quantities on the horizontal and vertical axis of the displayed coordinate system.

50	**S4:**	[*sets the vertical axis to 'Position (y) and then the horizontal axis to Position (x)*]
51	**S5:**	[*drags the corner of the graph window to make it smaller and then moves the window to the left so they can watch the circle's motion as it rolls down the ramp*]
52	**S5:**	Something like that.
53	**S4:**	And start?
54	**S5:**	Yeah.

Excerpt Summary In the first part in of the excerpt, the students look for a way to quantify the movement of the puck, in particular, to put a numerical value on the distance the puck travels off the edge of the table. By exploring the options provided by *Algodoo*, the students discover the Show Plot tool. Student 4 then interacts with the plotting tool to select the appropriate axes labels (the *x*-position and *y*-position of the virtual puck), and Student 5 positions the graph window in such a way that they can simultaneously observe both the virtual experiment and the plot.

(continued from above)

55	**S4:**	[*presses the play button and they watch the puck move with the data being drawn in the graph window simultaneously*] (Fig. 16.8)
56	**S5:**	And let's see. If we look closer at this . . . [*leans in to examine the graph*]
57	**S4:**	Here. [*points to the point on the graph corresponding to where he thinks the circle hit the ground*]
58	**S5:**	Yeah there. [*pointing to the same point as S4*]
59	**S5:**	We can see that we have to look at it from here. [*touches the point on the graph which he interprets as where the circle left the table*] to there. [*touching the point on the graph corresponding to where they agreed the circle hit the ground*]
60	**S4:**	Hits the ground there. That's what we need to get.
61	**S5:**	Yeah, we want to know the distance here? [*gestures to show the length from the end of the physical table in the room and looks to the interviewers for confirmation*]
62	**In:**	Mhm.
63	**S5:**	Yeah. Uh . . . [*pauses for a long time to examine the graph*]

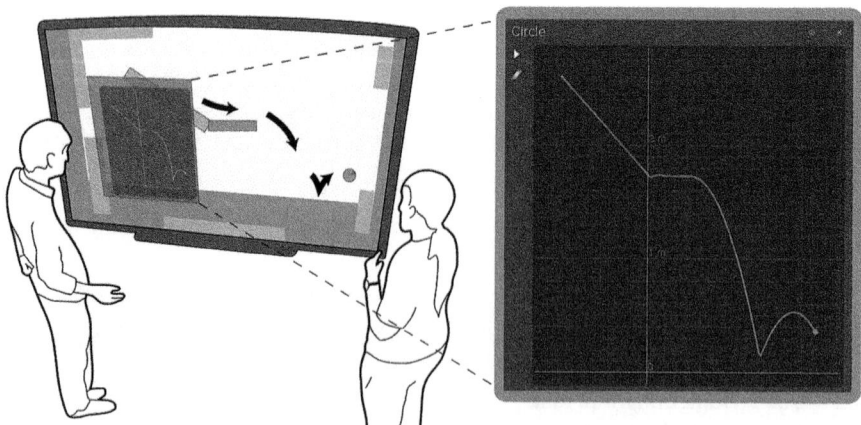

Fig. 16.8 The students examine the scene after watching the circle roll down the ramp and off the table. The graph displays a plot of the circle's motion

Excerpt Summary In the second part of the episode, the students have just run the simulation and noted its outcome by observing the movement of the puck, as well as the self-drawing graph in parallel. They continue by interpreting the graph. They start to relate characteristic points on the graph to spatial locations in the *Algodoo* scene, as well as in the physical experiment that is set up in the room next to the IWB. They identify the distance of interest and point out what they interpret as the corresponding distance on the graph.

(continued from above)

64 **S5:**   I'm trying to figure out why is there a zero here? [*points along the y-axis of the graph*] 'Cause we started way up here [*points to the upper left corner of the graph*], and where does this graph place the zero? How does this software determine where the origin is?

65 **In:**   Mhm. Is there a question?

66 **S5:**   Uh, I think so, I'm not... [*drags the corner of the graph window to make it larger*], I don't really know how to look at this graph to determine... I mean here it says 10 meters, there. [*points to the rightmost label of the x-axis*]

67 **In:**   So, what is this graph displaying really?

68 **S5:**   The y-position [*gestures up and down the IWB*] and the x-position. [*gestures left and right along the IWB*] (Fig. 16.9)

69 **In:**   Mhm.

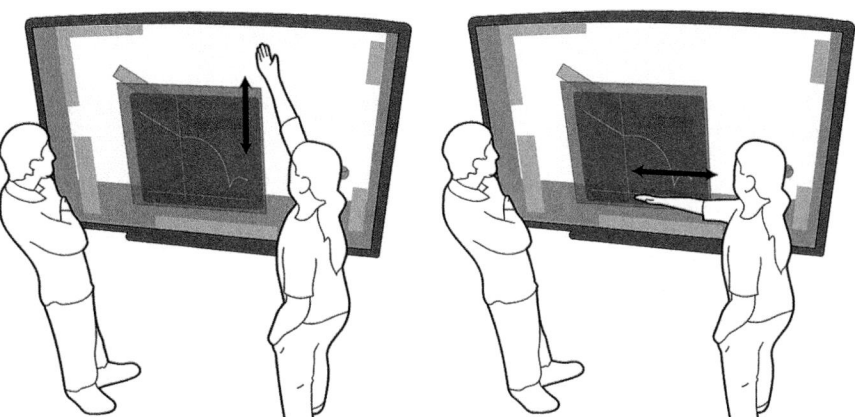

Fig. 16.9 Student 5 gestures to describe what each of the axes is displaying. He describes that the y-axis displays the y-position of the circle (gesturing up and down), while the x-axis displays the x-position of the circle (gesturing left and right)

70   **S5:**   But what I can't really see is where the *x*-position zero point is.
              That should be there. [*points to the origin in the graph window*]
              But it doesn't show much more [*taps around in the graph space to see
              what selecting the axes does then traces the graphed path of the ball in
              the plot to select various data points*]

71   **In:**   Can you say from the graph where the *x*-position zero is? [*pauses*]
              So, this graph, what does this graph represent? Like in other words,
              what would you say this graph represents? 'Cause you can have
              velocity versus time graphs. You can have *x* versus time graphs, but
              this is a *y* versus *x* graph

72   **S5:**   Yeah it describes exactly where the ball has been. It shows the path
              of the ball.

73   **In:**   Mhm! So, in space, right?

74   **S5:**   In space, yes.

75   **In:**   So, I think you can actually see where the *x*-zero is then.

76   **S5:**   Yeah when it starts rolling on the other one . . . [*grabs the graph
              window and drags it out of the way of the ramp*] When it starts rolling
              on that one. [*points to the intersection of the ramp rectangle and the
              table rectangle*] (Fig. 16.10)

77   **In:**   And where would you like it to be?

78   **S4:**   Here. [*points to the top right corner of the tilted rectangle*] (Fig. 16.11)

79   **S5:**   No [*drags the graph window out of the way*]. We want it on the end
              there [*points to the end of the horizontal rectangle*]

Fig. 16.10 Student 5 points to the intersection of the ramp rectangle (the tipped rectangle) and the table rectangle (the horizontal rectangle) to indicate the location in the scene which he interprets as the position of the $x = 0$ line of the graph

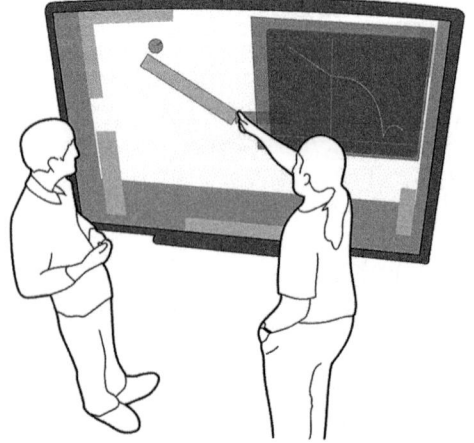

Fig. 16.11 Student 4 points to the position he thinks would be best for the $x = 0$ position at the top of the tilted rectangle (left image). Student 5 disagrees and points to the point he thinks they should place the $x = 0$, at the end of the horizontal rectangle (right image)

Excerpt Summary In this exchange, the students try to make sense of the position of the origin of the coordinate system used to describe the position of the puck. The instructors encourage them to interpret from the existing plot of the puck's motion, where the origin (zero) is currently placed and where they would like it to be. Student 4 proposes that the desired placement of zero for the x-coordinate would be the edge of the table (due to the convenience of reading off the distance from the edge of the table at which the puck first hits the ground). After line 79, with some technical help from the instructors, the students reposition the objects in *Algodoo* so that the right edge of the horizontal rectangle (the virtual table) is positioned at $x = 0$. This is done since *Algodoo* does not allow the user to move the origin of the built-in reference frame, which is fixed to the background of the scene.

(after positioning the virtual setup as desired)

80 **S4:** [*presses start and watches as the ball rolls down again, tracing a path on the graph similar to the one before, but with the axes reposition as they wanted*]

81 **S5:** [*presses pause*] Then we can find . . . [*traces finger along the data in the graph from top left to bottom right, stopping where the circle hit the ground*] the x-position! Point 75 meters (Fig. 16.12)

Excerpt Summary In this last excerpt from Case 2, the students manage to assign a numeric value to the horizontal distance the rolling puck travelled before it first hits the ground. They do this by touching the location in the graph where the tracked object (the virtual puck) appears to have first bounced and then reading the x-value of its position from the built-in graph examining tool (Fig. 16.12).

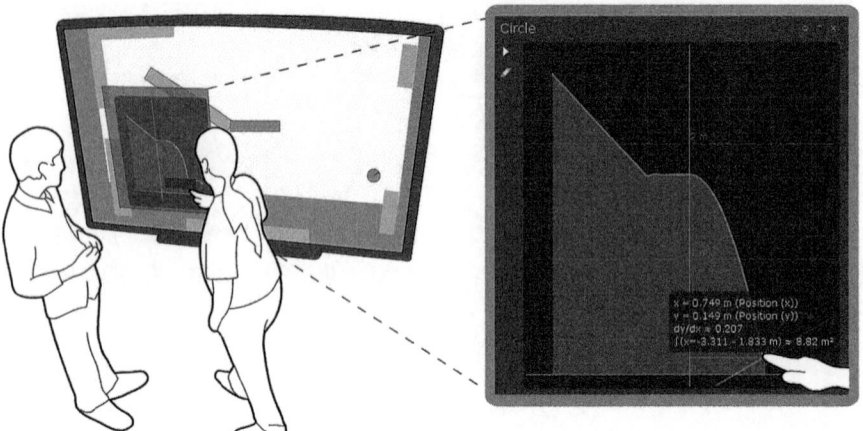

Fig. 16.12 shows Student 5 tracing the data in the graph (of the shifted setup where the y-axis is more conveniently placed) with his finger until he finds the point where the circle hits the ground. He is then able to read off the value for the horizontal distance from the x-coordinate of the dynamic graph label

In the student dialogue from Case 2, we see that the students stumbled upon the Show Plot tool in *Algodoo* and then tried to figure out how to place the origin of their graph in a useful position for their measurement purposes. In order to figure out how to move the axes to where they wanted, the students first had to interpret what the graph was showing so that they could understand how *Algodoo* had placed the origin for them (the origin is fixed by default to the background in *Algodoo*, and they had to move their virtual setup so that axes were aligned with the desired part of their ramp-puck model).

16.5.2.1   Analysis of Case 2

In Case 2, the students engaged with the *Algodoo*-IWB setup to mathematize the motion of a puck in a graph. Despite the physics content being different from that in Case 1, we use the students' interaction in Case 2 to highlight how *Algodoo* appeared to act as a microworld for the students by providing them with mathematical material to draw upon in a meaningful, if slightly unconventional, exploration of a physics phenomenon.

With the Show Plot tool in *Algodoo*, the two students in Case 2 made use of a graph in a somewhat atypical manner: that is, to *measure* the horizontal distance travelled by the puck after leaving the table within their *Algodoo* scene. As they might have used a meter stick to measure the physical distance that the puck travels in the non-virtual ramp-puck setup, the students used a graph within *Algodoo* to plot the position of their virtual puck (the circle) and read off the x-value from this graph as the x-component of its plotted motion. This implementation of the graphical representation is interesting in that the students measured a quantity *with* the graph rather than populating the graph with data measured by another tool. This is made possible in digital environments like *Algodoo* due to the fact that these programs are necessarily built up from mathematics. *Algodoo* was already tracking the position of the circle in relation to the background of the scene, so, for the students in Case 2, it was simply a matter of finding a way to display the position of the circle in a graph for their use.

However, their imaginative use of the Show Plot tool still required them to employ the mathematical representation *correctly*. In order for their graph to display the position of the circle, the students first had to select the appropriate quantities for each of the axes. Student 4 chose axes labels of Position (y) and Position (x), changing them from the default labels of Speed and Time. Even though *Algodoo* generated an option for graphical mathematical representation for the students, they were still required to engage with the representation enough to responsibly select an appropriate version of the graph for their given situation. The students had to tailor the mathematical representation so that it could be used in their unconventional implementation. This is our first example from Case 2 of how the microworldiness of *Algodoo* allowed the students to use mathematical representations in a creative yet meaningful manner: the software provided the students with mathematical materials in the form of a graphical tool, which they implemented in their own creative problem-solving.

The other way in which *Algodoo*'s microworld-like behaviour appears to have afforded unique opportunities to the students is in how it constrained their actual construction a model of the ramp-puck setup. While the transcript above focuses on the students' use of the Show Plot tool, the students' mathematization within *Algodoo* began even before the excerpts of line 50, when the students *geometrized* the ramp-puck setup into the virtual space. The students first had to interpret the parts of the physical experiment (the ramp, the table and the puck) as simple geometrical entities, spatially organized in the *Algodoo* scene so as to result in a simple geometrical model of the experiment. This meant that the students needed to make creative, physicist-like decisions about how to simplify the three-dimensional problem into a two-dimensional collection of simple shapes.

Furthermore, as the students overlaid the graph of the circle's motion in the *Algodoo* scene, they then needed to *interpret* the interactions of the objects within their model in terms of how they related to the mathematical representation. In his choice to plot the horizontal and vertical position of the circle in a graph, Student 4 effectively produced an abstract, mathematized version of the puck's trajectory. However, since the graph did not display some of the main visual features of the scene itself (i.e. the ramp rectangle, the table rectangle, the circle or the ground), the students were presented with the challenge of interpreting how the plotted data related to the virtual ramp-puck model. For example, the location of the edge of the table, which was particularly important for determining the distance of interest, was not explicitly represented in the graph itself. This led the students to explore the connection between the mathematical representation and the phenomenon which it represented. They do this by first running the simulation and then realizing that the axes of their plot were not where they wanted. Eventually, the students were likely able to relate specific points of the graph to places in the virtual setup in part due to the proximity and simultaneity of the representations (topics discussed in depth in work such as Ainsworth 2006).

We see in Case 2 how the Show Plot tool, while being used as a quantification tool for measuring horizontal distance, also involved the students in a purposeful coordination of a geometrical representation (the virtual ramp-puck model) and mathematical representation (the graph) of a physical experiment (the real ramp-puck setup). As we saw with Case 1, the student activity in Case 2 around the given prompt showcases how users of *Algodoo* can make creative yet meaningful use of the representations within the digital environment. The students were creatively engaged not only as they explored a novel physics phenomenon but also as they generated a geometrical model of a real experiment. They were involved in the tailoring of a mathematical representation of motion, and, by creatively leveraging the affordances of the *Algodoo*-IWB setup, they were able to determine the desired distance and continue with their task. This suggests that such *Algodoo*-IWB setups might be used for a variety of tasks, by a variety of students, to support student creativity and fluency in formal and mathematical representations of physics phenomena.

16.6 Discussion and Implications for Instruction

By appropriately encouraging and guiding students in environments such as *Algodoo*, those software that are rich in the mathematical materials with which users can build and have experiences, it may be possible for instructors to help students attain a better conceptual understanding of physics and to help them relate those conceptual understandings to mathematical formalisms. The cases presented in this chapter are used to show how the open structure of *Algodoo* inspired students to informally create and explore with formal mathematical representations.

16.6.1 Algodoo as a Tool for Conceptual Learning

We recognize *Algodoo* as a potentially valuable tool for expanding the possible ways in which students can engage with mathematics in physics contexts. The software allows the object of learning to be presented to students as something around which they can safely and inventively build an understanding of physics phenomena. Especially when paired with large touchscreen displays such as an IWB, students using *Algodoo* may be able to experience physics phenomena through mathematical representations in much the same way that they can begin to experience velocity and acceleration in our speedometer-rich culture. By bringing mathematical representations to life within the dynamic system of a virtual world, digital learning environments like *Algodoo* might better construe representations as part of – and intrinsically related to – observable phenomena, thereby also making representations and the phenomena they represent available to students as objects of inquiry. In a way, students using *Algodoo* can observe how mathematical representations behave much like one observes an experiment.

Put in another way, *Algodoo* seems to function as conceptual stepping stone between physical phenomena and mathematical formalisms. This is an idea that we have explored in previous research (Euler and Gregorcic 2018), wherein we specifically examined how students transitioned between using a physical laboratory setup, a digital model they created in *Algodoo*, and mathematical representations related to both the physical and digital environments. In this work, while building on Hestenes's (1992) mathematical modelling games and diSessa's (1988) discussion of the functions of educational technology, we interpret the role of *Algodoo* as one of a *semi-formalism*: that is, a conceptual intermediary between the experiences of the physical world and the formal models used in physics.

Furthermore, while much of Papert's work – and the well-known work of his colleague, Piaget – focused on learning in young children, we argue that *Algodoo* and other open-ended software have the potential to be a learning tool for a wide variety of students spanning many age groups. By providing a creative arena that adapts to the exploration and creativity of each user, *Algodoo* not only provides novice learners with alternative means for accessing physics but also allows more experienced learners to further develop, assess and/or verify their understanding of the interplay of physics and mathematics concepts. We suggest that *Algodoo* can be useful for physics learners from primary school through university.

16.6.2   Student Motivation and Interest

The processes discussed in this chapter are relevant not only from a conceptual learning perspective but also as a way of providing students with nonthreatening opportunities to approach problems in self-directed ways. The studied setup seems to have fostered exploratory behaviour even in novice users. This suggests that Algodoo and similar software could have potential for engaging learners in the early stages of mathematization through novel and less threatening ways than traditional instruction. In both cases presented here, we see that by giving students control to create and choose among the many available mathematical representations within *Algodoo*, as opposed to insisting that they use the 'most appropriate' representation for the task, the activity that results can be student-directed and playful in nature while at the same time meaningful from the perspective of conceptual learning.

This aligns well with prior research on the use of *Algodoo* in physics education, which has shown that it can be used in ways that promote the engagement and interest of students who do not consider themselves particularly savvy with physics (Gregorcic et al. 2017a). There is also a growing body of examples of *Algodoo* use wherein students from various backgrounds seem to consistently explore conceptual physics content in playful ways (Euler and Gregorcic 2018; Gregorcic et al. 2017b; Rådahl 2017). In addition to these published reports, our extensive anecdotal experience with the use of the software with a geographically diverse population of preservice and in-service physics teachers suggests that the visual clarity, user-friendliness and open-endedness of the software are usually met with great enthusiasm, particularly by preservice teachers.

16.6.3   Concluding Remarks

It is sometimes easy to be impressed with a new technology to the point of overestimating its utility in the classroom. It is worth noting that there exists much debate around the usefulness of open-ended technologies which align with the constructionist ideas of microworlds. This is especially the case among cognitivists who claim that exploratory learning places too much of a load on the cognitive processes of students (see such arguments as Hmelo-Silver et al. 2007; Kirschner et al. 2006; Sweller et al. 2007). Nonetheless, we hold that if the learning activities are appropriately framed (e.g. as playful inquiry with instructor guidance and specified tasks), the microworldiness of *Algodoo* seems to provide students with meaningful opportunities to engage with mathematical concepts, which they might have found as prohibitively challenging or uninspiring in traditional classroom or laboratory circumstances. As we stressed earlier, an open-ended software may not be sufficient on its own to ensure that powerful ideas are learned. However, as the cases in this chapter illustrate, when software such as *Algodoo* is paired with some intentional structure in the activity, students can still be creative – and self-directed – in their activities to the extent that they make meaningful use of mathematical representations. While further research is needed to find the optimal use of technologies, such as *Algodoo*, which may function as microworlds for students in the learning of physics, these digital tools seem to be potentially valuable in the way that they can (1) help students to coordinate physics concepts with mathematical representations and (2) foster student motivation in inviting avenues for playful exploration. As such, it represents a unique and exciting class of digital resources for the teaching and learning of physics across many levels of education.

References

Abelson, H., & Disessa, A. A. (1980). *The computer as a medium for exploring mathematics.* Cambridge: MIT Press.

Ainsworth, S. (2006). DeFT: A conceptual framework for considering learning with multiple representations. *Learning and Instruction, 16*(3), 183–198. https://doi.org/10.1016/j. learninstruc.2006.03.001.

Arnone, S., Moauro, F., & Siccardi, M. (2017). A modern Galileo tale. *Physics Education, 52*(1), 1–5.

Bezemer, J., & Mavers, D. (2011). Multimodal transcription as academic practice: A social semiotic perspective. *International Journal of Social Research Methodology, 14*(3), 191–206. https://doi.org/10.1080/13645579.2011.563616.

Clements, D. H. (1986). Effects of Logo and CAI environments on cognition and creativity. *Journal of Educational Psychology, 78*(4), 309–318. https://doi.org/10.1037//0022-0663.78.4.309.

Clements, D. H. (1990). Metacomponential development in a Logo programming environment. *Journal of Educational Psychology, 82*(1), 141–149. https://doi.org/10.1037/0022-0663.82.1.141.

Dancy, M., Christian, W., & Belloni, M. (2002). Teaching with Physlets: Examples from optics. *The Physics Teacher, 40*(8), 494–499. https://doi.org/10.1119/1.1526622.

diSessa, A. A. (1988). Knowledge in pieces. In G. E. Forman & P. B. Pufall (Eds.), *Constructivism in the computer age* (pp. 49–70). Hillsdale: Lawrence Erlbaum. https://doi.org/10.1159/000342945.

Etkina, E. (2015). Millikan award lecture: Students of physics—Listeners, observers, or collaborative participants in physics scientific practices? *American Journal of Physics, 83*(8), 669–679. https://doi.org/10.1119/1.4923432.

Euler, E., & Gregorcic, B. (2018). Exploring how physics students use a sandbox software to move between the physical and the formal. In *2017 physics education research conference proceedings* (pp. 128–131). American Association of Physics Teachers. https://doi.org/10.1119/perc.2017.pr.027.

Goodwin, C. (2007). Environmentally coupled gesture. In S. D. Duncan, J. Cassell, & E. T. Levy (Eds.), *Gesture and the dynamical dimension of language: Essays in honor of David McNeill* (pp. 195–212). Amsterdam: John Benjamins Publishing Company.

Gregorcic, B. (2015a). Exploring Kepler's laws using an interactive whiteboard and Algodoo. *Physics Education, 50*(5), 511–515. https://doi.org/10.1088/0031-9120/50/5/511.

Gregorcic, B. (2015b). *Investigating and applying advantages of interactive whiteboards in physics instruction*. Ljubljana: University of Ljubljana.

Gregorcic, B. (2016). Interactive whiteboards as a means of supporting students' physical engagement and collaborative inquiry in physics. In L. Thoms & R. Girwidz (Eds.), *Proceedings from the 20th international conference on multimedia in physics teaching and learning* (pp. 245–252). Mulhouse: European Physical Society.

Gregorcic, B., & Bodin, M. (2017). Algodoo: A tool for encouraging creativity in physics teaching and learning. *The Physics Teacher, 55*, 25–28.

Gregorcic, B., Etkina, E., & Planinsic, G. (2015). Designing and investigating new ways of interactive whiteboard use in physics instruction. In P. V. Engelhardt, A. D. Churukian, & D. L. Jones (Eds.), *2014 physics education research conference proceedings* (pp. 107–110). American Association of Physics Teachers. https://doi.org/10.1119/perc.2014.pr.023.

Gregorcic, B., Etkina, E., & Planinsic, G. (2017a). A new way of using the interactive whiteboard in a high school physics classroom: A case study. *Research in Science Education*, 1–25. https://doi.org/10.1007/s11165-016-9576-0.

Gregorcic, B., Planinsic, G., & Etkina, E. (2017b). Doing science by waving hands: Talk, symbiotic gesture, and interaction with digital content as resources in student inquiry. *Physical Review Physics Education Research, 13*(2), 1–17. https://doi.org/10.1103/PhysRevPhysEducRes.13.020104.

Hestenes, D. (1992). Modeling games in the Newtonian world. *American Journal of Physics, 60*(8), 732–748.

Hestenes, D. (1995). Modeling software for learning and doing physics. In C. Bernardini, C. Tarsitani, & M. Vicentini (Eds.), *Thinking physics for teaching* (pp. 25–65). Boston: Springer US. https://doi.org/10.1007/978-1-4615-1921-8_4.

Hmelo-Silver, C. E., Duncan, R. G., & Chinn, C. A. (2007). Scaffolding and achievement in problem-based and inquiry learning: A response to Kirschner, Sweller, and Clark (2006). *Educational Psychologist, 42*(2), 99–107. https://doi.org/10.1080/00461520701263368.

Jewitt, C., Bezemer, J., & O'Halloran, K. (2016). *Introducing multimodality (First)*. New York: Taylor & Francis.

Jimoyiannis, A., & Komis, V. (2001). Computer simulations in physics teaching and learning: A case study on students' understanding of trajectory motion. *Computers & Education, 36*(2), 183–204. https://doi.org/10.1016/S0360-1315(00)00059-2.

Kirschner, P. A., Sweller, J., & Clark, R. E. (2006). Why minimal guidance during instruction does not work: An analysis of the failure of constructivist, discovery, problem-based, experiential, and inquiry-based teaching. *Educational Psychologist, 41*(2), 75–86. https://doi.org/10.1207/s15326985ep4102_1.

Klahr, D., & Carver, S. M. (1988). Cognitive objectives in a LOGO debugging curriculum: Instruction, learning, and transfer. *Cognitive Psychology, 20*(3), 362–404. https://doi.org/10.1016/0010-0285(88)90004-7.

Laurillard, D. (2002). *Rethinking university teaching: A conversational framework for the effective use of learning technologies* (2nd ed.). New York: Routledge.

Lehrer, R., Randle, L., & Sancilio, L. (1989). Learning preproof geometry with LOGO. *Cognition and Instruction, 6*(2), 159–184.

Mayer, R. E., Dow, G. T., & Mayer, S. (2003). Multimedia learning in an interactive self-explaining environment: What works in the design of agent-based microworlds? *Journal of Educational Psychology, 95*(4), 806–812. https://doi.org/10.1037/0022-0663.95.4.806.

McDermott, L. C., Rosenquist, M. L., & van Zee, E. H. (1987). Student difficulties in connecting graphs and physics: Examples from kinematics. *American Journal of Physics, 55*, 503.

Mellingsæter, M. S., & Bungum, B. (2015). Students' use of the interactive whiteboard during physics group work. *European Journal of Engineering Education, 40*(February), 115–127. https://doi.org/10.1080/03043797.2014.928669.

Miller, C. S., Lehman, J. F., & Koedinger, K. R. (1999). Goals and learning in microworlds. *Cognitive Science, 23*(3), 305–336. https://doi.org/10.1016/S0364-0213(99)00007-5.

National Research Council. (2011). In M. A. Honey & M. L. Hilton (Eds.), *Learning science through computer games and simulations*. Washington, DC: The National Acadamies Press.

Papert, S. (1980). *Mindstorms: Children, computers and powerful ideas*. New York: Basic Books, Inc. https://doi.org/10.1016/0732-118X(83)90034-X.

Pea, R. D., & Kurland, D. M. (1984). On the cognitive effects of learning computer programming. *New Ideas in Psychology, 2*(2), 137–168. https://doi.org/10.1016/0732-118X(84)90018-7.

Perkins, K., Adams, W., Dubson, M., Finkelstein, N., Reid, S., Wieman, C., & LeMaster, R. (2006). PhET: Interactive simulations for teaching and learning physics. *The Physics Teacher, 44*(1), 18–23. https://doi.org/10.1119/1.2150754.

Plass, J. L., & Schwartz, R. N. (2014). Multimedia learning with simulations and microworlds. In R. E. Mayer (Ed.), The Cambridge handbook of multimedia learning (2nd Edi, pp. 729–761). Cambridge: Cambridge University Press.

Rådahl, E. (2017). *Responsive teaching using simulation software: The case of orbital motion*. Uppsala University.

Rieber, L. P. (1996). Seriously considering play: Designing interactive learning environments based on the blending of microworlds, simulations, and games. *Educational Technology Research and Development, 44*(2), 43–58. https://doi.org/10.1007/BF02300540.

Rieber, L. P. (2005). Multimedia learning in games, simulations, and microworlds. In R. E. Mayer (Ed.), *The Cambridge handbook of multimedia learning* (pp. 549–568). New York: Cambridge University Press.

Rosengrant, D., Van Heuvelen, A., & Etkina, E. (2009). Do students use and understand free-body diagrams? *Physical Review Special Topics – Physics Education Research, 5*(1), 010108. https://doi.org/10.1103/PhysRevSTPER.5.010108.

Roth, W.-M., & Lawless, D. (2002). Scientific investigations, metaphorical gestures, and the mergence of abstract scientific concepts. *Learning and Instruction, 12*, 285–304. https://doi.org/10.1016/S0959-4752(01)00023-8.

Sweller, J., Kirshner, P. A., & Clark, R. E. (2007). Why minimally guided teaching techniques do not work: A reply to commentaries. *Educational Psychologist, 42*(2), 115–121. https://doi.org/10.1080/00461520701263426.

Trowbridge, D. E., & McDermott, L. C. (1980). Investigation of student understanding of the concept of velocity in one dimension. *American Journal of Physics, 48*(12), 1020–1028. https://doi.org/10.1119/1.12298.

Trowbridge, D. E., & McDermott, L. C. (1981). Investigation of student understanding of the concept of acceleration in one dimension. *American Journal of Physics, 49*(3), 242–253. https://doi.org/10.1119/1.12525.

Van Heuvelen, A. (1991). Learning to think like a physicist: A review of research-based instructional strategies. *American Journal of Physics, 59*(10), 891–897.

Volkwyn, T. S., Airey, J., Gregorcic, B., Heijkensköld, F., & Linder, C. (2018). Physics students learning about abstract mathematical tools when engaging with "invisible" phenomena. In *2017 physics education research conference proceedings* (pp. 408–411). American Association of Physics Teachers. https://doi.org/10.1119/perc.2017.pr.097.

White, B. Y. (1984). Designing computer games to help physics students understand Newton's laws of motion. *Cognition and Instruction, 1*(1), 69–108.

Wieman, C. E., Adams, W. K., & Perkins, K. K. (2008). PhET: Simulations that enhance learning. *Science, 322*(5902), 682–683. https://doi.org/10.1126/science.1161948.

Printed by Printforce, the Netherlands